The growth and decay of ice

Studies in Polar Research

This series of publications reflects the growth of research activity in and about the polar regions, and provides a means of disseminating the results. Coverage is international and interdisciplinary: the books are relatively short and fully illustrated. Most are surveys of the present state of knowledge in a given subject rather than research reports, conference proceedings or collected papers. The scope of the series is wide and includes studies in all the biological, physical and social sciences.

Other titles in this series:

The Antarctic Circumpolar Ocean
Sir George Deacon

The Living Tundra*
Yu. I. Chernov, transl. D. Love

The Antarctic Treaty Regime
edited by Gillian D. Triggs

Antarctica: The Next Decade
edited by Sir Anthony Parsons

Antarctic Mineral Exploitation
Francisco Orrego Vicuna

Transit Management in the Northwest
  Passage
edited by C. Lamson and D. Vander
  Zwaag

Canada's Arctic Waters in
  International Law
Donat Pharand

Vegetation of the Soviet Polar Deserts
V. Aleksandrova, transl. D. Love

Reindeer on South Georgia: The
  Ecology of an Introduced
  Population
N. Leader-Williams

The Biology of Polar Bryophytes and
  Lichens
R. Longton

Microbial Ecosystems in Antarctica
Warwick F. Vincent

The Frozen Earth
P. J. Williams and M. W. Smith

A Chronological List of Antarctic
  Expedition and Related Historical
  Events
R. Headland

* Also available in paperback

The growth and decay of ice

G.S.H. LOCK

Department of Mechanical Engineering, University of Alberta
Edmonton, Alberta, Canada

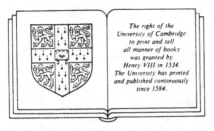

The right of the
University of Cambridge
to print and sell
all manner of books
was granted by
Henry VIII in 1534.
The University has printed
and published continuously
since 1584.

CAMBRIDGE UNIVERSITY PRESS

Cambridge
New York   Port Chester
Melbourne   Sydney

CAMBRIDGE UNIVERSITY PRESS
Cambridge, New York, Melbourne, Madrid, Cape Town, Singapore, São Paulo

Cambridge University Press
The Edinburgh Building, Cambridge CB2 2RU, UK

Published in the United States of America by Cambridge University Press, New York

www.cambridge.org
Information on this title: www.cambridge.org/9780521331333

First published 1990
This digitally printed first paperback version 2005

A catalogue record for this publication is available from the British Library

Library of Congress Cataloguing in Publication data
Lock, G. S. H.
The growth and decay of ice / by G.S.H. Lock.
p.   cm. – (Studies in polar research)
Includes bibliographical references.
ISBN 0 521 33133 1
1. Ice.   2. Freezing points.   I. Title.   II. Series
GB2403.2.L63   1990
551.3'1–dc20   89–25441   CIP

ISBN-13 978-0-521-33133-3 hardback
ISBN-10 0-521-33133-1 hardback

ISBN-13 978-0-521-02193-7 paperback
ISBN-10 0-521-02193-6 paperback

To the memory of R. R. Gilpin

Contents

Preface   ix

Acknowledgements   xiii

Nomenclature   xv

1   Introduction   1
1.1   Ice in the heavens   1
1.2   Ice ages   7
1.3   The incidence of ice   12
1.4   The response to ice   20

2   Thermodynamics of ice   24
2.1   Classical laws and principles   25
2.2   The Gibbs function and phase change   28
2.3   The Gibbs function and chemical potential   34
2.4   Diffusion in ice, water and water vapour   40
2.5   Metastability and the freezing point   44
2.6   Morphology and microenergetics   49
2.7   Thermal properties   53
2.8   Nucleation   55
2.9   Interfacial conditions and growth   60

3   The Stefan problem   63
3.1   The simplified model and its variations   64
3.2   The Neumann solution   71
3.3   Physico-mathematical analysis   75
3.4   Analytic techniques   79
3.5   Numerical techniques   86

4   Ice and water   102
4.1   Freezing of solutions and suspensions   102
4.2   Formation of an ice cover   112
4.3   Growth of an ice cover   121
4.4   The effect of natural convection on planar growth   134

4.5   The effect of forced convection on planar growth   140
4.6   Freezing inside cavities and conduits   151
4.7   Freezing on submerged bodies   161

5   Ice and air   **167**
5.1   Deposition of water vapour   167
5.2   Accretion   178
5.3   Atmospheric ice   192
5.4   Icing of airborne structures   205
5.5   Icing of waterborne and offshore structures   213
5.6   Icing of land-based structures and equipment   220

6   Ice and Earth   **229**
6.1   Surface energy exchange   229
6.2   Freezing on the Earth   242
6.3   The pore system: its architecture, hydrodynamics and
      thermodynamics   255
6.4   Freezing in the Earth   272

7   Ice and life   **298**
7.1   Water in the biosphere   298
7.2   Freezing of biofluids   302
7.3   The cell   312
7.4   Freezing of unicellular organisms   317
7.5   Freezing in plants   329
7.6   Freezing in animals   336
7.7   Freezing of foods   344

8   Decay of ice   **355**
8.1   Reversibility and symmetry   356
8.2   Melting of single ice crystals   358
8.3   Melting of polycrystalline ice   361
8.4   Decay of an ice cover   364
8.5   Ablation of atmospheric and structural icing   370
8.6   Thawing of soil   374
8.7   Thawing of organisms   380

      Notes   386
      Selected bibliography   394
      Index   422

Preface

This book was a long time in the making. It began with a lecture entitled the Growth and Decay of Ice (Lock, 1974) presented to the Fifth International Heat Transfer Conference in Tokyo, 1974. In retrospect, this lecture seems hopelessly inadequate, and must be viewed merely as an early attempt to grapple with the immensity of the subject during the search for a sense of unity. The subsequent suggestion that the lecture might be expanded into a monograph was taken seriously but just as I was about to embark on a sabbatical leave in 1975 to undertake that task I acquired administrative responsibilities heavy enough to prevent extensive work in glaciology. Only in the early 1980s was I able to return to the task, finally getting down to the library work during a sabbatical leave in Cambridge during 1985 and 1986.

I would like to claim that the intervening decade was a period of gestation, but I must confess it to be no more than a delay. It was a period during which my knowledge broadened and deepened but my overall views changed little, save through lessons in the sociology of scholarship learned at interdisciplinary boundaries. It has been a most interesting exercise reviewing the literature and attending conferences concerned, in one way or another, with the growth and decay of ice. I have found that glaciology is in one sense a single discipline, but in another it may be viewed as a microcosm of science itself: possessing a common purpose expressed in significantly different ways; unified, yet divided, by mathematics; uneasy in the balance between traditional and emerging views; scattered over the scales of length and time; and moving more readily across international, rather than interdisciplinary, boundaries.

I have chosen not to follow the more traditional approach to glaciology so well articulated by Shumskii (1964). Such an approach is well suited to the treatment of ice in a geophysical or geotechnical context, and expresses the central ideas in what I have termed *geoglaciology*: i.e., the study of ice in and on the earth. However, the petrological framework has to be bent somewhat if it is to extend to ice in the presence of all the many forms of water, and it is perhaps more natural to treat this material in a different context. Accordingly, I have taken *hydroglaciology* to mean the study of ice in and on water. The same line of reasoning suggests, even more strongly, two further contextual subdivisions: *aeroglaciology* is defined as the study of ice in the atmosphere, while *bioglaciology* describes the relationships between ice and living systems. Curiously enough, this contextual taxonomy (Lock, 1986) is organized on the ancient Greek elements, if life is substituted for fire, although the order has been rearranged in this book. After introductory and foundational chapters dealing with perspective, thermodynamics and mathematics, the first context is water. Despite its ultimate complexity, the interface between ice and pure water represents an excellent point of departure for the examination of various physical and chemical effects. This lays the groundwork for situations where the fluid side of the interface is not occupied solely by water, and thus provides entry to the context of air; hydrodynamics is exchanged for aerodynamics. Snow provides the natural link between ice in the air and ice on the ground. In particular, metamorphosis of snow heralds the ensuing discussion of ice in pores and porous media. Finally, soil and water link the physical significance to the biological significance when the context is life in its many forms.

I decided at the outset, as I had done in 1974, that decay would not be treated alongside growth. This is partly because each of the four main chapters would then be more difficult to balance, with examples of decay being either too numerous, and therefore repetitive, or not numerous enough, and therefore incomplete. However, the main reason for treating decay separately was the opportunity it presented for the general exploration of similarities and differences. I was thus able to review the factors common to both growth and decay while examining the circumstances where the difference between them was much more than a minus sign. In addition, the importance of time, particularly in cyclic sequences, could then be more readily pointed up.

This is a work of synthesis in which the four chosen contexts are presented separately but are woven together into a broader view of glaciology. Within each context, fundamentals are accompanied by pragmatism; theories are juxtaposed with implications and applications. While this is designed to produce self-contained chapters, each of which will be of particular interest to selected groups of specialists, it is also intended to incorporate the mathematical and physical principles common to all aspects of growth and decay. Furthermore, in addition to this 'vertical' integration, the organization also encourages 'horizontal' integration by allowing the experienced and inexperienced reader alike to enter any discipline or subdiscipline at an appropriate level. Each unidisciplinarian is therefore free to explore at will the realms of seemingly distant colleagues, and thus reap the benefits – conceptual, economic and practical – which accrue to an interdisciplinarian. Above all, it is hoped that a stronger and more unified view of glaciology, *sui generis*, may emerge.

The book is based upon the conventional wisdom. While I make no claim to be expert in every aspect of glaciology, in the interest of developing a comprehensive picture I have occasionally been forced to speculate on relationships hitherto unnoticed or events apparently unexplained. Should I have done so incorrectly or unnecessarily I hope the reader will be kind enough to provide me with an appropriate reference. A great many colleagues have already been good enough to give me the benefit of their knowledge or have been generous enough to read chapters or parts of chapters of the first draft. I take this opportunity to thank sincerely, and absolve from any blame in the final draft, the following: in Cambridge, Dr Terence Armstrong, Dr William Block, Dr David Drewry, Dr Felix Franks, Dr David Pegg, Sir Brian Pippard, Dr Gareth Rees, Dr Gordon Robin, Dr Murray Stewart and Dr Peter Wadhams; in Edmonton, Dr David Chanasyk, Dr K. C. Cheng, Dr Tom Forest, Dr Edward Gates, Dr Larry Gerard, Dr Edward Lozowski, Dr Locksley McGann, Dr Norbert Morgenstern, Dr David Sego, Dr Robert Tait and Dr Stanley Thompson; in Calgary, Dr Derick Nixon; in Fairbanks, Dr John Zarling; in Boulder, USA, Dr Charles Knight; in Kamloops, Dr Darryl Stout; in Victoria, Dr Richard Ring; in Vancouver, Dr William Powrie; in Austin, Dr Kenneth Diller; and in Dartmouth, USA, Dr Steven Daley.

The serious writing began in Cambridge when I was a visiting scholar at the Scott Polar Research Institute. Then director, Dr David Drewry, made me most welcome and his staff, Robert Headland, Valerie Galpin, Ailsa MacQueen and Rosemary Graham, were very accommodating. I also had the benefit of open exchange with members of the Sea Ice Group, and enjoyed several discussions with Dr Hilda Richardson, Secretary-General of the International Glaciological Society. During the same period, I was fortunate enough to spend several months in Clare Hall as a visiting fellow. To then President, Sir Michael Stoker, the fellows of Clare Hall, and especially to Senior Fellow, Dr Terence Armstrong, I am very grateful for the hospitality and congeniality which my wife and I enjoyed so much.

It was Peter Wadhams who visited me one morning in the gallery of the

Scott library and suggested that the Studies in Polar Research series of Cambridge University Press might be an appropriate publication vehicle for what was then little more than a brief outline. It is therefore fitting, and for me most fortunate, that he should be charged with the responsibility for technical editorship. His reading of the entire manuscript led to improvements for which I am grateful. The illustrations were prepared in the Graphics Division of the University of Alberta where Mr Hiroshi Yokota did the drawings. Pre-production assistance by Gareth Lock, Verity Hunter and Shamim Datoo is also appreciated, as is the bibliographical assistance of John Miletich of the University library and the help of staff of Cambridge University Press, Maureen Storey in particular. Above all, I am indebted to my wife, Edna, in so many ways: for preparing the bibliography; for the production of three successive drafts; for the supervision of copyright clearances; and for the production of the index.

G.S.H.L.
30th June, 1989
Edmonton.

Acknowledgements

Thanks are extended to the following for their courtesy and kind permission to publish, in whole or in part, diagrams and photographs identified in the text and more fully in the Bibliography:

Academic Press, D. Hillel, R.J.M. DeWiest and R.B. Duckworth; American Geophysical Union; American Institute of Aeronautics and Astronautics; American Institute of Chemical Engineers; American Meteorological Society; American Physiological Society and K.E. Zachariassen; American Society of Mechanical Engineers; Annual Reviews Inc; K. Arakawa; M.J. Ashwood-Smith; Association of American Geographers and J.R. MacKay; Australian Academy of Science; H.R. Bardarson; Bibliothek des Deutschen Wetterdlenstes; M.I. Budyko; Cambridge University Press; Dover Publications Inc; Elsevier Science Publishers; F. Franks; Geological Society of America; J.R. Gunderson; Harvard University Press; Hemisphere Publishing Corporation; Holt, Rinehart and Winston; International Glaciological Society; A.R. Jumikis; K. Kikuchi; L. Makkonen; McGraw-Hill Inc and H. Kohnke; Marcel Dekker Inc; R. Massom; National Academy of Sciences; National Aeronautics and Space Administration; National Research Council of Canada; Oxford University Press; Pergamon Press Inc, T. Hirata, W.S.B. Paterson and J. Gutschmidt; Plenum Publishing Corporation, W.F. Weeks and D.E. Pegg; Rockefeller University Press; Royal Society of Canada and P.V. Hobbs; Scripta Technica Inc; Springer-Verlag and N. Untersteiner; V.A. Squire; M. Sroka; Stanford University Press; St. Martin's Press, J.P. Clancy and A.W. Washburn; Taylor and Francis Ltd and B.J. Mason; N.A. Tsytovich; U.S. Army Cold Regions Research and Engineering Laboratory; University of Toronto Press; The University of Minnesota Press and S.C. Porter. The cover photograph is from the cover of Science, Vol 166, 17th Oct, 1969, and is used with the kind permission of Dr W. Dansgaard and the American Association for the Advancement of Science (copyright 1969).

Nomenclature

Symbols

a lattice parameter

A area, advective flux

B cooling rate

Bi Biot number

c thermal capacity per unit volume

c_p isobaric specific heat

c_v isometric specific heat

C concentration, capacity

d diameter, gap

D diffusivity, diameter

$\mathcal{D}$ diffusion coefficient, dispersion coefficient

e specific energy, emissive power density, void ratio

E energy, radiant emission, collision efficiency

Ec Eckert number

$\mathcal{E}$ electric charge

f extensive property per unit volume, fraction, function

F Helmholtz free energy, intensive variable

Fo Fourier number

g gravitational acceleration, Gibbs free energy per unit mass

G Gibbs free energy, irradiation, acceleration, geothermal heat flux

h specific enthalpy, thickness, heat transfer coefficient

H thickness, piezometric head

i electric current density, Van't Hoff's factor

I short wave radiation

j flux density

J flux, radiosity

$\mathcal{J}$ rate per unit volume

k conductivity, Boltzmann's constant, intrinsic permeability

K curvature, inertial parameter, hydraulic conductivity

l mean free path

L length

Le Lewis number

m mass fraction, molecular mass, mass per unit volume/area/length, degree of wetting

M mass, molecular weight, Mach number

$\mathcal{M}$ molality

n number, freezing fraction, number of moles

$\mathcal{N}$ porosity

N number per unit volume

Nu Nusselt number

$O()$ order of ()

P pressure, interpolation function

Pe Peclet number

Pr Prandtl number

q heat per unit area

Q heat

r radius, error

R gas constant, resistance, radius, residual

Ra Rayleigh number

Re Reynolds number

$\mathcal{R}$ universal gas constant, thaw–consolidation ratio

s specific entropy, volumetric source

S entropy, saturation ratio, slenderness ratio, salinity, area

Sc Schmidt number

Sh Sherwood number

Ste Stefan number

S_p Segregation potential

t absolute time

T absolute temperature

u specific internal energy

U internal energy, velocity

v specific volume, volume fraction, velocity

V volume, velocity

$\mathcal{V}$ velocity

w specific content

W work, thickness, velocity

x displacement, fraction

X displacement, extensive variable

y displacement

Y displacement

Z displacement

Greek letters

α angle, absorptance, dispersivity, matrix compressibility

β thermal expansion coefficient, local collision efficiency, fluid compressibility

γ extinction coefficient

Γ Gamma function

δ thickness, infinitesimal, Dirac delta

Δ finite difference, interval

ϵ emittance, superheat ratio, charge per unit mass

ζ extensive property

η similarity variable, ledge energy, recovery factor, thermal efficiency

θ temperature difference

$\varkappa$ thermal diffusivity

λ wavelength, latent heat

μ chemical potential, kinematic viscosity

ν frequency, momentum diffusivity

ξ interface location

π Peltier coefficient

Π osmotic pressure

ϱ density, reflectance (albedo)

σ Stefan number, Stefan–Boltzmann constant, surface tension

Σ summation

τ transmittance, time

T tortuosity

ϕ normalized temperature difference, potential

Φ normalized temperature difference

Subscripts

a air

A advective, atmosphere

b black, base, body

c coolant, cell

C characteristic, convective, conductive

cr critical

d droplet

D drag, dried

e embryonic, external, equilibrated

E eutectic, evaporative, electric

E energy

f fluid, film, frozen, final

ff freezing fringe

F film, frozen

g generation per unit volume

H horizontal

i ice

I incident, interface

k k-th component

K kinetic

l lens

L evaporative

m   mean, metastable, maximum, matrix, melting
M   mass
n   normal, nucleus
n   *n*-th mode
O   overburden
p   pipe, pore
P   isobaric
Q   heat
r   retained
r   radius *r*, *r*-direction
R   radiative, reflected, reference
s   substrate, solid, solute, snow, shed, soil matrix
s   *s*-direction
S   surface, sensible
S   entropy, isentropic

SP   separation
sat   saturated
t   tip, terminal
T   triple point, tangential
T   isothermal
TH   thermal
uf   unfrozen
v   vapour
v   per unit volume
V   vertical
w   water
W   work
x,y,z   *x*-, *y*- and *z*-directions
∞   bulk, distant
λ   wavelength
0   reference

Superscripts

·   per unit time
0   standard datum
*   spontaneous nucleation
i   stable equilibrium
V, L   vapour, liquid
+   transport
∧   interface
'   apparent, modified

−   mean
(e)   element *e*
f   steady state
m   molecular diffusion
d   dispersion
Q   heat
G   glass (vitrification)
A   approximation

1

Introduction

The purpose of this introduction is to develop a broad perspective within which to view ice. Accordingly, the chapter deals with origin, evolution, incidence and impact. It begins with ice in an astronomical context and an astronomical time scale, bringing us gradually to the point when the Earth attained the form it has today. Then follows a brief outline of the variable presence of ice, particularly on a large scale, in an attempt to reveal its two opposite faces: the great beauty and the immense danger. The chapter ends with some of the recorded responses of mankind to ice.

1.1   Ice in the heavens

It is now generally accepted by astronomers that the Universe evolved from a cataclysmic explosion which took place at least 10^{10} years ago. Ever since that moment, the energy and matter so released have been expanding radially outwards, eventually cooling at an asymptotic rate which, according to the tenets of thermodynamics and relativity theory, is expressed in the relation

$$T \propto 1/t^{\frac{2}{3}} \tag{1.1}$$

where T is the absolute temperature and t is the 'absolute' time. At present, the temperature in deep space is about 2.7 K thus implying that for the first 10^7 years the magnitude was in excess of 273 K: i.e., for all but the first 0.1% of its existence the Universe has been cold enough to tolerate the formation of ice, assuming that H_2O molecules were available in sufficient numbers. (Strictly speaking $T \propto t^{\frac{1}{2}}$ during the first 10^6 years but the conclusion remains the same.)

The primordial medium in the early Universe consisted largely of the first two elements, hydrogen and helium, produced from nucleosynthesis, but it was not until the medium had given rise to galactic formations in which stars were born, evolved, and died that the helium nuclei began

Table 1.1. *Materials of the Universe*
(following Allen (1973) and Illingworth (1985))

Relative cosmic abundances (by mass)

H	7.3×10^{-1}	Na	4×10^{-5}
He	2.5×10^{-1}	Mg	5×10^{-4}
Li	4×10^{-8}	Al	6×10^{-5}
Be	8×10^{-10}	Si	7×10^{-4}
B	6×10^{-9}	P	5×10^{-6}
C	3.0×10^{-3}	S	4×10^{-4}
N	1.0×10^{-3}	Cl	3×10^{-6}
O	8.0×10^{-3}	A	7×10^{-5}
F	5×10^{-7}	K	2×10^{-6}
Ne	1.0×10^{-3}	Ca	4×10^{-5}

Some components of the interstellar medium

OH^-	CN
H_2O	HCN
CO	NH_3
H_2CO	HN_2^+
CH_3OH	CH_4

interacting to produce other elements, notably ^{12}C and ^{16}O. Table 1.1 lists some of the elements created and then distributed by supernova explosions and stellar winds (Allen, 1973; Illingworth, 1985). Also listed in Table 1.1 are some of the molecules and ions subsequently formed as the interstellar medium cooled. Cosmic dust is an important component of the interstellar medium. It consists of carbon, silicate and iron particles which swarm into dense clouds, at which point they are observed to possess icy mantles, the most primitive and most widespread form of glaciation in the Universe.

After gravity had generated the galaxy, and thence a variety of stars, the gas and dust remaining in the vicinity of the Sun gradually transformed itself by accretion into the solar system we know today. In the condensation process, the materials with the highest evaporation temperatures (refractories) formed first and may have grown into fairly large masses before the later volatile components formed grains; turbulence kept them mixed. The mechanism by which the pre-planetary grains grew into planetesimals and eventually coalesced into the present planets is unknown, but it appears to have led to planetary structures in which refractories were concentrated in the core while volatiles remained closer to the surface.

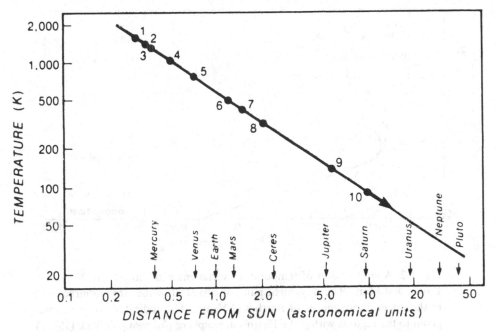

Fig. 1.1. Indicated in this diagram are the temperatures and locations at
which major planetary constituents would be expected to condense from the
primordial solar nebula: 1, refractory minerals like the oxides of calcium,
aluminium, and titanium, and rare metals like tungsten and osmium; 2,
common metals like iron, nickel, cobalt, and their alloys; 3, magnesium-rich
silicates; 4, alkali feldspars (silicates rich in sodium and potassium); 5, iron
sulphide; 6, the lowest temperature at which unoxidized iron metal can exist;
7, hydrated minerals rich in calcium; 8, hydrated minerals rich in iron and
magnesium; 9, water ice; and 10, other ices, (following Lewis (1982)).

Important to the understanding of planetary composition is the relation
between the condensation temperatures of various constituents and the
corresponding distance from the sun (Lewis, 1982). Fig. 1.1 illustrates this
relation from which it is evident that structural planetary ice in any abundance may not be expected within the orbit of Jupiter. Equally important,
at least from the point of view of the terrestrial planets, is the development
and evolution of an atmosphere (Pollack, 1982a). It seems likely that
outgassing of volatile-laden minerals was the primary atmospheric source,
although it is also possible that the solar nebula, the solar wind and
collision with volatile rich comets may have made contributions. Initially,
the gases vented were chiefly water vapour, carbon dioxide, carbon monoxide, hydrogen and nitrogen, as Table 1.1 might suggest, but their continued participation in the energy exchange processes at the surface of the
planet gradually produced substantial changes which varied from planet to

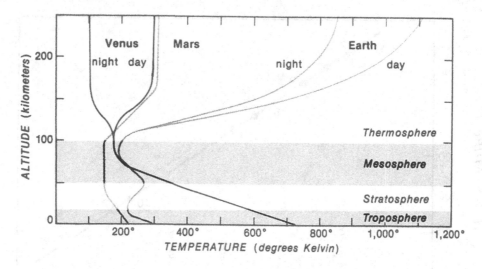

Fig. 1.2. A comparison of temperature variations with altitude for Venus, Earth, and Mars. The day–night pairs of curves show the strong diurnal cycle in the upper atmospheres of Earth and Venus. Also indicated are the names given to the regions within the Earth's atmosphere (following Pollack (1982a)).

planet. Volcanism, the loss of hydrogen to space, the accumulation of inert nitrogen, the formation and condensation of water vapour and the chemical reactions of carbon dioxide and carbon monoxide with water vapour and rock, all led to atmospheric evolution during approximately 5×10^9 years of planetary existence.

The atmospheres of the Earth and its two nearest neighbours currently contain many gases in common but their proportions differ (Pollack, 1982a). Perhaps not surprisingly they all contain H_2O, though in very small amounts: about 1% for Earth, and much less for Mars and Venus. Fig. 1.2 shows the variation of temperature with altitude for these planets, and makes it abundantly clear that H_2O will not remain in vapour form at every altitude; this is particularly true of the low altitudes on Mars and Earth. Some of the condensed water vapour is, or was, present as water but much of it exists as ice, especially in the polar regions of these planets. Fig. 1.3 shows the spiral form of Mars' north polar ice cap on which wind erosion and vapour deposition appear to have alternated in the production of layers of ice and particulates. The Martian regolith, the surface soil and rock, is also a likely location for condensed water vapour, probably in the form of extensive permafrost; water may have flowed in the past, and may still exist beneath and within the permafrost (Masursky, 1982; Krass, 1984).

Further out in the solar system are Jupiter and Saturn which, although

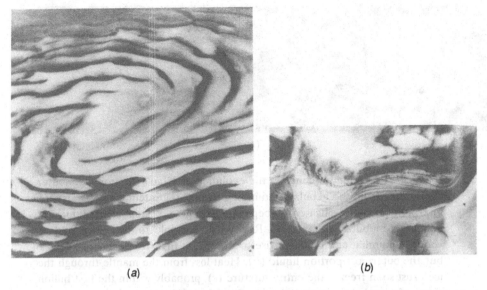

(a)   (b)

Fig. 1.3. From its high-inclination orbit, the Viking 2 orbiter photographed the involved spiral of Mars' north polar water-ice cap seen in (a). Visible in (b) is a cliff in the Martian north polar cap; strikingly layered deposits of dust and ice have been eroded, overlain by fresher deposits of ice, then eroded again (following Pollack (1982a)).

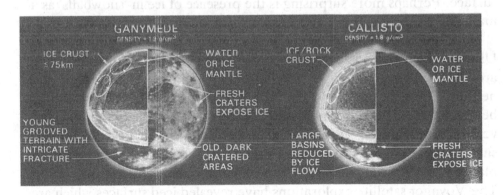

Fig. 1.4. These schematic illustrations portray the satellite interior as presently understood (following Johnson (1982)).

possessing no ice themselves, are in close proximity to it (Johnson, 1982; Krass, 1984). Two of the moons of Jupiter, Ganymede and Callistro, are not only covered in ice but also have a deep mantle of ice or water: over one-third of their volume is H_2O. As Fig. 1.4 indicates, an ice/rock crust has developed above the mantle as a result of meteoric impact over an extended period of time. Fig. 1.5 shows a possible sequence of events in

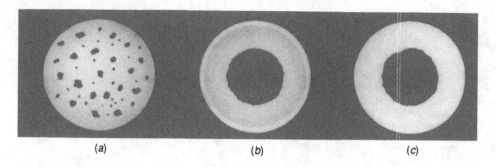

(a)   (b)   (c)

Fig. 1.5. A possible sequence in the evolution of Titan's interior. Soon after condensing from the nebula surrounding Saturn, the satellite's heterogeneous mixture of ices and silicates (*a*) begins to segregate, as heat created during formation mobilizes the interior. Rocky material sinks to the centre, and water, ammonia, and methane rise to the top; at first, residual heat keeps all but the outermost portion liquid (*b*). Heat loss from the mantle through the ice crust soon freezes the entire mixture (*c*), probably within the first billion years of Titan's existence (following Pollack, (1982b)).

the formation of Titan, Saturn's largest moon (Pollack, 1982b). The density of the ice crust may be expected to increase with depth, so much so that ices I, II, V and VI[1] are likely to exist in concentric rings beneath the surface. Perhaps more surprising is the presence of ice in 'snowballs' as a major constituent in the A and B rings surrounding Saturn (Burns, 1982).

Near the outer edge of the solar system are the planets Neptune and Uranus which also possess rocky cores surrounded by 'ice' mantles; they are covered in a 'crust' of gravitationally compressed hydrogen and helium. The state of the H_2O in these mantles is not yet known but it is believed to be a highly-compressed, ionized fluid capable of thermal convection and therefore capable of generating a magnetic field. Icy worlds such as these are common in the outer solar system, as evidenced by the planet Pluto and the moons of Uranus (Morrison and Cruikshank, 1982). The Voyager satellite explorations have revealed iced surfaces which are not only pitted by meteoric impact but are still in the process of geologic re-surfacing, even by water volcanoes.

Perhaps the most spectacular icy bodies in the solar system are the comets of which there are well over 500 (Brandt, 1982). They are best known for their long tails of plasma or dust, but their head, or coma, which consists of a sphere of gas and dust often 10^5–10^6 km in radius, contains a nucleus. This nucleus, which is believed to be the source of all cometary gas and dust, is composed mainly of ice, as suggested in Fig. 1.6. As the Sun is approached, heating of the nucleus establishes a sublimation

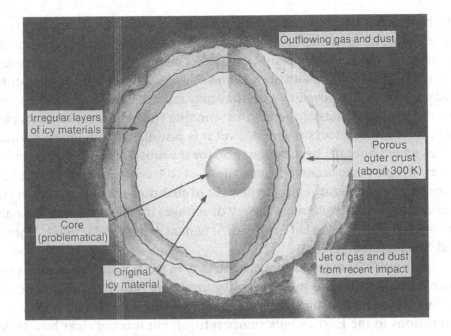

Fig. 1.6. Fred Whipple's 'icy conglomerate' nucleus model, as extended by Armand Delsemme (following Brandt (1982)).

process just below the accumulated dust of the crust, thereby generating the coma and ultimately the tail. Since H_2O is believed to be the principal ingredient of the nucleus it is estimated that the sublimating surface, which may have a radius of 10^2–10^4 m, must have a temperature of about 215 K when it is near the Earth's orbit.

1.2 Ice ages

An elementary heat balance on any body in the solar system reveals that

$$T_b = T_s \left(\frac{R_s}{2R_b} \right)^{\frac{1}{2}} \tag{1.2}$$

where T_b, T_s are the uniform[2] surface temperatures of the body and the Sun, respectively, and R_b, R_s are the distances of the body and the Sun's surface, respectively, from the centre of the Sun. This simple expression indicates that the Earth's average equilibrium temperature should be about 280 K, which is a surprisingly good estimate carrying with it the implication that water should be the prevalent form of H_2O on this planet. Such a prediction is too coarse to accommodate the spatial effects of geometry and atmosphere and the transient effects of spin, orbit and axis tilt, but it does lead naturally to the question of what would happen to the huge amounts of planetary water should the surface temperature over substantial regions of the planet ever fall a mere 10 K below the average.

Everyone is familiar with diurnal and seasonal variations in temperature and the freeze–thaw cycles which they may induce. Less familiar, perhaps, are the precise details of the Earth's spinning and orbital movements and their fluctuating effect on climate; yet it is just such astronomical period-icities which offer the best explanation for the substantial alterations in ice cover which have occurred over the past 10^9 years and, in particular, during the last million years. In the light of such time scales it is sobering to realize that the astronomical theory of ice ages has been formulated only during the past 150 years (Croll, 1875; Milankovitch, 1941; Hays, Imbrie and Schackleton, 1976).

Equation (1.2) suggests that eccentricity in the Earth's orbit could pro-duce annual variations in the equilibrium temperature, at least near the surface (a period of one year is not long enough to permit substantial alterations in the Earth's bulk temperature), but it is not clear how these might lead to a longer term change. Yet climatic records of the past 500 000 years indicate a close correlation between climate and eccen-tricity, measured as the interfocal distance divided by the major axis (Hays *et al.*, 1976). It thus appears that the influence of other bodies in the solar system produces a cyclic variation in eccentricity, the amplitude being about 6% and the period being of the order of 10^5 years, but the precise mechanism that fosters the widespread growth of ice remains uncertain.

Also important is the Earth's spin or, more precisely, the axis about which the spinning occurs. If this axis lay normal to the ecliptic plane i.e., the plane in which the Earth rotates about the Sun, its daily effect would be periodic and its seasonal effect would be absent. The fact that the axis is currently tilted about 23.5° from the normal to the ecliptic creates differential heating in relation to the plane in which the tilt takes place. At the two orbital positions where the plane is intersected by the Sun, the solstices[3], the differential is greatest: days are shorter and colder at the pole further from the Sun. The solstice plane does not coincide with the major axis of the orbit and is currently displaced from it by about 13 days.

The seasonal effect of axis tilt is plainly evident each year; it is in fact greater than the annual effect of eccentricity (currently about 1%). Obser-vations of tilt reveal that it has not remained constant over the years; thus other bodies in the solar system once again make their presence felt (Berger, 1977), by altering the amplitude of the axis tilt and by rotating the axis about the normal to the ecliptic. The first of these cyclic effects is a

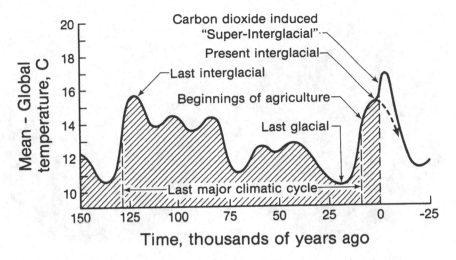

Fig. 1.7. Past and future climate. According to the astronomical theory of ice ages, the natural course of future climate (shown by the dashed line) would be a cooling trend leading to full glacial conditions 23 000 years from now. The warming effect of carbon dioxide, however, may well interpose a 'superinterglacial.' The next ice age would be delayed until the warming had run its course, perhaps 2000 years from now (following Mitchell (1977)).

variation of about 1.5° over a period of approximately 42 000 years. The second, the precession, has a period of approximately 23 000 years (Hays *et al.*, 1976).

Evidently movements elsewhere in the solar system have a predictable, if complex, effect on the density of the radiative flux incident at each point on the Earth's surface, but it is clear that although the radiative flux may trigger climatic change, and will continue to play an important part in it, the climate itself is not likely to respond passively. With respect to water in particular, alterations in the heat balance, locally and globally, lead to variations in circulation and evaporation, and these in turn will combine with altered atmospheric temperatures to vary winds and precipitation rates. The precipitation of snow with its high albedo alters the radiative balance yet further and thus augments the very cooling effect which produced the snow. Such dynamic, interactive effects may be seen every year and it is not difficult to understand how the annual growth and decay of massive planetary ice could shift its balance in response to astronomical perturbations. Fig. 1.7 uses the Earth's mean global temperature to reflect climatic change over the past 1.5×10^5 years (Mitchell, 1977) and shows how superposition of the three principal disturbances produces periodic glacial conditions.

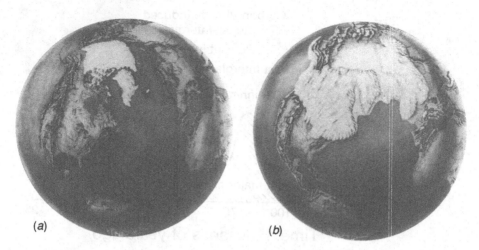

(a)   (b)

Fig. 1.8. Earth today during (*a*) a hot northern summer and (*b*) a summer of
the last ice age. Twenty thousand years ago, great ice sheets covered parts
of North America, Europe, and Asia; surface waters of the arctic and parts of
the North Atlantic Oceans were frozen; and the sea level was 350 ft lower than
it is today. Many parts of the continental shelf, including a corridor between
Asia and North America, became dry land. (Drawing by Anastasia
Sotiropoulos, based on information compiled by George Denton and other
members of the CLIMAP project.) (Following Imbrie and Imbrie (1979).)

During most of the Earth's 5 billion years of existence the astronomical
theory of ice ages may not apply because of the enormous structural
changes which have occurred in our planet. There is, however, some
geological evidence of glacial ages during the last billion years (Imbrie and
Imbrie, 1979). These evidently occurred 250–320 million years ago, in the
Permo-Carboniferous period of the Paleozoic era, and 600–800 million
years ago in the Pre-Cambrian era; the present (late Cenozoic) glacial age
began some 10 million years ago. The best current explanation of these
glacial ages relies on the theory of continental drift from which it appears
that only during certain geological periods has there been a substantial
portion of the Earth's land surface located near the poles (Imbrie and
Imbrie, 1979). Then, as now, astronomical variations have a greater effect
on climate which, as mentioned earlier, may be expected to respond in a
non-linear manner, particularly at high latitudes and high elevations. During the Permo-Carboniferous glacial age a huge section of the Earth's
surface was covered by the supercontinent of Gondwana, whose southern
tip included the South Pole. It was then that glaciation occurred in what
are now South America, Africa, India, Australia and Antarctica. The
withdrawal of the substantial land mass from the South Pole brought an

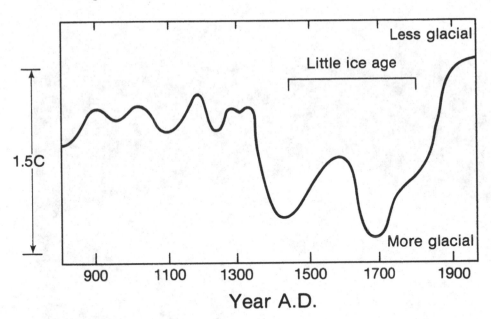

Fig. 1.9. Climate of the past 1000 years. The graph is an estimate of winter conditions in Eastern Europe, as compiled from manuscript records. During the Little Ice Age (AD 1450–1850), mountain glaciers all over the world advanced considerably beyond their present limits (following Lamb (1969)).

end to that glacial age, the Earth then being much warmer until 55 million years ago when, in the declining Cenozoic era, Gondwana evidently broke up into the continents of today and thus spawned the Antarctic land mass at the South Pole and the North American and Eurasian land masses encroaching on the North Pole.

The present glacial age has been in existence since the Miocene period (which began about 20 million years ago) when the Antarctic ice sheets first formed and glaciers began to appear high in the northern hemisphere. In the Pliocene period, which immediately followed, ice sheets were also formed near the North Pole (about 3 million years ago), and appear to have been there ever since, fluctuating in extent in accordance with the three principal frequencies. During the recent past, the climatic record abounds in these astronomical periodicities which reveal, among other things, that the most recent period of glaciation occurred just 20 000 years ago; its effect is illustrated in Fig. 1.8. This suggests that we are currently at the centre of an interglacial period, as Fig. 1.7 indicates. The very recent past is more difficult to analyse because additional minor effects become relatively more important; the response of the biosphere may indeed be more than minor. Geological records enable us to construct the most

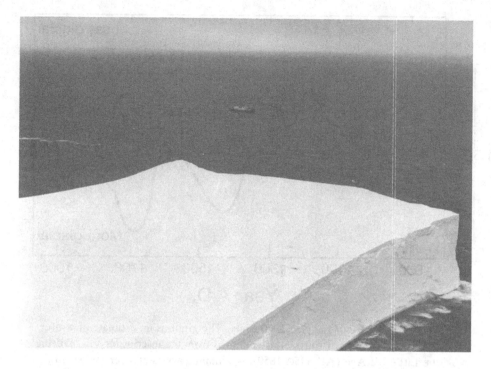

Fig. 1.10. Overlooking one edge of a tabular iceberg in the Southern Ocean. Beyond is the research vessel *Endurance* (courtesy V. Squire).

recent 10000 years of climate (Imbrie and Imbrie, 1979), and the written historical record may be used to study the latest millennium (Lamb, 1969). The latter record, shown in Fig. 1.9, reveals the so-called Little Ice Age with its cold spells early in the fifteenth century and late in the seventeenth century; it also indicates the subsequent warming trend later documented in detailed meteorological records.

1.3   The incidence of ice

The Saracen knight in Scott's tale *The Talisman* listened in profound disbelief as the Christian knight told of riding his steed across a lake. But the Saracen knight had never seen or heard of a northern lake during winter. Readers of this book are unlikely to need convincing that such a crossing is not only possible but not in the least extraordinary. Even so, before entering upon a scientific discussion of ice, it is worthwhile giving an overview of its widespread occurrence, and attempting to reveal, albeit in small compass, how variable are its forms, how beautiful they may be and how threatening.

The surface of this planet is well endowed with water, most of it frozen.

Huge tracts of land in the polar and subpolar regions, and even in more temperate zones, are underlain by frozen soil and rock. Almost two-thirds of the area of ice covering the surface at any given time lies on top of the seas, the remaining fraction consisting of ice caps and ice sheets; it is the latter which cover about 10% of the Earth's land surface and contain over 75% of the Earth's fresh water. Every year a layer of snow is deposited on these ice covers and on contiguous regions which are at least as large again. The snow, of course, originates in the atmosphere which also generates and deposits other forms of ice such as frost and hail. This great amount of ice, so variable in its presence and form, has done much to shape the surface of the planet. It has also exerted a considerable influence on the adaptation of many forms of life, from the microscopic to the mammoth, and not least on human society itself.

The impact of ice is felt annually by many people, notably in the northern nations of the American, European and Asian continents. Awareness of ice is obviously a function of season and lifestyle, and ranges from the immediate to the remote. In general terms, the most recurrent symbol of ice in water is perhaps the iceberg with its majestically sculpted flanks and towering appearance. Fig. 1.10 is but one illustration of this sea giant. Other massive forms of frozen water may be equally impressive and among these the frozen spray of the Gullfoss Waterfalls, shown in Fig. 1.11, is particularly spectacular.

Ice in and on water, saline or fresh, may also take forms which are not enjoyed for their appearance; more likely they will be regarded as an inconvenience, a nuisance, a threat and even an extreme hazard. The effect of a burst domestic water pipe is familiar enough, and it is not difficult to imagine how a particularly severe winter might affect the water supply of thousands of inhabitants in a northern town. Less familiar, perhaps, is the potential role of the tiny ice crystals – frazil – which may appear in enormous numbers in open, turbulent water. By choking intakes, frazil has been known to shut down a 30 MW hydro-electric station in less than one hour (Schaefer, 1950). On a grander scale, ice covers on lakes, rivers and seas create a sinister and swiftly changing threat to transportation. There are many instances of ships being holed or crushed by ice. Only specially designed vessels may venture safely in such waters. Freezing of a large body of water imposes obvious restrictions but these may be no greater than those which occur during melting, which is equally beyond human control. The freeze up and break up periods in many northern rivers, for example, may both be accompanied by massive ice jams.

Ice in the atmosphere seldom occurs in such massive forms, its origins

Fig. 1.11. Frozen spray at Gullfoss Waterfalls (following Bardarson (1974)).

being found in surface evaporation, diffusion and advection, all of which ensure a widespread distribution of water vapour. Condensation of this water vapour, either directly into ice crystals or into water droplets which subsequently freeze, leads to a variety of ice forms. Most of these are either eye-catching in their own right or produce atmospheric phenomena which send photographers, amateur and professional alike, in search of their cameras. The most familiar examples are provided by snow crystals, which have been so painstakingly recorded (Bentley and Humphreys, 1962). Fig. 1.12 shows several photographs of snow crystals and provides a glimpse of the infinite number of variations which are possible on such a simple geometric theme, no two snow crystals being identical. Ice crystals themselves are beautifully manifest either through an individual sparkle, or collectively in phenomena such as the parhelion (Minnaert, 1954). Almost everyone is familiar with grotto-like icicle hangings, and with the

Fig. 1.12. Some regular snow crystal shapes (following Bentley and Humphreys (1962)).

Fig. 1.13. Shipboard icing (courtesy L. Makkonen, US Coast Guard and W. P. Zakrzewski).

Fig. 1.14. Upper part of ablation zone of Gilkey Glacier (Alaska), a main
trunk made up of contributions from over a dozen different tributaries and
many medial moraines. Ogives, arch-shaped dirt bands, also known as
'Forbes Bands', can be seen on several of the ice streams pointing
down-glacier. (Photo taken during Juneau Icefield Research Project.)

foliage engravings which occasionally appear on a window pane as compen-
sation for the underheating of a room in winter.

There is, however, another side to atmospheric icing which may be just as
threatening as hoarfrost is enchanting. There is no need to emphasize the
dangers of a snow blizzard, of driving sleet, or of a severe hail storm, but the
impact of ice accretion is perhaps less well known. This form of icing arises
when supercooled droplets of water collide with and then freeze upon a solid
object. In supercooled clouds, for example, engine carburettors have been
choked and aircraft wings have had their aerodynamic characteristics
dramatically altered, all to the frustration of the pilot and the jeopardy of
passengers. On the ground, fog and drizzle may produce similar effects on
road and rail vehicles; nor are building structures immune. Icing of overhead power lines may be heavy enough to snap them. Perhaps most dramatic of all is accretion which occurs on a ship's superstructure as a result of
spray tossed up in heavy weather. This is illustrated in Fig. 1.13. Many a

Fig. 1.15. Icing on a road at Norman Wells caused by water issuing from the ground between the freezing active layer and the permafrost, and then freezing in low air temperatures (following Brown (1970)).

vessel has accumulated so much ice that, listing badly and losing manoeuvrability, it has finally foundered (Lock, 1972).

Ice in and on the earth also reveals the two familiar faces. There are many photographs of snow drifts which, like sand dunes in a desert, exhibit the many beautiful sculpting effects of wind. Less easy to capture are the prismatic refractions of individual snow crystals, which are almost lost in the reflected glare of the drift. On a very much larger scale, the edge of an ice sheet may become the calving ground of icebergs, thus providing an awe-inspiring sight, while a large glacier seen from a distance imposes its own majesty. Fig. 1.14 is a representative example of the latter.

Within the earth itself, the presence of ice is not always obvious, but its effects are frequently seen. Alternation of freezing and thawing produce an unsymmetrical response in the terrain and all that it supports. On a large scale, these effects have led to such land forms as patterned ground and pingos, while on a smaller scale they have given rise to such uneven distributions of ice that local variations in surface elevation have become quite marked. Heaving and slumping beneath buildings have an all-too-obvious effect on structural integrity, but such an effect is equally signifi-

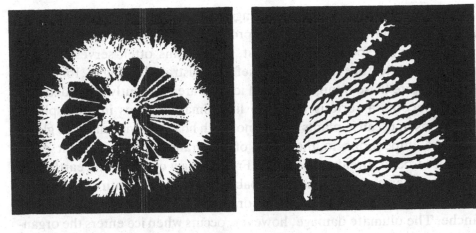

Fig. 1.16. Architecture seen through hoarfrost (following Bentley and Humphreys (1962)).

cant under a road, a rail bed, or a runway. It is also important beneath a pipeline carrying oil, gas or water. However, the thermal regime is not the only threat to transportation and communications, as Fig. 1.15 illustrates. This shows the effect of partial blockage on groundwater which was thereby forced to seep out and freeze during winter (Brown, 1970). The effect may be produced naturally, through unusual climatic variations or, as illustrated, through human alterations to the surface cover.

As mentioned earlier, the presence and variability of ice causes life forms to adapt. The precise form of this adaptation depends upon the particular type of life, the particular type of ice and the circumstances under which they meet; it is therefore impossible to generalize, except perhaps to distinguish between ice on the outside of an organism and ice on the inside. Ice in general, and snow in particular, are an important and integral part of the northern ecology. The insulating property of snow, for example, enables many creatures, ranging in size from insect to polar bear, to spend the winter safely under its protection. More passive, perhaps, is an organism's response to hoarfrost, though this often serves to bring intricate architecture into sharp relief, as illustrated in Fig. 1.16.

But despite these benign effects of ice, we are constantly reminded that it may become a serious threat to life. Small changes in atmospheric conditions may cause plants to coat not with hoar but with rime, its glassy structure impervious to the gases of respiration and its weight great enough to bend, break and flatten. Freeze–thaw cycles in the earth alter the root systems of trees and are capable of moving the entire tree bodily, producing a drunken forest. More dramatic still is the effect of an avalanche. The ultimate damage, however, occurs when ice enters the organism; in higher animals generally, and man in particular, the effect is usually lethal.

1.4   The response to ice

As a whole, human beings have always lived with ice but they have not always regarded its existence and meaning in the same way. No human records of the last glacial age remain but the ancient world made many pronouncements. Aristotle, for example, laying the groundwork for atmospheric science, noted that '. . . from the clouds there fall three bodies formed by cooling: water, snow and hail . . .' and '. . . snow is the same as frost . . .' (*Meteorologica*); and to him must be attributed the words which began a debate that still continues: 'If the water has been previously heated, this contributes to the rapidity with which it freezes: for it cools more quickly.' About the same time, Han Ying in China observed that '. . . (flowers) of snow are six-pointed.' Much later, Kepler turned away from his telescope to study the mystery of the snow crystal's hexagonal shape, and published a detailed, though inconclusive, account of his thoughts (Kepler, 1611). More recently, ice has attracted the attention of such eminent scientists as Tyndall, Kelvin, Faraday, Stefan, Langmuir, Pauling, Onsager and Shumskii.

Scientists alone do not provide a very accurate assessment of the significance of ice in our lives: theirs is the dispassionate view driven by curiosity. Poets are usually closer to the mark. 'Blow, blow, thou winter wind,' wrote Shakespeare in *As You Like It*; and in the same passage:

Freeze, freeze, thou bitter sky
That dost not bite so nigh
　As benefits forgot:
Though thou the waters warp,

Thy sting is not so sharp
As friends remember'd not.

Lament not the coming of ice.

In the late nineteenth century, American William Cullen Bryant (Bentley and Humphreys, 1962) wrote this poem called *Snow*:

Here delicate snow-stars, out of the cloud,
Come floating down in airy play,
Like spangles dropped from the glistening crowd,
That whiten by night the Milky Way.

Contrast that romantic view with this extract from a poem of the same title written by an unknown medieval Welsh poet (Clancy, 1965).

No sleep, no leaving my house:
This causes me discomfort.
No world, no ford, no hillside,
Nowhere clear, no land today.
No girl's promise will lure me
Out of my house into snow.
A plague, these plumes cling on clothes
As if they played dragon.
My clothing is my excuse:
It looks much like a miller's.
Isn't it true, after New Year's,
Everyone's dressed in ermine?
In the first month of the year
God's busy making hermits.

Whether we play with snow or are forced to live with it makes all the difference.

The Canadian poet Sir Charles Roberts (*Selected Poems*, 1974) penned this exquisite description of hoarfrost on a window.

One night came Winter noiselessly, and leaned
Against my window pane.
In the deep stillness of his heart convened
The ghosts of all his slain.

Leaves, and ephemera, and stars of earth
And fugitives of grass,
White spirits loosed from bonds of mortal birth,
He drew them on the glass.

This perception of ice and snow was not always shared by the explorers in search of the Northwest Passage. In the narrative poem *John Franklin, His Enterprise*, St Maur writes of starving, exhausted trekkers:

... the freezing wind
          that whines its long complaint,
matching our every effort
          with rebuff and stopping
our slow deliberate steps
          with each capricious breath that swirls
the gentle snowflakes
           tiny crystals of ice
until they pile up
           millions high
in walls of solid white.

From one period of history to another, and from one country to another, the existence and meaning of ice has brought forth a full spectrum of human response. E. J. Pratt, in 'The Titanic' (*Collected Poems*, 1958), starkly gives us the iceberg;

... 'nothing but the brute
and paleolithic outline of a face ...
... waiting a world-memorial hour, its rude
Corundum form stripped to its Greenland core.

While the Japanese poet Shunzei, writing centuries earlier in the traditional tanka form (Brower and Miner, 1961), finds a state of mind in a stream freezing over:

Now here frozen over
Now there just fleeting from the grip of ice,
    The stream between the hills
Is choked within its rocky channel
And sobs its suffering in the winter dawn.

But it was none other than Robert Frost who declared in 'Fire and Ice' (*The Poetry of Robert Frost* (1969)) that:

Some say the world will end in fire,
some say in ice.
From what I've tasted of desire
I hold with those who favour fire.
But if it had to perish twice,
I think I know enough of hate
To say that for destruction ice
Is also great
And would suffice.

It is fitting to leave the last word on ice to those who, like their ancestors, have lived almost their entire lives in its presence. Until recently, their lan-

guages did not have a written form and therefore their poetic heritage has
been in the oral tradition. It is too early to judge Inuit poetry but there is
every reason to believe that it will produce the finest of human responses to
ice. It should also generate the broadest response, and within it we may
expect the greatest subtlety. A sampling of a Kangiryuarmiut–English
dictionary (Lowe, 1983), for example, reveals the following entries on
snow:

first snow	sugar snow
falling snow	fine sugar snow
falling powered snow	snowdrift
fallen snow	long snowdrift
snowflake	snowdrift shaped like a duck's head
snow to make water	snow blowing along a surface
light soft snow	whiteout
fresh soft snow	melting snow
frosty sparkling snow	

In general, the English description is only possible with the heavy use of
adjectives or other types of qualification. The Inuit have the natural language, formed through daily experience, first hand, and evolved over
many centuries (Freeman, 1984).

The chapters that follow deal only with scientific and technological aspects
of ice. Treatment is at the graduate level, although no attempt has been
made to provide exhaustive surveys of research activity. Wherever possible, emphasis has been placed on conventional wisdom in order to provide a synthesis. After two chapters on the thermodynamics and mathematics of ice formation, there are four specialized chapters, each dealing
with ice in a particular context. In keeping with the outline given in Section
1.3, they deal with water, air, earth and life. The book ends on a warming
note.

The main purpose of the four principal chapters – 4–7 – is to introduce
the reader, research student and practitioner alike, to various types of ice
growth. This is done in terms of the mathematics, physics and biology
appropriate to the context, thus permitting the development of models
consistent with our current understanding. Completion of any of these
chapters should enable the reader to apply the knowledge immediately in
that particular field or, in the context of graduate work, to begin research
on a specialized but advanced topic. To read the entire book is to embark
on what I have found to be an enlightening journey.

2

Thermodynamics of ice

Growth and decay of ice are both governed by the laws of nature and, in particular, by the laws of thermodynamics. This chapter begins with an outline of classical thermodynamics as it applies to ice, water and water vapour when treated as continua. The discussion continues with a brief review of the most important thermodynamic variables, the Gibbs function in particular. Later sections on diffusion and metastability extend the discussion beyond the limitations of reversibility and stable equilibrium.

Despite the power and versatility of the continuum approach it offers limited insight into many theoretical and practical matters. To offset this disadvantage, several sections are devoted to microscopic or molecular analysis. Each has the aim of contributing to our understanding of interface growth or decay; collectively they cover a wide range of topics: the structure of the H_2O molecule and its ensembles; thermal properties; the nature of nucleation; and the process of crystal growth.

The use of vectors enables many of the expressions developed in this chapter to be written in succinct forms which readily lend themselves to a general interpretation. In the interest of brevity, therefore, free use has been made of vector calculus along with the elements of differential and integral calculus. While this is common practice in many branches of science and engineering, it may pose some difficulties in others. In the event, the reader may choose between two paths: to review the rudiments of mathematics in an appropriate textbook (e.g., Aris (1962)); or to accept the statement given and proceed to its application later in the book. Wherever feasible, key mathematical statements are accompanied by verbal statements describing the corresponding physical truth, and it should thus be possible to follow the gist of the argument without detailed mathematical knowledge.

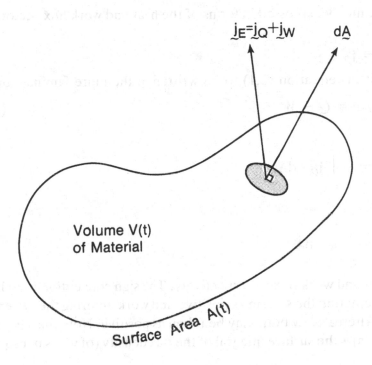

Fig. 2.1. Energy leaving a volume of material: energy leaves the volume $V(t)$ through the surface area $A(t)$. The energy flux consisting of heat and work is

$$\int_{A(t)} \mathbf{j}_E \cdot d\mathbf{A} = \int_{A(t)} \mathbf{j}_Q \cdot d\mathbf{A} + \int_{A(t)} \mathbf{j}_W \cdot d\mathbf{A}$$

2.1 Classical laws and principles

The First Law of Thermodynamics may be stated in a number of ways, each of which either implies or leads to a definition of energy. Perhaps the most important consequence of the law, here stated as a principle, is that energy is conserved. Consider the energy leaving the volume $V(t)$ of material shown in Fig. 2.1. Since energy is conserved, the rate of depletion is equal to the exiting flux. Written mathematically for a continuum, the *energy principle* is

$$dE/dt = - \int_{A(t)} \mathbf{j}_E \cdot d\mathbf{A} \qquad (2.1)$$

where E is the energy of the body of given material bounded by a surface $A(t)$, and $\mathbf{j}_E$ is the flux density of the energy leaving the body. This latter term is commonly expressed in terms of the heat and work flux vectors by taking

$$\mathbf{j}_E = \mathbf{j}_Q + \mathbf{j}_W$$

thus permitting equation (2.1) to be written in the more familiar form

$$dE/dt = \dot{Q} - \dot{W} \tag{2.2}$$

where

$$\dot{Q} = -\int_{A(t)} \mathbf{j}_Q \cdot d\mathbf{A}$$

and

$$\dot{W} = \int_{A(t)} \mathbf{j}_W \cdot d\mathbf{A}$$

are the heat and work fluxes, respectively. The sign convention used here takes heat entering the system as positive and work entering the system as negative. Alternatively both may be taken as positive. Note that the total flux (in watts) is the surface integral of the flux density (of watts per square metre.)

The Second Law of Thermodynamics has also been formulated in a variety of ways each of which leads to a definition of entropy. The corresponding *entropy principle* is

$$dS/dt = -\int_{A(t)} \mathbf{j}_S \cdot d\mathbf{A} + \int_{V(t)} \dot{s}_g dV \tag{2.3}$$

where S is the entropy of the body and $\mathbf{j}_S$ is the outward entropy flux density. This statement reveals that the rate of depletion is not generally equal to the exiting flux and therefore entropy is not conserved unless the generation rate $\dot{s}_g$ is identically zero; in general, $\dot{s}_g \geq 0$. Generation here is represented as a volumetric source. The entropy flux is related to the heat flux at a point by the expression

$$\mathbf{j}_Q = T\mathbf{j}_S \tag{2.4}$$

where T is the absolute temperature.

Paralleling the description of energy and entropy is the description of matter itself. This may be embodied in the *species principle*

$$dM_k/dt = \int_{V(t)} \dot{m}_{gk} dV \tag{2.5}$$

in which M_k is the mass of species k and $\dot{m}_{gk}$ is its volumetric generation rate resulting from chemical reactions etc. Since the total mass $M = \sum_k M_k$ is taken to be fixed, the surface flux term is absent from the right hand side of equation (2.5), thus implying that

$$\frac{dM}{dt} = \frac{d}{dt} \sum_k M_k = \int_{V(t)} \sum_k \dot{m}_{gk} \, dV = 0 \qquad (2.6)$$

In other words the total mass of the given body is conserved but the mass of any particular species within it may increase, though only at the expense of the others.

It is often useful to re-write the above principle using Reynolds' transport theorem

$$\frac{d}{dt} \int_{V(t)} f \, dV = \int_{V(t)} \frac{\partial f}{\partial t} \, dV + \int_{A(t)} f \mathbf{V} \cdot d\mathbf{A} \qquad (2.7)$$

where f is any extensive material property expressed per unit volume. For a fixed volume, this has the interpretation that the total change of the property within the volume is the sum of two separate contributions: internal changes of f with respect to time; and changes attributable to material (having f) crossing the boundary surface. When the property is mass, $f = \rho$, and therefore equations (2.6) and (2.7) dictate that

$$\frac{dM}{dt} = \int_{V(t)} \left(\frac{\partial \rho}{\partial t} \right) dV + \int_{A(t)} \rho \mathbf{V} \cdot d\mathbf{A} = 0$$

Hence

$$\int_{V(t)} \left(\frac{\partial \rho}{\partial t} + \nabla \cdot \rho \mathbf{V} \right) dV = 0$$

or, as the volume shrinks to a point,

$$\frac{\partial \rho}{\partial t} + \nabla \cdot \rho \mathbf{V} = \frac{d\rho}{dt} + \rho \nabla \cdot \mathbf{V} = 0 \qquad (2.8)$$

This is the continuity equation in which

$$d\rho/dt = \partial \rho/\partial t + \mathbf{V} \cdot \nabla \rho$$

is differentiation[1] following the motion.

This result may now be combined with equation (2.7). When $f = \rho e$, where e is the energy per unit mass, equations (2.7) and (2.8) may be used to convert equation (2.1) into the point form

$$\rho(\partial e/\partial t + \mathbf{V} \cdot \nabla e) = -\nabla \cdot \mathbf{j}_E \qquad (2.9)$$

Similarly, equation (2.3) converts to

$$\rho(\partial s/\partial t + \mathbf{V} \cdot \nabla s) = -\nabla \cdot \mathbf{j}_S + \dot{s}_\text{g} \tag{2.10}$$

where s is the specific entropy. Equations (2.8), (2.9) and (2.10), together with the corresponding equation of motion, form the basis of the continuum approach to ice formation and decay.

2.2   The Gibbs function and phase change

For an interval of time dt, equation (2.2) may be re-stated[2] as

$$dE = đQ - đW$$

while equations (2.3) and (2.4) combine to give

$$dS/dt = \dot{Q}/T + \dot{S}_\text{g}$$

where T is the boundary temperature, taken to be uniform, and $\dot{S}_\text{g}$ is the total entropy generation rate within the boundary. Hence

$$đQ = TdS - T\dot{S}_\text{g}\,dt$$

which, together with the above expression for dE, reveals that

$$dE + đW - TdS = -T\dot{S}_\text{g}\,dt \leqslant 0$$

because T, $\dot{S}_\text{g}$ and dt are all assumed to be positive. This inequality defines the domain of spontaneously occurring (unstable) states, as opposed to the domain of other allowed states which do not occur spontaneously but which may be reached through stable, equilibrium processes. The result thus leads to a general criterion for stable equilibrium: namely, that for any infinitesimal changes δ in the stable states

$$\delta E + \delta W - T\delta S \geqslant 0 \tag{2.11}$$

in which the equality sign corresponds to the limiting 'surface' of stable equilibrium. Equation (2.11) is often stated in the particular forms $(\delta S)_E \leqslant 0$ or $(\delta E)_S \geqslant 0$ for isolated systems, which are denied work interactions.

For the special case of a phase change during which $đW = PdV$ and $dE = dU$, this reduces to

$$\delta U + P\delta V - T\delta S \geqslant 0$$

Now the Gibbs function is defined by

$$G = U + PV - TS$$

and hence

$$dG - VdP + SdT = dU + PdV - TdS$$

so that for stable equilibrium

$$\delta G - V\delta P + S\delta T \geqslant 0$$

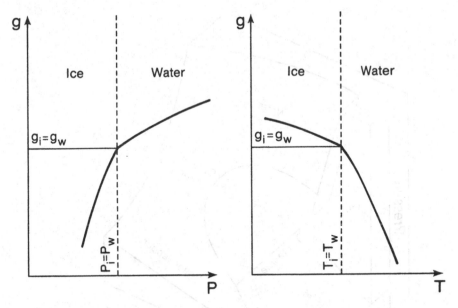

Fig. 2.2. Specific Gibbs function dependence on temperature and pressure.

It thus follows that when P and T are held constant

$$\delta G \geqslant 0$$

is the particular criterion for stable equilibrium. This situation may occur, for example, when pure ice is in stable equilibrium[3] with the liquid or vapour which it formed, or from which it was formed.

For a mixture of ice and water in stable equilibrium, with total mass M and ice fraction x,

$$G = Mxg_i + M(1 - x)g_w$$

where g_i, g_w are the corresponding specific Gibbs functions. An infinitesimal change in the ice fraction of this system is a change (between two stable states) for which $dG = 0$. Hence

$$dG = (\partial G/\partial x)\,dx = M(g_i - g_w)\,dx = 0$$

and consequently the requirement

$$g_i = g_w \tag{2.12}$$

must be satisfied.

Representative plots of the specific Gibbs function for pure ice and water are shown in Fig. 2.2 from which it is evident that the partial differential coefficients in the expansion

$$dg = \left(\frac{\partial g}{\partial P}\right)_T dP + \left(\frac{\partial g}{\partial T}\right)_P dT$$

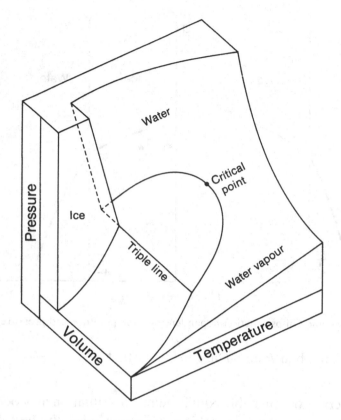

Fig. 2.3. Stable equilibrium *P–V–T* surfaces for ice, water and water vapour.

are discontinuous at the equilibrium condition. For either phase, the definition of the Gibbs function given earlier requires that

$$dg = vdP - sdT$$

under stable, equilibrium conditions, when it is clear that the specific volume and specific entropy are different for ice and water. Moreover, for an infinitesimal shift between the stable states at which equation (2.12) applies,

$$dg_i = dg_w$$

or

$$v_i dP_i - s_i dT_i = v_w dP_w - s_w dT_w$$

Therefore, since $P_i = P_w$ and $T_i = T_w$ under these conditions, it follows that

$$\frac{dP}{dT} = \frac{s_i - s_w}{v_i - v_w}$$

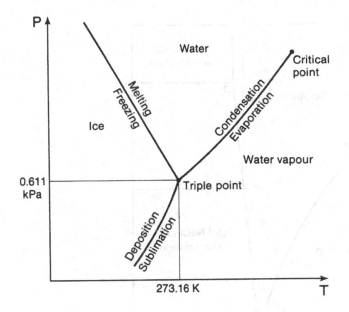

Fig. 2.4. Rules surfaces of bulk H_2O: end view.

or

$$\frac{dP}{dT} = \frac{-\lambda_{iw}}{T(v_i - v_w)} \qquad (2.13)$$

where λ_{iw} is the latent heat of fusion. This is the Clausius–Clapeyron equation for a first[4] order phase change between ice and water. Similar equations apply to the evaporation and sublimation phase changes.

Fig. 2.3 shows the equilibrium P–V–T surface for bulk H_2O which displays the characteristic decrease in density upon solidification. Fig. 2.4 is another view of the same surface looking along the volume axis. This clearly shows the ruled two-phase surfaces of Fig. 2.3 as three curves intersecting at the triple 'point' where $T = 273.16$ K and $P = 0.611$ kPa. As dictated by the Clausius–Clapeyron equation, dP/dT is positive for sublimation and evaporation but negative for melting.

The equilibrium behaviour of ice, water and water vapour is changed by the presence of solutes. Sea water, for example, is an aqueous solution of many substances, principal among which is sodium chloride. The concentration of sodium chloride is thus an additional variable converting the two-phase *curves* of a single component (Fig. 2.4) into two-phase *surfaces*. Fig. 2.5 shows a typical section through the sodium chloride–H_2O surface. The point A is the freezing point for pure H_2O at atmospheric pressure: i.e., the *ice point*, at which $T = 273.15$ K, slightly lower than the triple

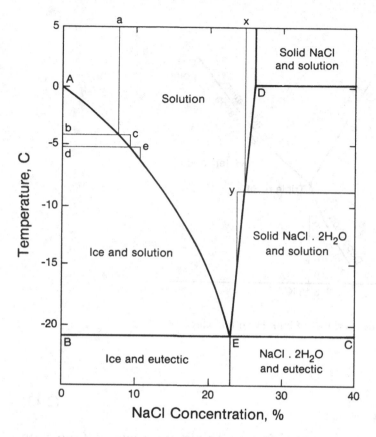

Fig. 2.5. Stable equilibrium phase diagram for sodium chloride and H_2O.

point temperature. The freezing temperature of a solution whose initial concentration is given by the point a is indicated by the point b. Since sodium chloride is essentially insoluble in ice, the solid deposited is very pure ice, thus reducing the H_2O concentration in the solution and raising the sodium chloride concentration in the direction of the point c. A further increment in cooling of the solution along the liquidus AE produces more ice, at temperature d, and a further increase in concentration towards the point e. Such small changes may be replaced by infinitesimals which, in the limit, generate a smooth excursion along AE provided stable conditions prevail. If this process continues, more and more pure ice will be produced until the sodium chloride concentration of the solution has reached the eutectic point at E and the entire solution has solidified. Subsequent cooling produces no further change in concentration.

A similar sequence of events occurs if the solution has an initial sodium chloride concentration indicated by x, the principal difference being the

Table 2.1. *Eutectic temperatures of various aqueous solutions*
(following Michel (1971))

	Substance	Eutectic temperature (°C)		Substance	Eutectic temperature (°C)
CaO	Calcium oxide	−0.15	NaOH	Sodium hydroxide	−28.0
K_2SO_4	Potassium sulphate	−1.52	$Ca(NO_3)_2$	Calcium nitrate	−28.0
$NaHCO_3$	Sodium bicarbonate	−2.33	$NaClO_4$	Sodium chlorate	−32.0
$NiSO_4$	Nickel sulphate	−3.40	$MgCl_2$	Magnesium chloride	−33.6
NaF	Sodium fluoride	−5.6	K_2CO_3	Potassium carbonate	−36.5
BaCl	Barium chloride	−7.8	$CaCl_2$	Calcium chloride	−55.0
KCl	Potassium chloride	−10.7	$ZnCl_2$	Zinc chloride	−62.0
$MgSO_4$	Manganese sulphate	−11.40	KOH	Potassium hydroxide	−65.2
NH_4Cl	Ammonium chloride	−16.0	H_2SO_4	Sulphuric acid	−74.5
NaCl	Sodium chloride	−21.2	HCl	Hydrochloric acid	−86.0

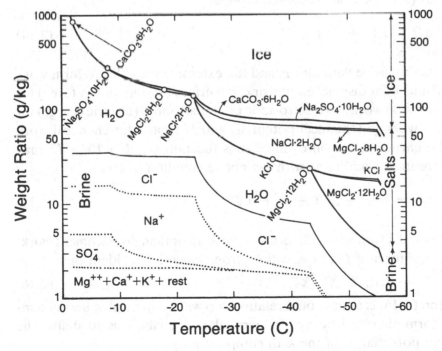

Fig. 2.6. Phase relations for 'standard' sea ice. Circles on the brine-salt curve indicate points at which salts precipitate (following Weeks and Ackley (1986)).

freezing mixture in which the solution is no longer in equilibrium with pure ice but with the dehydrate $NaCl.2H_2O$. As cooling continues along the liquidus DE the salt solution becomes less and less concentrated until the eutectic point is again reached. Once more, cooling below the solidus BEC occurs with a fixed (eutectic) composition: a finely divided mixture of pure ice and $NaCl.2H_2O$. Table 2.1 lists the eutectic temperatures of various aqueous solutions.

A more comprehensive phase diagram for sea ice, based on the work of Assur (Weeks and Ackley, 1986), is shown in Fig. 2.6. From this it is evident that ions other than those of sodium and chlorine are not only present but precipitate out as salts over a wide range of temperatures in a variety of hydrate forms.

2.3   The Gibbs function and chemical potential

Continuing the discussion of equilibrium thermodynamic surfaces, consider now the application of the energy and entropy principles to a stationary body of material under stable conditions. Taking the intensive variables T, P etc. to be uniform throughout the body, equations (2.2), (2.3) and (2.4) may be combined to give

$$dU = TdS + \sum_i F_i dX_i \qquad (2.14)$$

where the intensive variables F_i and the extensive variables X_i form work pairs: P and V for displacement work; σ (surface tension) and A (area) for surface tension work; ϕ (electrostatic potential) and $\mathscr{E}$ (electric charge) for electrical work; μ (chemical potential) and M (mass) for chemical work etc. The earlier definition of the Gibbs function $G = U + PV - TS$ may be differentiated and used with the above equation to give

$$dG = VdP - SdT + \sum_i F_i' dX_i' \qquad (2.15)$$

where F_i' and X_i' represent all forms of work other than displacement work. This is one form of *Gibbs equation*, from which it is evident that

$$G = G(P, T, X_1', X_2', \ldots) \qquad (2.16)$$

Equation (2.16) is a functional relationship widely applicable in problems of ice formation and decay. In particular, it enables us to define the chemical potential μ_k of the k-th component by

$$\mu_k = \left(\frac{\partial G}{\partial M_k}\right)_{PTM_{j \neq k} X_1' X_2' \ldots}$$

That is, the chemical potential is the partial Gibbs function.

An alternative[5] definition of the Gibbs function is given by

$$G = \sum_i F_i' X_i'$$

from which it follows that

$$dG = \sum_i F_i' dX_i' + \sum_i X_i' dF_i'$$

Combining this with equation (2.15) gives

$$-VdP + SdT + \sum_i X_i' dF_i' = 0 \qquad (2.17)$$

which is known as the *Gibbs–Duhem equation*. The Gibbs–Duhem equation is often useful in determining the relationship between the chemical potential and other variables, as may be demonstrated by factoring out the chemical work terms so that equation (2.17) then becomes

$$-VdP + SdT + \sum_k M_k d\mu_k + \sum_k X_i'' dF_i'' = 0$$

in which the pairs F_i'' and X_i'' represent all forms of work other than displacement or chemical work. Thus, for example, an infinitesimal change in the chemical potential of a pure substance subject only to displacement and chemical work is given by

$$d\mu = -sdT$$

under isobaric conditions, and

$$d\mu = vdP$$

under isothermal conditions. The latter circumstances applied to pure water vapour behaving like an ideal gas thus lead to the result

$$\mu = \mu^\circ + RT \ln P/P^\circ \qquad (2.18)$$

whereas for an incompressible condensed phase, either ice or water, it is evident that

$$\mu = \mu^\circ + v(P - P^\circ) \qquad (2.19)$$

The superscript refers to the pressure datum P° (usually 100 kPa) at which

$$\mu^\circ = \mu(T, P^\circ)$$

From the definition of chemical potential it is apparent that $\mu = g$ for any system with a single component. More generally, μ_i is the partial Gibbs function of component i in a system which may not only contain several components but several phases. In the absence of surface and body force effects, for example, the Gibbs equation (2.15) for a single phase system with m components reduces to

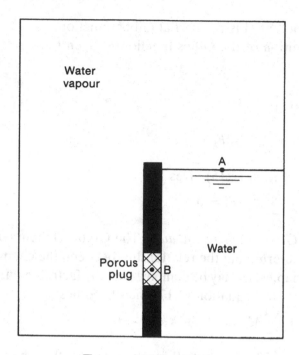

Temperature uniform

Fig. 2.7. Equilibrium of water and water vapour at a curved interface.

$$dG = \sum_{i=1}^{m} \mu_i \, dM_i$$

under conditions of fixed temperature and pressure. Hence for n phases in stable equilibrium

$$dG = \sum_{i=1}^{m} \mu_i^1 \, dM_i^1 + \ldots + \sum_{i=1}^{m} \mu_i^j \, dM_i^j + \ldots + \sum_{i=1}^{m} \mu_i^n \, dM_i^n = 0$$

But since

$$M_i = \sum_{j=1}^{n} M_i^j$$

then

$$dM_i = \sum_{j=1}^{n} dM_i^j = 0$$

which can only be satisfied together with the Gibbs equation if

$$\mu_i^1 = \mu_i^2 = \ldots = \mu_i^j = \ldots = \mu_i^n,$$

a more general result than equation (2.12). In fact this result is not limited
to systems which are uniform in pressure, but also applies, in particular, to
those in which pressure non-uniformity is created through surface tension
effects. Thus for particulate and porous systems containing ice, water and
water vapour, stable equilibrium implies that the chemical potential of
each component is uniform throughout the various phases unless impermeable partitions or body forces within the system prevent it. In general,
stable *thermodynamic* equilibrium, requiring thermal, mechanical and
chemical equilibrium, implies uniformity in the intensive properties of
temperature, pressure and chemical potential, if surface and body forces
and internal partitions are absent.

Consider the implications of chemical potential uniformity across a
boundary, beginning with stable equilibrium at the curved interface
formed between water and water vapour at the same temperature. Fig. 2.7
shows a vessel in which pure water vapour is partly separated from pure
water by a solid wall containing a porous plug. At the point A on the flat
interface, $\mu^V = \mu^L$, while at the point B on a curved interface inside the
porous plug, $\mu^{\circledV} = \mu^{\circledL}$, where the circled superscript indicates curvature.
However, because of capillarity and gravity $\mu^V \neq \mu^{\circledV}$ and $\mu^L \neq \mu^{\circledL}$, as may
be demonstrated by imagining hypothetical isothermal processes under-
gone by both the water and the vapour in moving between stable states on
flat (A) and curved (B) interfaces. As noted earlier, $d\mu_T = (v\,dP)_T$ and
hence

$$\mu^{\circledV} - \mu^V = \int_A^B v\,dP = RT\ln P^{\circledV}/P^V$$

if the vapour is ideal. Similarly,

$$\mu^{\circledL} - \mu^L = \int_A^B v\,dP = v^L(P^{\circledL} - P^L)$$

if the water is incompressible. But stable equilibrium requires that
$$\mu^{\circledV} - \mu^V = \mu^{\circledL} - \mu^L$$
and therefore
$$v^L(P^{\circledL} - P^L) = RT\ln(P^{\circledV}/P^V)$$
or
$$P^{\circledV}/P^V = \exp[v^L(P^{\circledL} - P^L)/RT]$$

which reveals how the equilibrium vapour pressure increases with capillary-induced water pressure or, if the water is frozen, with capillary-
induced ice pressure.

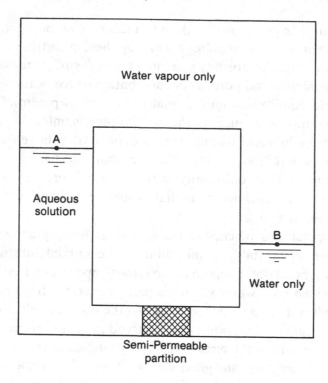

Fig. 2.8. Equilibrium of water and an aqueous solution at a semi-permeable partition.

This result is often described as the Poynting effect, and is typically quite small, as may be seen by writing

$$\ln P^{\unicode{x24CB}}/P^{V} \simeq \frac{P^{\unicode{x24CB}} - P^{V}}{P^{V}} = \frac{P^{\unicode{x24CB}} - P^{V}}{P^{L}}$$

so that

$$\frac{P^{\unicode{x24CB}}}{P^{V}} \simeq 1 + \frac{v^{L}}{v^{V}} \left(\frac{P^{\unicode{x24C1}} - P^{L}}{P^{L}} \right)$$

recalling that $P^{V}v^{V} = RT$. Since $v^{L} \ll v^{V}$, it follows that $P^{\unicode{x24CB}} \simeq P^{V} = P^{L}$, which is an important fact because the pressure difference across the curved interface is given by

$$P^{\unicode{x24C1}} - P^{\unicode{x24CB}} = 2\sigma/r$$

where r is the mean[6] radius of curvature, and hence

$$\ln(P^{\unicode{x24CB}}/P^{V}) \simeq 2v\sigma/RTr \tag{2.20}$$

in which v and σ are chosen according to whether the condensed phase is water or ice. This result was first derived by Kelvin.

Now consider uniformity of the chemical potential across a partition (permeable to water only) which separates pure water from an aqueous solution. Stable equilibrium demands a pressure difference, the osmotic pressure, as illustrated in Fig. 2.8 for a single, non-volatile solute. For an ideal, dilute solution, Raoult's law states that the water vapour pressure for a given temperature and solution pressure is given by

$$P^V = x_w P^1$$

where x_w is the mole fraction of the water, and P^1 is the vapour pressure when $x_w = 1$. Now since $\mu^V = \mu^L$ at the liquid–vapour interfaces, it follows from equation (2.18) that

$$\mu_A^V = RT \ln P^V + F(T) = \mu_A^L \qquad\qquad \text{at A}$$

and

$$\mu_B^V = RT \ln P^1 + F(T) = \mu_B^L \qquad\qquad \text{at B}$$

The difference in the chemical potential of the vapour between A and B is $\Delta\mu^V = RT \ln x_w$, using Raoult's law, and therefore, using equation (2.19),

$$- v^L \Delta P^L = RT \ln x_w$$

or

$$\Pi = -(RT/v^L) \ln x_w \qquad\qquad (2.21)$$

which defines the osmotic pressure Π. Since $1 = x_w + x_s$, where x_s is the solute mole fraction, this result may also be written in the approximate form

$$\Pi \simeq (RT/v^L) x_S$$

It is sometimes convenient to write the equation in terms of molality (the number of solute moles per unit mass of solvent).

As a third example of the consequence of uniformity in the chemical potential consider ice in stable equilibrium with the aqueous solution from which it was formed. Since $\mu_i = \mu_w$, it follows that

$$\mu_i = \mu_{w0} + R_w T \ln x_w \qquad\qquad (2.22)$$

in which the first term represents the chemical potential of pure water (at the given pressure and temperature) and the second term corresponds to the change produced by solute concentration. As earlier, it is assumed that this change may be described using Raoult's law. Now the specific enthalpy of pure ice or water is defined by

$$h = u + Pv$$

and consequently $dh = Tds + vdP$, using equation (2.14). Hence

$$(\partial h/\partial T)_P = T(\partial s/\partial T)_P$$

Furthermore, $\mu = g = h - Ts$ for a pure substance, so that

$$\left[\frac{\partial}{\partial T}\left(\frac{\mu}{T} \right) \right]_P = -\frac{h}{T^2}$$

and hence equation (2.22) may be divided through by T and differentiated to yield

$$\frac{\partial}{\partial T}(\ln x_w) = \frac{h_w - h_i}{R_w T^2} = \frac{\lambda_{iw}}{R_w T^2}$$

Integrating between T^0 and T, we obtain

$$\ln x_w = \frac{-\lambda_{iw}(T^0 - T)}{R_w TT^0}$$

or, approximately,

$$x_s = \frac{\lambda_{iw}(T^0 - T)}{R_w (T^0)^2}$$

thus yielding the freezing point depression as

$$T^0 - T = \frac{R_w (T^0)^2}{\lambda_{iw}} x_s$$

For a solute which is insoluble in ice, this expression represents the tangent of the $T - x_s$ curve at $x_s = 0$. It is a convenient approximation for most dilute aqueous solutions (see Fig. 2.5, for example) and is commonly written in the form

$$\Delta T_{fp} = K_{fp} \, \mathcal{M} \tag{2.23}$$

where $\mathcal{M}$ is the molality and

$$K_{fp} = \frac{\mathcal{R}(T^0)^2}{\lambda_{iw}} = 1.86 \text{ kg K mol}^{-1}$$

where T^0 is the ice point temperature and $\mathcal{R}$ is the universal gas constant. Should the solute form several distinct ions, the number of moles in solution must be modified, roughly in proportion to the number of ions. The freezing point depression is altered accordingly. This and other restrictions on equation (2.23) are discussed in Fennema, Powrie and Marth (1973).

2.4   Diffusion in ice, water and water vapour

Diffusion is not within the province of equilibrium thermodynamics because it implies non-uniformity in the intensive thermodynamic properties, the temperature and chemical potential in particular. It is an irreversible process for which the full form of the entropy principle, equation (2.3), is required. Growth and decay of ice are intimately associated with the diffusion of heat to, or from, the ice surface. The heat flux density is represented by Fourier's law

$$\mathbf{j}_Q = -k_Q \, \nabla T \tag{2.24}$$

where k_Q is the thermal conductivity. Thus, in the X-direction,

$$j_X = -k_Q \, \partial T / \partial X$$

The flow of heat is often accompanied by a diffusive mass flux represented by the analogous Fickian form

$$\mathbf{j}_M = -k_M \, \nabla \mu \tag{2.25}$$

which may be applied to a variety of phenomena: self-diffusion, diffusion of water vapour, electron diffusion, etc.

There are many circumstances where the diffusion of heat and mass may be treated independently of each other, in which case the thermal balance at the ice surface may be adequately described in terms of equation (2.24) accompanied by a latent heat flux. In general, however, the diffusive fluxes are coupled and must be written in the form

$$\left. \begin{array}{l} \mathbf{j}_{TH} = K_{QQ} \nabla (1/T) - K_{Q1} \, \nabla \mu_1 / T \\ \mathbf{j}_{M1} = K_{1Q} \nabla (1/T) - K_{11} \, \nabla \mu_1 / T \end{array} \right\} \tag{2.26}$$

in which the first equation represents the thermal energy flux (strictly the flux density) and the second represents the flux (density) of a single component labelled 1. This particular choice of fluxes and gradients is one of several which satisfy *Onsager's reciprocal relation*

$$K_{rt} = K_{tr}$$

the criterion for which is that $\mathbf{j}$ and $\mathbf{X}$ in the flux equations

$$\mathbf{j}_r = \sum_t K_{rt} \mathbf{X}_t$$

must be such that

$$\dot{s}_g = \sum_r \mathbf{j}_r \cdot \mathbf{X}_r$$

where $\dot{s}_g$ is the volumetric entropy generation rate introduced in equation (2.3). The Onsager reciprocal relation $K_{1Q} = K_{Q1}$ is the first of four identities which must be introduced if the coefficients K in equations (2.26) are to be related to the more familiar phenomenological coefficients contained in equations (2.24) and (2.25).

The remaining identities are taken as follows. The *isothermal mass diffusion coefficient* k_M is defined by

$$(\mathbf{j}_{M1})_{\nabla T=0} = -k_M(\nabla \mu_1)_{\nabla T=0}$$

thus requiring that

$$K_{11} = k_M T$$

The *thermal conductivity*, on the other hand, is defined under stagnation conditions by

$$(\mathbf{j}_{TH})_{j_{m1}=0} = -k_Q(\nabla T)_{j_{m1}=0}$$

which yields

$$K_{QQ} = \frac{K_{1Q}^2 + k_Q K_{11} T^2}{K_{11}}$$

Thirdly, the *isothermal heat of transport* Q_1^+ is defined by

$$(\mathbf{j}_{TH}/\mathbf{j}_{M1})_{\nabla T=0} = K_{Q1}/K_{11} = Q_1^+$$

Hence

$$K_{Q1} = K_{1Q} = k_M T Q_1^+$$

Equations (2.26) may now be re-written in the phenomenological form

$$\left.\begin{array}{l} \mathbf{j}_{TH} = -[k_Q + k_M(Q_1^+)^2/T]\nabla T - k_M Q_1^+ \nabla \mu_1 \\ \mathbf{j}_{M1} = [(-k_M Q_1^+)/T]\nabla T \quad\quad - k_M \nabla \mu_1 \end{array}\right\} \tag{2.27}$$

Eliminating μ_1, yields

$$\mathbf{j}_{TH} = -k_Q \nabla T + \mathbf{j}_{M1} Q_1^+ \tag{2.28}$$

revealing that the thermal flux consists of the Fourier flux and a transport flux dependent upon Q_1^+ which has the alternative definition

$$Q_1^+ = -T(\nabla \mu_1 / \nabla T)_{j_{M1}=0} \tag{2.29}$$

Applied to the coupled flow of heat and electricity, for which $\mathbf{i} = \epsilon_1 \mathbf{j}_{M1}$ is the current density, and ϵ_1 is the charge per unit mass, equations (2.27) become

$$\left.\begin{array}{l} \mathbf{j}_{TH} = [k_Q + k_E T(s_1^+)^2]\nabla T - k_E T s_1^+ \nabla \phi \\ \mathbf{i} \quad = -k_E s_1^+ \nabla T \quad\quad\quad - k_E \nabla \phi \end{array}\right\} \tag{2.30}$$

where k_E is the isothermal electrical conductivity, and

$$s_1^+ = \frac{S_1^+}{\epsilon_1} = \frac{Q_1^+}{T\epsilon_1} = -\left(\frac{d\phi}{dT}\right)_{i=0} \tag{2.31}$$

is the Seebeck coefficient. The chemical potential (sometimes called the electrochemical potential) in these equations has been replaced by $\mu_1 = \phi \epsilon_1$, following the Gibbs equation (2.15) and the definition of μ applied to the electrons: i.e., $dG_1 = \mu_1 dM_1 = \phi d\mathscr{E}_1$, where ϕ is the electrostatic potential and $\mathscr{E}_1 = M_1 \epsilon_1$ is the electric charge. The Seebeck coefficient, as the entropy per unit charge, is related to the Peltier coefficient π by

$$\pi_1 = (\mathbf{j}_{TH}/\mathbf{i})_{\nabla T=0} = T s_1^+ \tag{2.32}$$

When applied to the coupled flow of heat and mass, equations (2.27) become

$$\left.\begin{aligned}
\mathbf{j}_{TH} &= -\left(k_Q + \frac{\rho_1 D_1 Q_1^{+2}}{R_1 T^2}\right)\nabla T - \frac{\rho_1 D_1 Q_1^{+}}{R_1 T}\nabla\mu_1 \\
\mathbf{j}_{M1} &= -\frac{\rho_1 D_1 Q_1^{+}}{R_1 T^2}\nabla T \qquad\quad - \frac{\rho_1 D_1}{R_1 T}\nabla\mu_1
\end{aligned}\right\} \tag{2.33}$$

in which equation (2.17) has been used along with the definition of the isothermal mass diffusivity D_1 given by

$$(\mathbf{j}_{M1})_{\nabla T=0} = -\rho D_1(\nabla C_1)_{\nabla T=0} = -D_1(\nabla\rho_1)_{\nabla T=0}$$

which assumes a low concentration C_1. The Soret effect (thermo-diffusion) is represented by the first term on the right hand side of the second of equations (2.33), and the Dufour effect (diffusion-thermo) by the second term on the right hand side of the first equation. Both contain the heat of transport Q_1^{+}.

It is instructive to examine the stagnation condition for binary diffusion in an aqueous solution under isothermal conditions (de Groot and Mazur, 1962). For a solute flux density $\mathbf{j}_s$ and a water flux density $\mathbf{j}_w$, the coupled flow equations are most conveniently expressed by

$$\mathbf{j}' = -K_{ww}(\nabla\mu_w/T) - K_{ws}(\nabla P/T)$$
$$\mathbf{j}'' = -K_{sw}(\nabla\mu_w/T) - K_{ss}(\nabla P/T)$$

where

$$\mathbf{j}' = \mathbf{j}_w - (c_w/c_s)\mathbf{j}_s$$

and

$$\mathbf{j}'' = v_w\,\mathbf{j}_w + v_s\mathbf{j}_s$$

Here c represents concentration and v specific volume. When stable equilibrium exists across a membrane permeable only to the water, the above equations generate the following special results:

(1) The solute flux is prevented ($\mathbf{j}_s = 0$), and hence
$$\mathbf{j}'/\mathbf{j}'' = 1/v_w$$

(2) The mobile water is in stable equilibrium across the membrane ($\nabla\mu_w = 0$), and therefore
$$\mathbf{j}'/\mathbf{j}'' = K_{ws}/K_{ss}$$

(3) The volume flux is zero ($\mathbf{j}'' = 0$), consequently
$$(\nabla P/\nabla\mu_w)_{j=0} = (dP/d\mu_w)_{j=0} = -K_{ws}/K_{ss}$$

since $K_{ws} = K_{sw}$. These three results reveal that the solute induced

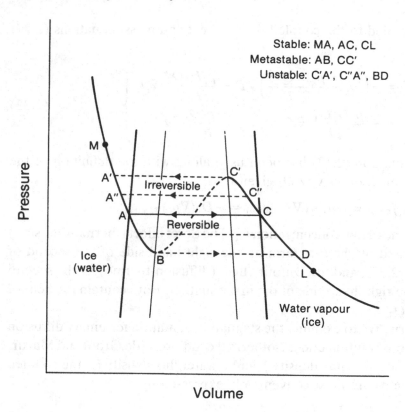

Fig. 2.9. Isotherm traversing the two-phase region.

pressure difference across the membrane, as reflected in the vertical difference in chemical potential, is given by

$$\Pi = \Delta P = -\Delta\mu_w/v_w = (-RT/v_w)\ln x_w$$

if the solution is dilute. This provides an alternative derivation, based on an irreversible formulation, of the osmotic pressure equation (2.21). Under these conditions, the entropy generation rate $\dot{s}_g$ is identically zero, confirming that stable equilibrium does indeed exist. Should the membrane be permeable to the solute, $\Delta P = 0$. In general, the flow of the components influences the pressure gradient which is then no longer purely osmotic.

2.5   Metastability and the freezing point

The idea of equilibrium, so commonly used in thermodynamics, carries with it a sense of balance between the system of interest and its environment. This balance is seldom static. The difference between unstable equilibrium and stable equilibrium is that the former implies a

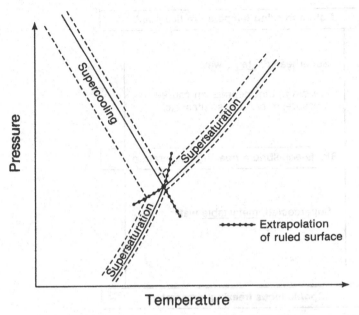

Fig. 2.10. Metastable extensions of the P–V–T surfaces of H$_2$O.

spontaneous change of state of the system without any change in the environment, whereas the latter implies that system changes may occur only as the result of, and in exact correspondence to, environmental changes. Metastable equilibrium occupies a middle ground in which small changes reflect stable behaviour whereas large changes produce instability. The principal limitation of the metastable concept is found in the interpretation of 'large' and 'small'. A further problem lies with the rate at which changes to the system actually take place.

Metastable behaviour is often important during a change of phase involving ice. For stable behaviour, the entropy extremes embodied in the inequality (2.11) have implications with respect to the slope of the *P–V–T* surface shown in Fig. 2.3. Stable equilibrium requires that $(\partial P/\partial V)_T \leqslant 0$, which is seen to be satisfied everywhere on the surface. On the other hand, $(\partial P/\partial V)_T > 0$ would imply unstable equilibrium. Fig. 2.9 shows an isotherm traversing the two-phase region between a more dense phase M on the left and a less dense phase L on the right; Fig. 2.3 reveals that this may refer to ice (M) and water vapour (L) or to water (M) and ice (L). If stable conditions exist, nucleation of the more dense phase takes place at the end of the isothermal compression process LC; further reductions in volume simply alter relative amounts of the two phases in a reversible process which ends at A. Similarly, nucleation of the less dense phase occurs after isothermal expansion along MA.

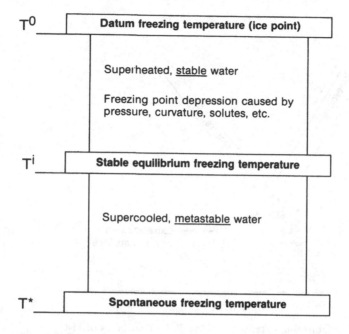

Fig. 2.11. Freezing temperatures of water: definitions.

In general, stable behaviour may be extended into a limited zone inside the two-phase region on both sides, although too great an excursion from the edge of the region will generate an irreversible change of phase. Considering ice and water vapour, for example, if the equilibrium saturation vapour pressure at C is exceeded during compression, the state of the resulting supersaturated vapour will be represented by a point lying on CC′. Nucleation occurring at any point C″ on this curve produces an irreversible change from C″ to the point A″ where the ice is once again in stable equilibrium. The limit of the metastable condition of supersaturation is indicated by the point C′ at which spontaneous nucleation occurs. Between B and C′ the positive slope of the curve prohibits stable or metastable behaviour. The latter may thus be viewed as an extension of the unruled equilibrium surfaces, as indicated by the dashed lines in Fig. 2.10 which is based on Fig. 2.4. Small changes of state within the metastable zones exhibit stable equilibrium behaviour and therefore the methods and results of previous sections also apply there. It is worth noting that excursions on these extended surfaces may be produced by an isothermal pressure variation, e.g., supersaturation, or by an isobaric temperature variation, e.g., supercooling; or by a combination. Isothermal and isobaric processes are both useful in the analysis of ice formation and decay.

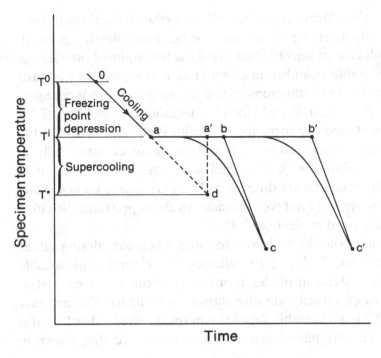

Fig. 2.12. Freezing behaviour during a constant rate of heat loss.

In Section 2.3, solutes and interface curvature were both discussed as sources of departure from the stable equilibrium states appropriate to pure, planar H_2O interfaces. In particular, they result in the displacement of the P–V–T equilibrium surface relative to that indicated earlier in Fig. 2.3, but this occurs without a departure from stable equilibrium. It is therefore evident that actual pressures and temperatures not located on the surfaces of Fig. 2.3 may be attributed to either (or both) of two very different causes: a metastable extension of the surface, or a stable displacement of it. This distinction is important.

Fig. 2.11 illustrates the need for precision in the term *melting point* or, to use a more common term, *freezing point*. The *datum freezing temperature*, $T^0 = 0\,°C$, or 273.15 K, will be taken as a fixed reference. (The datum freezing temperature is sometimes called the ice point. The triple point 273.16 K is an alternative datum; the difference is negligible for most purposes.) In practice, ice may nucleate at a lower temperature T^i because of stable alterations in pressure, interface curvature, solute concentration, etc. Above T^i, ice will not exist if stable equilibrium conditions prevail. At T^i, the *stable equilibrium freezing temperature* (sometimes called the saturation temperature) ice forms so long as stable equilibrium conditions exist. Should ice thus form, as it often does, further reductions in temperature occur only after the latent heat has been completely removed; subsequent withdrawal of sensible heat from the ice produces *subcooling*. Almost as often, stable equilibrium of water at T^i is replaced by metastable equilibrium; further reductions in temperature then produce *supercooling* of the water which, if and when the threshold temperature T^* is reached, will spontaneously form ice. This threshold will be called the *spontaneous freezing temperature*, and will be discussed more fully in section 2.8. Metastability is thus gauged not in relation to T^0, but in relation to T^i. The temperature difference $T^0 - T^i$ is usually known as the *freezing point depression* (whatever the cause of the depression: the effect of solutes was discussed in Section 2.3).

These distinctions should be borne in mind when considering either freezing or deposition. As Fig. 2.10 indicates, $T < T^i$ implies metastable equilibrium on an extension of the appropriate unruled surface. Extrapolations of the ruled surfaces are also shown on the figure. On the other hand, $T^i < T < T^0$ implies stable equilibrium on the fluid side of a saturation curve which may have been displaced relative to that shown in Fig. 2.3.

Fig. 2.12 offers another interpretation of freezing temperature, as observed in an experiment in which an isothermal (spatially uniform) specimen containing water undergoes a constant rate of heat loss: i.e., when

$$\dot{Q} = M c_p \, dT/dt = \text{constant}$$

so that

$$dT/dt \propto 1/c_p$$

in the absence of phase change. Under stable equilibrium conditions, a pure specimen will nucleate at $T^i \leqslant T^0$, releasing latent heat steadily from a to b at T^i until the entire specimen is frozen; subsequently the specimen is subcooled (b–c) at a cooling rate usually greater than that of the unfrozen specimen, as revealed by the expression immediately above. Should the material freeze over a temperature range, the cooling curve takes the continuous form a–c lying beneath the discontinuity at b. Now if the ice does not nucleate at T^i, cooling continues under metastable conditions, as suggested by the broken line a–d, until T^* is reached, at which point the spontaneous release of latent heat restores the specimen temperature to the stable equilibrium value; cooling then continues as before, depending upon whether there is a single freezing temperature (a'–b') or a freezing temperature range (a'–c').

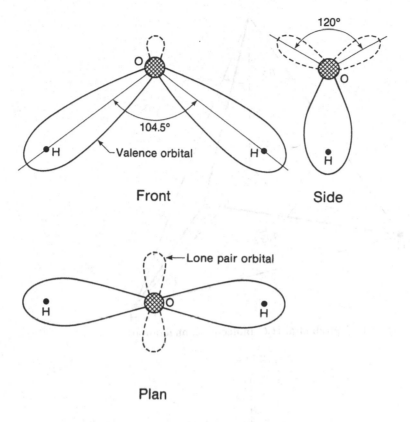

Front   Side

Plan

Fig. 2.13. Electron orbitals for H_2O molecule: three views.

In all of the above, it has been assumed that nucleation and subsequent interfacial growth are unimpeded. In practice, diffusive and molecular kinetic processes reduce the rate of change to a finite level, and indeed may reduce it to a negligible value. A glass transition, for example, may occur at a rate so low as to be virtually unmeasurable: the glass continues to exist in unstable equilibrium. Many aqueous solutions undergo a similar transition which is usually referred to as vitrification. The freezing point of such solutions is difficult to define because the freezing process may take a long time, during which, strictly speaking, the solution is not in metastable equilibrium and is not governed by the dictates of stable equilibrium. As a consequence, vitreous solutions may be erroneously regarded as super-cooled (metastable) liquids or subcooled (stable) solids.

2.6   Morphology and microenergetics
The microstructure of ice, in any of its many forms, is clearly related to the architecture of the water molecule which has been

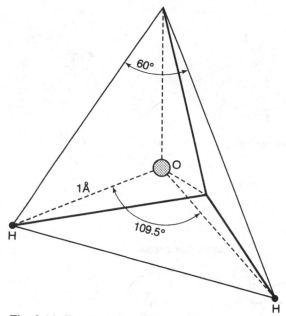

Fig. 2.14. Tetrahedral H_2O molecule configuration.

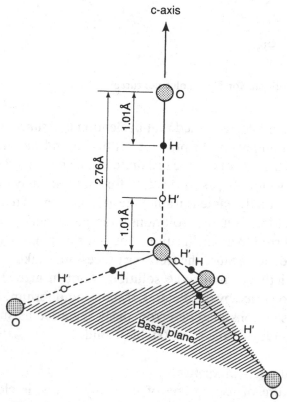

Fig. 2.15. Valence and hydrogen bonding.

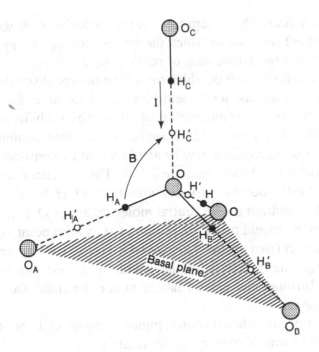

Fig. 2.16. Propagation of defects.

extensively studied. It is illustrated in Fig. 2.13 which shows two pairs of electron orbitals lying in mutually-perpendicular planes: two valence orbitals, and two lone pair orbitals. As indicated, the included angles are 104.5° and about 120°, thus revealing a molecule readily adaptable to a tetrahedral structure in which the included angles would both be 109.5°. It is also apparent that the molecule will have three normal modes of vibration: one distortional, in which the smaller included angle varies; and two longitudinal, one symmetric and the other unsymmetric. The wavelengths corresponding to these fundamental frequencies are all found to lie in the infra-red region of the electromagnetic spectrum.

With some adjustment to the orbitals, notably in the lone pairs, the water molecule can conform closely, though not exactly, to the tetrahedral ice arrangement shown in Fig. 2.14. In this crystalline array, neighbouring oxygen atoms are 2.76 Å apart. The *basal plane* contains three oxygen atoms as shown in Figs. 2.15 and 2.16, the *c*-axis being defined normal to it. Fig. 2.15 indicates that there are two possible positions for the protons in this configuration: 1.01 Å or 1.76 Å from an oxygen atom, each implying a different characteristic bonding energy. The full lines O–H represent intramolecular *covalent* bonds (valence orbitals) which require slightly less than 5 eV each to form or break, about half the energy of dissociation or formation; the broken lines O–H' represent intermolecular *hydrogen* bonds (lone pair orbitals) for each of which the corresponding energy is about 0.29 eV, about half the latent heat of sublimation.

The proton's two locations, both of which cannot be occupied simultaneously under stable conditions, lend themselves to protonic mobility which gives ice its electric and thermoelectric properties. This mobility has a number of sources (dislocations, free surfaces and intergranular surfaces), but its intrinsic mechanism may be attributed to two particular defects: the *Bjerrum* defect and the *ionization* defect. The Bjerrum defect occurs when the H–O–H molecule rotates about an O–H bond. In Fig. 2.16, for example, rotation of the central molecule about O–H_B, as indicated by the arrow B, would bring the proton at H_A to the point H_C' thus generating a D-defect (two protons) along O–O_C and an L-defect (no protons) along O–O_A. These defects are not stable arrangements and therefore propagate through the crystal lattice under the influence of thermal and electric fields.

An ionization defect occurs when a proton jumps along an O–O bond, thus generating a pair of ions according to the relation

$$2H_2O \rightleftharpoons OH^- + H_3O^+$$

This is illustrated in Fig. 2.16 by the arrow I in which the jump of the proton from H_C to H_C' leaves the proton-deficient O_C molecule negatively charged (OH$^-$) and the proton-enriched O molecule positively charged (H$_3$O$^+$). Both Bjerrum and ionization defects may be present in ice crystals where the energies associated with them are about 0.68 eV and 0.74 eV, respectively.

Two other defects, which have comparable energies and densities, are the *interstitial* defect and the *vacancy* defect. The former refers to molecules located in the lattice interstices (i.e., not in lattice sites) and the latter refers simply to the absence of molecules from lattice sites. Such defects are believed to form the basis of molecular (or vacancy) diffusion in ice, although the precise mechanism is not fully understood.

Before leaving this short section it is important to note that the discussion has been limited to the most common form of ice, designated Ih. Unless otherwise qualified, the word *ice* throughout this book refers only to ice Ih. Ices II through IX have different microstructures and therefore cannot be expected to exhibit the same behaviour as Ih, beyond reflecting the properties of the H$_2$O molecule. The remarks of this section therefore apply to pressures less than about 200 MPa. Strictly speaking, they are also limited to temperatures above $-100\,°C$ in the vicinity of which the cubic form of ice Ic is transformed to Ih, but since the reverse transformation has not been observed for temperatures ranging down as low as $-140\,°C$, it is reasonable to treat Ih as the stable form over a fairly wide range of temperatures. Below $-140\,°C$ a vitreous form of ice exists. In ice Ic, despite its cubic lattice, the oxygen atoms are still arranged tetrahedrally, and therefore we might expect to find many of the characteristics of Ih in Ic, but this could not be expected of the lower-temperature amorphous form.

2.7 Thermal properties

In natural circumstances the temperature of ice is seldom far from the datum freezing temperature T^0; much of the time, the temperature is above $-55\,°C$. The homologous temperature ratio T/T^0 is therefore often greater than 0.8, thus suggesting that ice should be regarded as a material close to its melting point where property variations may be substantial. This book is essentially limited to thermophysical properties, among the most important of which are density, specific heat, latent heat and thermal conductivity. Each of these will be considered in turn.

The lattice structure discussed in the previous section provides a basis for theoretical calculations of density. Given the mass of the water molecule as $2.992 \times 10^{-26}\,kg$, and the tetrahedral geometry with oxygen atoms spaced $2.76\,Å$ apart, the density of ice Ih is calculated to be $0.9167 \times 10^3\,kg\,m^{-3}$ which is in excellent agreement with measurements at the datum freezing temperature. There is no significant density difference between ices Ih and Ic, and the high pressure polymorphs exhibit densities ranging from 16% to almost 100% greater than ice Ih, the increases roughly paralleling the pressures. Ice VII, for example, has a density of $1.66 \times 10^3\,kg\,m^{-3}$ at $25 \times 10^5\,kPa$. Vitreous ice has a density of about $0.940 \times 10^3\,kg\,m^{-3}$.

At T^0 pure water has a density of $1.000 \times 10^3\,kg\,m^{-3}$. The hydrogen bonds in ice at this temperature evidently restrain its densification, so that the immediate effect of melting, in which only a small fraction of the hydrogen bonds are broken, is to allow a small (8%) shrinkage in volume. Further heating of the melt increases thermal fluctuations which eventually overshadow the initial contraction effect. The two effects are in balance near $4\,°C$, at which point they create a local maximum in density.

The isobaric and isometric specific heats for a single phase substance are defined, respectively, by

$$c_p = \left(\frac{\partial h}{\partial T}\right)_P = T\left(\frac{\partial s}{\partial T}\right)_P \qquad (2.34)$$

and

$$c_v = \left(\frac{\partial u}{\partial T}\right)_v = T\left(\frac{\partial s}{\partial T}\right)_V \tag{2.35}$$

They may be combined in the relation

$$c_p - c_v = -T\left(\frac{\partial P}{\partial v}\right)_T\left(\frac{\partial v}{\partial T}\right)_P^2 \tag{2.36}$$

As noted earlier, $(\partial P/\partial v)_T < 0$ for stable conditions, and hence $c_p > c_v$ except when $T = 0\,\mathrm{K}$. When $(\partial P/\partial v)_T = 0$ (a critical point) or $(\partial v/\partial T)_P = 0$ (a density maximum), $c_p = c_v$. Typically, $c_p \approx c_v$ for a solid phase, and since neither depends strongly on pressure or volume, it is reasonable to assume that the specific heat of ice may be described by the empirical form

$$c_p = \sum_n c_n \theta^n \tag{2.37}$$

where $\theta = T - T^0$; this is consistent with the known effect of temperature on interatomic forces and distances. Experiments (Giauque and Stout, 1936) indicate that $c_0 = 2.10\,\mathrm{kJ\,kg^{-1}K^{-1}}$ and $c_1 = 7.7 \times 10^{-3}\,\mathrm{kJ\,kg^{-1}K^{-2}}$ which together yield an accuracy of about 1% in equation (2.37) down to $-100\,°\mathrm{C}$. It might be expected that a similar expression would apply to supercooled water but in fact the specific heat increases dramatically as the temperature drops between $0\,°\mathrm{C}$ and $-40\,°\mathrm{C}$ (Angell, 1982).

The specific heats measure the capacity of the lattice to store thermal energy, essentially in the form of lattice vibrations. The latent heat, on the other hand, is the enthalpy difference of the two phases. Thus the latent heat of fusion is given by

$$\lambda_{iw} = h_w - h_i$$

and hence, using equation (2.34),

$$(\partial \lambda_{iw}/\partial T)_P = c_{pw} - c_{pi}$$

which, given the form of equation (2.37), suggests the empirical form

$$\lambda_{iw} = \sum_n \lambda_n \theta^n \tag{2.38}$$

Experimental data (Hobbs, 1974) reveal that $\lambda_0 = 334\,\mathrm{kJ\,kg^{-1}}$ and $\lambda_1 = 5\,\mathrm{kJ\,K^{-1}kg^{-1}}$. The latent heat of sublimation is about $0.53\,\mathrm{eV}$ per molecule at 273 K as compared with $0.06\,\mathrm{eV}$ for the latent heat of fusion. The molecular model of an ice crystal implies breakage of all the hydrogen bonds if each molecule is to be removed from the influence of the lattice (12% of the hydrogen bonds break during melting). Near the triple point the latent heat of sublimation of pure H_2O is $2838\,kJ\,kg^{-1}$, of which only about 5% is attributable to PdV work (Rossini *et al.*, 1952). The empirical relation

$$\lambda_{iv} = 2838 + 0.8\,\theta \qquad (2.39)$$

reveals that λ_{iv} is not too dependent on temperature.

Thermal conductivity in ice k_Q is a phonon diffusion phenomenon which, if treated analogously to diffusion of molecular kinetic energy, may be represented by the expression

$$k_Q = \tfrac{1}{3}C_v\mathcal{V}_s l_p \qquad (2.40)$$

where $\mathcal{V}_s$ is the phonon mean velocity, (roughly equal to the sonic velocity), and l_p is the phonon mean free path. The mutual interactions of phonons, unlike those of vapour molecules, produce a mean free path length which is inversely proportional to the absolute temperature, (Debye, 1914; Peierls 1929) and therefore equation (2.40), in which C_v and $\mathcal{V}_s$ have a weaker temperature dependence, suggests that $k_Q \propto 1/T$. Experimental data (Dillard and Timmerhaus, 1966) indicate that in the range $108\,K < T < 273\,K$.

$$k_Q = 0.468 + 488/T \qquad (2.41)$$

in $W\,m^{-1}\,K^{-1}$. A more common empirical curve, which also extends down to 108 K, is

$$k_Q = \sum_n k_n\,\theta^n \qquad (2.42)$$

in which $k_0 = 2.17\,W\,m^{-1}\,K^{-1}$, $k_1 = -3.4 \times 10^{-3}\,W\,m^{-1}\,K^{-2}$ and $k_2 = 9.1 \times 10^{-5}\,W\,m^{-1}\,K^{-3}$.

It is interesting to note that the thermal conductivity of pure water at $0\,°C$ is $0.57\,W\,m^{-1}\,K^{-1}$, only about one-quarter of that of ice. Since the specific heat of water is about twice that of ice, it follows that the thermal diffusivity $\varkappa = k/\varrho c_p$ of water is approximately one-eighth of that of ice. These orders of magnitude mark an important difference between ice and water when transient heat conduction is being considered.

2.8 Nucleation

The morphological evidence discussed earlier suggests that pure water near the freezing point has an open, ordered structure. Water vapour, on the other hand, is highly disordered, as a consequence of which

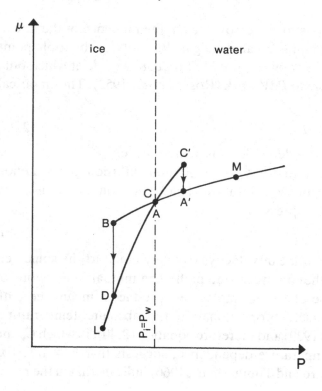

Fig. 2.17. Metastable excursions for ice and water.

vapour molecules only come together through chance encounters generated by thermal fluctuations. Such encounters may result in two or more molecules joining and remaining together as a cluster of water or ice. At the same time, further molecular collisions with the cluster, and collisions between clusters, may lead to fragmentation and a regeneration of single H_2O molecules, depending on the prevailing conditions. The size and number of clusters in the vapour must either increase or decrease, or, under extremum conditions, remain unchanged. The third possibility may correspond to an unstable equilibrium state in which the clusters have reached a critical radius r^*; above this radius they grow spontaneously while below it they decay spontaneously.

The structure of pure water is not easily described; nor is the liquid encounter process well understood. Given a quasi-crystalline model of water, nucleation may be interpreted as the alignment and stiffening of highly flexible lattice networks, whereas a hard sphere model would lend itself to an interpretation of nucleation similar to that described above for the vapour phase. The real situation is perhaps somewhere in between and may be interpreted, at least in part, by either model.

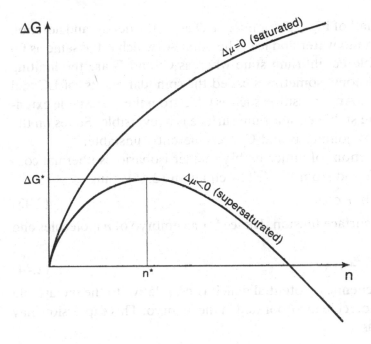

Fig. 2.18. Gibbs function for various embryo sizes.

Homogeneous nucleation, namely that occurring spontaneously throughout a supercooled fluid, may be understood in part by examining the behaviour of a single cluster or embryo of the embryonic phase. Under equilibrium conditions, the critical or threshold size of an embryo represents a stationary state. Before considering this behaviour, however, it is worth recalling previous discussions of the equilibrium g–P–T and P–v–T surfaces. For the latter, the individual surfaces for ice, water and water vapour are separated by ruled two-phase regions whereas in the former they intersect along three curves g_{iv} (P, T), g_{iw} (P, T) and $g_{vw}(P, T)$ which join at the triple point. It is known that $(\partial g/\partial P)_T$, which is equal to v, decreases when vapour is transformed into ice or ice becomes water; it is also known that $-(\partial g/\partial T)_P$ which is equal to s, increases when ice is transformed into either fluid phase. Two particular examples are given in Fig. 2.2 which also confirms another important fact: namely that the chemical potential (specific Gibbs function) of a pure substance is always greater in the supercooled or supersaturated phase than in the embryonic phase. Metastable extensions of the equilibrium g–P–T water (and water vapour) surfaces lie above the stable equilibrium ice surface, as may be illustrated by the ice–water transition shown in Fig. 2.17. This is a re-plot of the left hand half of Fig. 2.2 for pure H_2O using the details and notation of Fig. 2.9 applied to water and ice. The point A (which is the same as C) represents a stable equilibrium state whereas C' and B are the limiting metastable extensions, sometimes called the spinodal points, of LC and MA, respectively. Any transition such as C'A' from the metastable extension CC' onto the stable equilibrium surface is irreversible. States on the surface formed by joining B and C' are inherently unstable.

The Gibbs function[7] of an ice embryo under isobaric, isothermal conditions may be found from the Gibbs equation (2.15): thus

$$dG = \mu dM + \sigma dA \tag{2.43}$$

in which σ is the surface tension. Hence for an embryo of n molecules and surface area A,

$$\Delta G = n\Delta\mu + \sigma A \tag{2.44}$$

where $\Delta\mu$ is the chemical potential deficit (i.e., relative to the metastable phase) of each molecule incorporated in the embryo. This expression may also be written as

$$\Delta G = n\Delta\mu + Kn^{\frac{2}{3}} \tag{2.45}$$

where K is a geometrical coefficient. Fig. 2.18 shows how the addition of molecules to the embryo changes the Gibbs function. When $\Delta\mu = 0$ (saturated conditions) the curve is monotonically increasing, but when $\Delta\mu < 0$ (supersaturated or supercooled conditions) the curve exhibits a maximum at $n = n^*$. If the embryo is assumed to be spherical, this corresponds to a threshold radius given by

$$r^* = -2\sigma/\rho_n\Delta\mu \tag{2.46}$$

where ρ_n is the number of embryo molecules per unit volume. At this radius, the embryo is in *unstable equilibrium* (any change $dG < 0$) and is thus poised between spontaneous growth and spontaneous decay. The corresponding threshold Gibbs function is given by

$$\Delta G^* = \frac{4}{3}\pi\sigma r^{*2} = \frac{16}{3}\pi\sigma^3/\rho_n^2(\Delta\mu)^2 \tag{2.47}$$

The chemical potential deficit in the above equations is given by

$$\Delta\mu = \mu_e - \mu_m$$

where μ_e, μ_m are the chemical potentials of the embryonic and metastable phases, respectively. This deficit is readily found under isothermal or isobaric conditions, either of which may provide a useful representation, depending upon the circumstances. As noted in section 2.3, $d\mu = vdP$ for a pure substance under isothermal conditions, and hence

$$\Delta\mu = -kT \ln(P_m^V/P_e^V) \tag{2.48}$$

where P^V is the vapour pressure and k is Boltzmann's constant. This expression is convenient in the treatment of supersaturated vapour. For supercooled water, it is usually more convenient to use $d\mu = -s\,dT$ under isobaric conditions to determine

$$\Delta\mu = -\int_{T^i}^{T^*} (s_e - s_m)\,dT \approx -\frac{\lambda}{T}(T^i - T^*) \tag{2.49}$$

between the stable equilibrium freezing temperature T^i and the spontaneous freezing temperature T^*: the latent heat $\lambda = T(s_m - s_e)$.

The above analysis reveals how the spontaneous growth of an individual embryo may occur but it does not provide the rate at which nucleation events take place. Assuming that critical embryos, for which $\Delta G = \Delta G^*$ obey the Boltzmann distribution, their density at temperature T is represented by

$$n = n_v \exp(-\Delta G^*/kT)$$

where n_v is the number of fluid molecules per unit volume. Of these, only those embryos which successfully add to their bulk will nucleate. At the surface of each embryo, molecules are constantly joining and leaving, their net rate depending upon the activation energy barrier $\Delta G\ddagger$ through which they must pass. It may be shown (Fletcher, 1970) that the volumetric rate at which nucleation thus occurs is given by

$$\mathcal{J} = A \exp(-\Delta G\ddagger/kT) \exp(-\Delta G^*/kT) \tag{2.50}$$

In the freezing of water, for example, $A \simeq 10^{41}\,m^{-3}s^{-1}$ and $\exp(-\Delta G\ddagger/kT) \simeq 10^{-5}$, so that

$$\mathcal{J}_{iw} = O[10^{36} \exp(-\Delta G^*/kT)] \tag{2.51}$$

where O represents the order of magnitude and ΔG^* is given by equation (2.47) applied to fusion. Despite several theoretical limitations (Franks, 1985) in the derivation of the above equation[8], it does show consistency with experimental results. In particular, it reveals that the nucleation rate increases rapidly with decreasing temperature, thus heralding a lower bound for homogeneous nucleation close to $-40\,°C$.

A similar approach may be taken for *heterogeneous* nucleation in which a foreign substrate initiates embryo formation (Fletcher, 1970). The effect of the substrate, and of its molecular geometry and chemistry in particular, is to alter the threshold energy ΔG^* by a factor which depends upon the size of the substrate as well as the degree of wetting. Although difficult to apply in practice, the theory of heterogeneous nucleation does demonstrate the substantial reduction which foreign surfaces may produce in the tolerable supercooling. The above nucleation theories also provide a basis on which to study the effect of solutes and supercooling; this will be discussed further in Chapter 4.

2.9   Interfacial conditions and growth

The process of ice nucleus growth is complex and not fully understood. Numerous factors come into play but each may be viewed as an aspect of either of two[9] essential components of the problem, or to the balance between them. These components consist of factors which influence the transport of H_2O molecules through the fluid towards the interface, and factors which influence the accommodation of these molecules on the ice surface. In essence, the first component is a problem of diffusion in which the mass flux will, in general, be accompanied by other fluxes, as discussed in Section 2.4. If uncoupled, the mass flux at the interface is determined from the concentration field $C(\mathbf{X}, t)$ using an equation of the form

$$\partial C/\partial t = D\nabla^2 C$$

or, in normalized[10] form,

$$\frac{1}{Fo}\frac{\partial c}{\partial \tau} = \nabla^2 c$$

(2.52)

where Fo is a Fourier number. Equation (2.52)[11] must be solved subject to suitable initial and boundary conditions over the interface and distant from it. This topic will be discussed more fully in the next chapter. At this point it is sufficient to note that if the heat flux is also uncoupled an energy balance leads to an expression for the interface velocity $\mathbf{V}_I$ given by

$$\varrho_i \lambda \mathbf{V}_I = k_i(\nabla T_i)_I - k_f(\nabla T_f)_I \tag{2.53}$$

where the first term on the right hand side represents the heat flux withdrawn through the ice and the second term gives the heat flux supplied through the fluid. When the fluid is superheated, the first term on the right hand side must dominate the second if the ice is to grow; when the fluid is supercooled, the second term is usually the dominant. In either case, the solution of equations (2.52) must be such that

$$\rho_i \mathbf{V}_I = \rho D\nabla C = k_M \nabla \mu \tag{2.54}$$

at the interface, thus linking the heat and mass fluxes through a phase change even though they may be uncoupled as simultaneous flows. For pure water, mass diffusion has no role to play except through the effect of vacancy diffusion which influences molecular mobility in the condensed phases, as noted in Section 2.6.

The left hand side of equation (2.53) represents the latent heat flux which arises from the second component of the growth problem: namely, molecular accommodation and the accompanying interfacial energetics. Equations (2.48) and (2.49) both provide convenient expressions for $\Delta\mu$: these are necessary for the determination of r^* and ΔG^* which, as equations (2.46) and (2.47) indicate, are inverse functions of $\Delta\mu$. Thus whenever the supercooling or supersaturation is large, $|\Delta\mu|$ is large; the number of critical embryos is then large but their radius is small. Should this critical radius become less than or equal to the order of the molecular radius, the molecule may be adsorbed indiscriminately at any site; the 'crystal' surface will then be rough and diffuse. Accommodation at a rough interface is characterized by the absence of an energy barrier. Equation (2.54) suggests that the interface velocity is then proportional to the overall driving potential difference $|\Delta\mu|$, and thus, using equation (2.49), it is to be expected that

$$V_1 \propto (T^i - T) \tag{2.55}$$

for a rough surface. Growth is controlled by diffusion or self-diffusion alone. Diffusion dominance may be approached at high values of supersaturation or supercooling, even when the ice surface is not rough.

For perfectly smooth surfaces, accommodation is marked by an energy barrier. Interface growth in a supercooled fluid may then be treated in much the same way as nucleation, by modelling the embryo as a two-dimensional island of radius r and height a on the interface (Fletcher, 1970). Thus, in analogy with equation (2.45),

$$\Delta G = \rho_n \pi r^2 a \Delta\mu + 2\pi r a \sigma_1 \tag{2.56}$$

leading to

$$r^* = -\sigma_1/\rho_n\Delta\mu$$

and

$$\Delta G^* = -\pi\sigma_1^2 a/\rho_n\Delta\mu$$

$$\left.\right\} \tag{2.57}$$

In these expressions σ_1 is the line or edge tension (Defay, Prigogine, Bellemans and Everett, 1966). The interface velocity is proportional to the rate at which such critical embryonic islands appear and is therefore given by

$$V_1 \propto \exp\left(-\Delta G^*/kT\right) \tag{2.58}$$

Using equation (2.57) and (2.49) applied to the freezing of water, this may also be written approximately, for small supercoolings, as

$$V_I \propto \exp\left[-A/(T_i - T)\right] \tag{2.59}$$

where A is an empirically-determined constant.

An alternative approach to accommodation is to abandon the idea of a perfectly smooth surface and to treat growth as a surface dislocation phenomenon (Burton, Cabrera and Frank, 1951). The screw dislocation in particular has a step shaped like an Archimedean spiral which propagates radially at a rate V_s proportional to $|\Delta\mu|$. The growth of the interface is thus given by $V_I \propto V_s/d$, where $d = 4\pi r^*$ is the distance between successive turns of the spiral and r^* is the island critical radius. Since r^* is given by equation (2.57), we find that

$$V_I \propto (\Delta\mu)^2$$

or, using equation (2.49),

$$V_I \propto (T^i - T)^2 \tag{2.60}$$

for the freezing of water.

Experimental observations on freezing in supercooled water (Fletcher, 1970) provide some evidence in support of each of equations (2.55), (2.59) and (2.60). The majority of a-axis results approximate an empirical relation of the form

$$V_I \propto (T^i - T)^n \tag{2.61}$$

in which $1.3 < n < 2.2$: typically, $n \simeq 2$. Equation (2.59) appears to represent c-axis growth (Hillig, 1958) which is facetted and much slower than a-axis growth.

3

The Stefan problem

Ever since the pioneering work conducted by Stefan on sea ice formation, problems of freezing and melting have been described as Stefan problems. The essence of such problems, at least from the mathematical point of view, is that they involve a moving heat source whose position is unknown *a priori*. This source, the latent heat liberated or absorbed at the interface of two phases, arises out of the prevailing thermodynamic conditions and influences the temperature distribution. The temperature distribution, however, is influenced by the velocity of the interface, and hence the two are intimately related.

Under certain conditions it is possible to separate the determination of the temperature field from the prediction of interface movement. Section

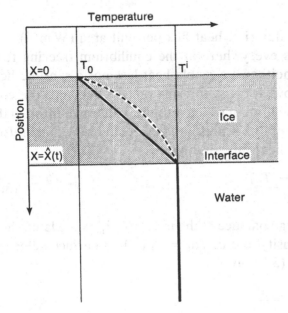

Fig. 3.1. The basic problem.

3.1 deals with such problems. More generally, a fuller analysis is required to incorporate the intrinsic coupling and to include the effect of sensible heat. Sections 3.2, 3.3 and 3.4 provide this analysis and show both the power and the limitation of analytic methods. The chapter concludes with a discussion of numerical methods for those wishing to apply them to problems of ice growth or decay.

3.1   The simplified model and its variations

Fig. 3.1 illustrates the problem considered by Stefan (1891). Lying over a large body of water at the stable equilibrium freezing temperature T^i, a layer of ice grows as the result of its upper surface temperature being held at some fixed value $T_0 < T^i$. The thickness $\hat{X}$ of the resultant ice layer is expected to be a monotonically increasing function of the time t.

Using the temperature–position coordinates shown in Fig. 3.1, the actual temperature distribution in the ice would look something like the dashed curve, but under quasi-steady conditions it would be represented by the straight line approximation to the gradient in Fourier's law

$$\mathbf{j}_Q = -k \nabla T$$

Thus, if the thermal conductivity is constant

$$j = \frac{k_i (T^i - T_0)}{\hat{X}} \tag{3.1}$$

is the density of the heat flux (i.e. heat flux per unit area ($W\,m^{-2}$).

Now since the water is everywhere at the equilibrium freezing temperature, it experiences neither a sensible heat loss nor a sensible heat gain. At the interface, however, the freezing process demands the continual removal of latent heat, all of which must be withdrawn through the ice by conduction. The heat flux represented by equation (3.1) therefore corresponds to the latent heat flux, and hence

$$\rho_i \lambda_{iw} \frac{d\hat{X}}{dt} = \frac{k_i (T^i - T_0)}{\hat{X}} \tag{3.2}$$

constitutes a thermal energy balance at the interface: λ_{iw} is the latent heat of fusion, ρ_i is the ice density and $d\hat{X}/dt = V_I$ is the interface velocity.

Re-arranging equation (3.2) gives

$$\hat{X} d\hat{X} = \frac{k_i (T^i - T_0) dt}{\rho_i \lambda_{iw}}$$

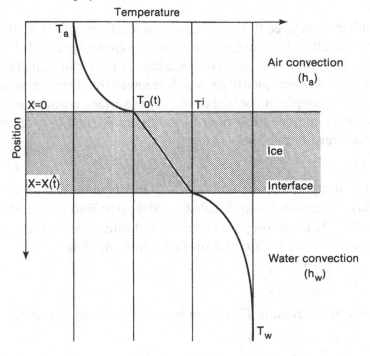

Fig. 3.2. Basic problem with convective boundary conditions.

which integrates to

$$\hat{X}^2 = \int_0^t \frac{2k_i(T^i - T_0)\mathrm{d}\tau}{\rho_i \lambda_{iw}} + \hat{X}^2(0),$$

where $\hat{X}(0)$ is the ice thickness when the upper surface temperature was first lowered to T_0. If $\hat{X}(0) = 0$, and the ice properties are constant, this result takes the simple form of the *Stefan solution*

$$\hat{X} = \left[\frac{2k_i(T^i - T_0)t}{\rho_i \lambda_{iw}}\right]^{\frac{1}{2}} \tag{3.3}$$

which has been in frequent use for almost a century.

The distinguishing feature of the above approach is that it only applies to quasi-steady growth, and is thus limited to situations where the heat capacity of the ice is negligible. This limitation is usually not very severe, and does not prevent the method from being applied to related problems which, as a class, are numerous (Lock, 1969a). Several important examples, each a variation on the basic theme, will now be treated in detail.

Firstly, consider a more realistic description of the upper boundary condition. Rather than specify the surface temperature, which is often unknown and difficult to measure, the heat flux might be specified. Very often, the heat flux *per se* is not known either, but it can usually be expressed in terms of other quantities which are available from empirical relationships. For example, if cooling of the upper surface is attributed to thermal convection in the atmosphere, as illustrated in Fig. 3.2, the surface heat flux density may be written

$$j = h_a (T_0 - T_a) \qquad (3.4)$$

where T_a is the temperature of ambient air, and h_a is a heat transfer coefficient. This expression could be substituted directly into the interface equation (3.2) but T_0 is no longer a constant and must first be found by combining equation (3.1) with equation (3.4), thus yielding

$$T_0 - T_a = \frac{k_i(T^i - T_0)}{h_a \hat{X} + k_i}$$

Using this result and equation (3.4), the interface equation becomes

$$\rho_i \lambda_{iw} = \frac{d\hat{X}}{dt} = \frac{k_i(T^i - T_a)}{\hat{X} + k_i/h_a}$$

which integrates to give

$$t = \frac{\rho_i \lambda_{iw}}{2k_i(T^i - T_a)} \left[\hat{X}^2 + \frac{2k_i \hat{X}}{h_a} \right] \qquad (3.5)$$

if the properties are again assumed to be constant and the water is initially free of ice. As the resistance $1/h_a$ of the convective system decreases to zero, this result reverts to the Stefan solution, as required.

Should the upper boundary condition be primarily one of radiative[1] emission, the linearity of the problem is lost because

$$j = \epsilon \sigma T_0^4 \qquad (3.6)$$

where σ is the Stefan–Boltzmann constant and ϵ is the surface emittance, often called the emissivity. Substituting this in equation (3.1) yields

$$T_0 = T^i - \alpha T_0^4$$

where $\alpha = \epsilon \sigma \hat{X}/k_i$. When $\alpha \ll 1$, a perturbation solution of this equation leads to

$$T_0 = T^i - \alpha(T^i)^4 + 4\alpha^2 (T^i)^7 - 22\alpha^3(T^i)^{10} + \cdots$$

Substituting this into the interface equation, and integrating, gives

$$t = \frac{k_i \rho_i \lambda_{iw}}{(\epsilon \sigma)^2 (T^i)^4} [\alpha + 2\alpha^2(T^i)^3 - 2\alpha^3(T^i)^6 + 7\alpha^4(T^i)^9 - \cdots] \qquad (3.7)$$

which, as $\alpha \rightarrow 0$, yields the simple radiative form

$$\hat{X}(t) = \epsilon\sigma(T^i)^4 t / \rho_i \lambda_{iw} \tag{3.8}$$

implying that when $T_0 \approx T^i$ the radiative heat loss and the interface velocity are virtually constant. Equations (3.5) and (3.7) are typical of many solutions to the Stefan problem: instead of obtaining $\hat{X}(t)$, the result is in the form $t(\hat{X})$. The latter function may be inverted, if necessary, but this seldom confers a computational advantage.

Similar modifications may be made to the lower boundary condition which affects the interface equation directly. As noted in the previous chapter, the presence of solutes in the water will influence the equilibrium freezing temperature, and will generally alter the shape of the interface (see Section 4.1). The effect of water superheat, however, is to introduce an additional heat flux into the interface equation (3.2) which must be extended to read

$$\rho_i \lambda_{iw} \frac{d\hat{X}}{dt} = \frac{k_i(T^i - T_0)}{\hat{X}} - h_w(T_w - T^i)$$

where h_w is the water heat transfer coefficient seen in Fig. 3.2. Re-arranging this,

$$\hat{X}d\hat{X} = \frac{k_i(T^i - T_0)}{\rho_i \lambda_{iw}} dt - \frac{h_w(T_w - T^i)\hat{X}}{\rho_i \lambda_{iw}} dt$$

which may be integrated formally to give

$$\hat{X}^2(t) = \frac{2k_i(T^i - T_0)t}{\rho_i \lambda_{iw}} - \frac{2h_w(T_w - T^i)}{\rho_i \lambda_{iw}} \int_0^t \hat{X} d\tau$$

if $\hat{X}(0) = 0$. This equation, in which the first term on the right hand side represents the Stefan solution, may be readily solved either by expanding $\hat{X}$ as a power series in t or by successive substitutions. More succinctly, the transformation

$$\hat{X}' = \frac{k_i(T^i - T_0)}{\rho_i \lambda_{iw}} - \frac{h_w(T_w - T^i)\hat{X}}{\rho_i \lambda_{iw}}$$

permits direct integration. For a body of water initially free of ice, the closed form solution is given by

$$t = \frac{\rho_i \lambda_{iw}}{h_w(T_w - T^i)} \left\{ \frac{k_i(T^i - T_0)}{h_w(T_w - T^i)} \ln\left[1 \Big/ \left(1 - \frac{h_w(T_w - T^i)\hat{X}}{k_i(T^i - T_0)}\right)\right] \right.$$

$$\left. - \hat{X} \right\} \tag{3.9}$$

As $t \rightarrow \infty$,

$$\hat{X} \to \frac{k_i (T^i - T_0)}{h_w (T_w - T^i)},$$

the steady state thickness which is, of course, independent of the initial thickness.

The restriction of constant properties may also be relaxed under certain circumstances. For example, the effective thermal conductivity of frozen material may often be represented solely as a function of temperature or position. If it is a position-dependent function $k_i(X)$ then the transformation to a new position variable

$$F(X) = \int_0^X \frac{d\zeta}{k_i(\zeta)}$$

leads to a temperature profile given by

$$\frac{T - T_0}{T^i - T_0} = \frac{\int_0^X d\zeta / k_i(\zeta)}{\int_0^{\hat{X}} d\zeta / k_i(\zeta)} = \frac{F(X)}{F(\hat{X})}$$

The temperature is no longer a linear function of X, and hence the interface equation must be written more precisely as

$$\rho_i \lambda_{iw} \frac{d\hat{X}}{dt} = k_i(\hat{X}) \left(\frac{\partial T}{\partial X} \right)_{\hat{X}} = \left(\frac{dT}{dF} \right)_{\hat{X}}$$

Substituting the temperature profile, and integrating, it is found that

$$t(\hat{X}) = \frac{\rho_i \lambda_{iw}}{(T^i - T_0)} \int_0^{\hat{X}} F(\zeta) \, d\zeta \tag{3.10}$$

the right hand side of which is readily determined once $k_i(X)$ has been specified. Similarly, for a temperature-dependent thermal conductivity, a new temperature variable

$$G(T) = \int_{T_R}^T k_i(\tau) \, d\tau$$

in which T_R is an arbitrary reference temperature, leads to the implicit temperature distribution

$$\frac{G(T) - G(T_o)}{G(T^i) - G(T_0)} = \frac{X}{\hat{X}}$$

Using this expression, the interface equation may be integrated to give

$$\hat{X}^2(t) = \frac{2}{\rho_i \lambda_{iw}} [G(T^i) - G(T_o)] t \tag{3.11}$$

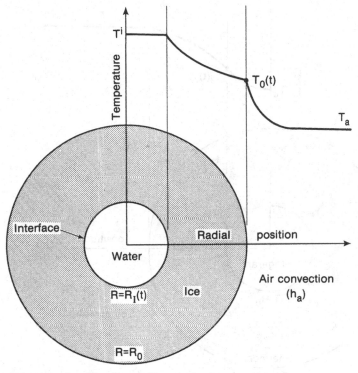

Fig. 3.3. Freezing in an air-cooled sphere

Again, the specification of the thermal conductivity's dependency makes the calculation simple.

Finally, it is worth noting that the quasi-steady approach is not limited to planar problems, but is readily extended to simple, common geometries for which a single space coordinate will suffice. It may also be used in multi-dimensional problems (see Section 3.3). Freezing inside an air-cooled sphere is represented in Fig. 3.3. If the water is initially at the temperature T^i, the quasi-steady temperature distribution in the ice is given by

$$\frac{T - T_0}{T^i - T_0} = \frac{1/R - 1/R_0}{1/R_I - 1/R_0}$$

where R_I is the radius of the ice–water interface. Hence the convective outer boundary condition may be written as

$$h_a(T_0 - T_a) = -k_i \left(\frac{\partial T}{\partial R}\right)_{R_0} = \frac{k_i(T^i - T_0)}{R_0^2(1/R_I - 1/R_0)}$$

and therefore

$$T^i - T_0 = \frac{h_a(T^i - T_a)}{h_a + \dfrac{k_i}{R_0^2(1/R_I - 1/R_0)}} .$$

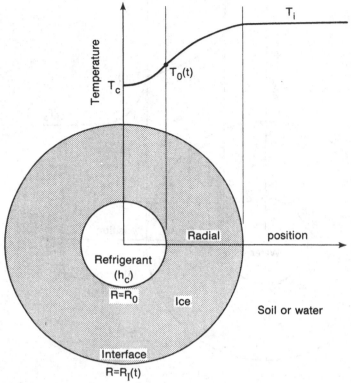

Fig. 3.4. Freezing around a pipe.

Substituting this into the interface equation given by

$$\rho_i \lambda_{iw} \frac{dR_I}{dt} = k_i \left(\frac{\partial T}{\partial R}\right)_{R_I} = \frac{-k_i(T^i - T_0)}{R_I^2(1/R_I - 1/R_0)}$$

re-arranging and integrating, gives

$$t = \frac{\rho_i \lambda_{iw} R_0^2}{2k_i(T^i - T_a)} \left[\frac{2}{3Bi_a}(1 - r^3) + \frac{1}{3}(1 + 2r^3) - r^2 \right] \qquad (3.12)$$

if t is measured from the moment when $R_I = R_0$, $r = R_I/R_0$ and $Bi_a = h_a R_0/k_i$ is a Biot number. The time taken to freeze the sphere completely ($r \to 0$) is thus given by

$$t = \frac{\rho_i \lambda_{iw} R_0^2}{6k_i(T^i - T_a)} \left(1 + \frac{2}{Bi_a}\right)$$

In view of the density difference between ice and water, freezing will cause an increase in water pressure until the tensile stress induced in the ice creates a fracture (e.g. in the bursting of droplets rapidly frozen in the atmosphere) or, if the freezing rate is low enough, leads to plastic yielding.

Another practical example is the freezing of soil or water surrounding a pipe carrying a refrigerant whose bulk temperature is T_c. This situation is illustrated in Fig. 3.4 where the thickness W_p of the pipe wall is not shown because the wall thermal resistance ($W_p/2\pi R_0 k_p$ per unit length) is usually small enough to be ignored. The quasi-steady temperature distribution within the ice is given by

$$\frac{T - T_0}{T^i - T_0} = \frac{\ln(R/R_0)}{\ln(R_I/R_0)}$$

from which the inner convective boundary condition may be satisfied to give

$$T^i - T_0 = \frac{h_c(T^i - T_c)}{h_c + k_i/R_0 \ln(R_I/R_0)}$$

This in turn may be substituted into the interface equation which, after integration, yields

$$t = \frac{\rho_i \lambda_{iw} R_0^2}{2k_i(T^i - T_c)}\left[\left(\frac{1}{Bi_c} - \frac{1}{2}\right)(r^2 - 1) + r^2 \ln r\right] \tag{3.13}$$

if the pipe is initially free of ice, $r = R_I/R_0$ and $Bi_c = h_c R_0/k_i$.

There are many other variations on the basic theme of quasi-steady analysis. Uncoupling of the ice temperature distribution from interface motion permits the use of all of the techniques used in linear heat conduction, except where they are specifically excluded, as for radiation boundary conditions or temperature-dependent properties. Together with those outlined above, these additional linear variations may be used separately or in combination, and thus embrace a large body of practical situations. When sensible heat effects are negligible quasi-steady analysis produces a ready, and simple, approximate solution to the Stefan problem. When sensible heat becomes important it is necessary to use other methods.

3.2   The Neumann solution

The first mathematical solutions of a solidification problem are attributable not to Stefan but to Lamé and Clapeyron (1831) and Neumann (Carslaw and Jaeger, 1959) whose work was undertaken in the 1860s. The classical solution of the latter is worth developing in its entirety.

The problem consists of a semi-infinite body of quiescent water whose initial temperature T_∞ is uniformly above the equilibrium freezing point T^i. At time $t = 0$, the free surface temperature of the domain is instantly lowered to, and subsequently maintained at, a fixed value of $T_0 < T^i$.

With X again being measured from the free surface, the equations governing heat conduction through the ice and water are, respectively,

$$\left.\begin{array}{l} \partial\theta_i/\partial t = \varkappa_i\, \partial^2\theta_i/\partial X^2 \\ \partial\theta_w/\partial t = \varkappa_w\, \partial^2\theta_w/\partial X^2 \end{array}\right\} \tag{3.14}$$

if the thermal properties are assumed to be constant and the two phases have the same density: $\theta_i = T_i - T_0$, $\theta_w = T_w - T_0$ and $\varkappa = k/\varrho c_p$ is the thermal diffusivity. These are to be solved subject to the initial condition

$$\theta_w(X,0) = T_\infty - T_0 = \theta_\infty$$

and the boundary conditions

$$\left.\begin{array}{l} \theta_i(0,t) = 0 \\ \theta_i(\hat{X},t) = \theta_w(\hat{X},t) = \theta_I \\ \theta_w(\infty,t) = \theta_\infty \end{array}\right\} \tag{3.15}$$

The position $X = \hat{X}(t)$ once more marks the ice–water interface at which $T_w = T_i = T_I = T^i$, and the full interface equation is given by

$$\rho_i\lambda_{iw}\frac{d\hat{X}}{dt} = k_i\left(\frac{\partial\theta_i}{\partial X}\right)_I - k_w\left(\frac{\partial\theta_w}{\partial X}\right)_I \tag{3.16}$$

the solution to which must satisfy the initial condition $\hat{X}(0) = 0$.

Solutions of the heat conduction equations (3.14) (Carslaw and Jaeger, 1959) which satisfy the conditions at $X = 0$ are

$$\left.\begin{array}{l} \theta_i(X,t) = A\operatorname{erf}\left[\dfrac{X}{2(\varkappa_i t)^{\frac12}}\right] \\[4mm] \text{for the ice, } 0 \leqslant X \leqslant \hat{X}, \text{ and} \\[4mm] \theta_w(X,t) = \theta_\infty + B\operatorname{erfc}\left[\dfrac{X}{2(\varkappa_w t)^{\frac12}}\right] \end{array}\right\} \tag{3.17}$$

for the water, $X \geqslant \hat{X}$. At the interface, $\theta_i = \theta_w = \theta_I$, and hence

$$A\operatorname{erf}\left[\frac{\hat{X}}{2(\varkappa_i t)^{\frac12}}\right] = \theta_I = \theta_\infty + B\operatorname{erfc}\left[\frac{\hat{X}}{2(\varkappa_w t)^{\frac12}}\right] \tag{3.18}$$

from which it follows that, since A and B are arbitrary constants, the arguments of the error functions must also be constants.

In particular, this implies that

$$\hat{X}(t) = 2\beta(\varkappa_i t)^{\frac12} \tag{3.19}$$

where β is a constant, yet to be determined. This form was anticipated in the Stefan solution, equation (3.3) of the previous section. It therefore appears that the inclusion of sensible heat in equations (3.14) does not change the essential structure of the solution.

From equations (3.18) it is evident that

$$A = \frac{\theta_I}{\operatorname{erf} \beta}$$

and

$$B = \frac{\theta_I - \theta_\infty}{\operatorname{erfc}\left[\beta(\varkappa_i/\varkappa_w)^{\frac{1}{2}}\right]}$$

and hence the temperature distributions are given by

$$\left.\begin{aligned}
\theta_i(X,t) &= \theta_I \operatorname{erf}\left[\frac{X}{2(\varkappa_i t)^{\frac{1}{2}}}\right]\Big/\operatorname{erf}\beta \\
\theta_w(X,t) &= \theta_\infty\left\{1 - \left(1 - \frac{\theta_I}{\theta_\infty}\right)\frac{\operatorname{erfc}[X/2(\varkappa_w t)^{\frac{1}{2}}]}{\operatorname{erfc}[\beta(\varkappa_i/\varkappa_w)^{\frac{1}{2}}]}\right\}
\end{aligned}\right\} \qquad (3.20)$$

The curvature in the error functions ensures that, at any given time, the temperature profiles in the ice and water are also curved. However, it is worth noting that

$$\operatorname{erf} x \to 2x/\pi^{\frac{1}{2}}$$

as $x \to 0$, and hence

$$\theta_i(X,t) \approx \theta_I X/(\pi\varkappa_i t)^{\frac{1}{2}}\operatorname{erf}\beta$$

when $X \ll 2(\varkappa_i t)^{\frac{1}{2}}$. Therefore the temperature profile in the ice tends towards a linear (Stefan) form as $\beta \to 0$.

To determine the coefficient β it is necessary to use the interface equation. Substituting equations (3.20) into equation (3.16) yields

$$\frac{\beta\pi^{\frac{1}{2}}}{Ste} = \frac{\exp(-\beta^2)}{\operatorname{erf}\beta} - \frac{k_w}{k_i}\left(\frac{\varkappa_i}{\varkappa_w}\right)^{\frac{1}{2}}\left(\frac{\theta_\infty - \theta_I}{\theta_I}\right)\frac{\exp(-\beta^2\varkappa_i/\varkappa_w)}{\operatorname{erfc}[\beta(\varkappa_i/\varkappa_w)^{\frac{1}{2}}]} \qquad (3.21)$$

where $Ste = c_p\theta_I/\lambda_{iw}$ is the ratio of the sensible heat in the ice to the latent heat. This is a key parameter which appears in Stefan's original paper and is now called the Stefan number (Lock, 1969a). Given the temperature ratio and property ratios, equation (3.21) indicates that β may be found for any particular Stefan number. For the special case of no superheat in the water, i.e., $\theta_\infty = \theta_I$, the equation evidently simplifies to

$$Ste/\pi^{\frac{1}{2}} = \beta\exp\beta^2\operatorname{erf}\beta \qquad (3.22)$$

Now if the exponential and error functions are expanded in series form, this expression may be written

$$\frac{Ste}{\pi^{\frac{1}{2}}} = \frac{2\beta}{\pi^{\frac{1}{2}}}(1 + \beta^2 + \cdots)\left(\beta - \frac{\beta^3}{3} + \cdots\right)$$

and hence

$$Ste = 2\beta^2\left(1 + \frac{2}{3}\beta^2 + \ldots\right)$$

which implies that as $Ste \to 0$, $\beta \to (Ste/2)^{\frac{1}{2}}$. Substituting this limit into equation (3.19) gives

$$\hat{X} = (2k_i\,\theta_1 t/\rho_i\lambda_{iw})^{\frac{1}{2}}$$

which is the Stefan solution. It is thus clear that the *relative* importance of sensible heat is dictated by the magnitude of the Stefan number. As $Ste \to 0$, the problems assume a quasi-steady form.

The planar growth of ice in supercooled water provides an interesting extension of Neumann's method. At the onset of nucleation, the material temperature often rises quickly from its initial value T_∞ to the equilibrium freezing temperature T^i. Assuming the ice contained in the region $0 \leqslant X \leqslant \hat{X}$ is neither heated nor cooled subsequently, its temperature will remain uniformly at T^i; the latent heat must then be removed entirely by conduction into the supercooled water for which the temperature distribution is given, for $X \geqslant \hat{X}$, by

$$\theta_w(X,t) = \frac{\theta_I\,\mathrm{erfc}\,[X/2(\varkappa_w t)^{\frac{1}{2}}]}{\mathrm{erfc}\,\beta} \tag{3.23}$$

where $\theta_w = T - T_\infty$. Note that this definition of θ differs from that used earlier in the Neumann solution, and β is now defined by $\hat{X} = 2\beta(\varkappa_w t)^{\frac{1}{2}}$. In the absence of conduction within the ice, the interface equation reduces to

$$\rho_i\lambda_{iw}\frac{d\hat{X}}{dt} = -k_w\left(\frac{\partial\theta_w}{\partial X}\right)_I$$

which again has a solution in the form of equation (3.19). Substitution of equation (3.23) leads to

$$Ste/\pi^{\frac{1}{2}} = \beta\exp\beta^2\,\mathrm{erfc}\,\beta \tag{3.24}$$

in which $Ste = c_{pw}\,\theta_I/\lambda_{iw}$, and the density difference between ice and water has been ignored. Again, as $Ste \to 0$, $\beta \to 0$, but the small degree of supercooling implied does not simplify the temperature profile in the water which is unbounded, and therefore $X/2(\varkappa_w t)^{\frac{1}{2}}$ may not be restricted to small values. However, the temperature field does tend towards a steady state solution in the moving coordinate X' measured from the interface: i.e. $X' = X - V_1 t$. This solution is given by

$$\theta_w(X') = \theta_I\exp\left(\frac{-V_I X'}{\varkappa_w}\right)$$

In any event, the solution is restricted by the requirement that the interface remain planar, a condition which is not satisfied for very long in pure

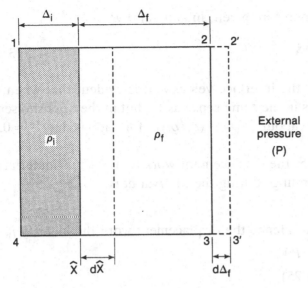

Fig. 3.5. Ice growth from fluid phase.

water or aqueous solutions unless the interface velocity is very low (Woodruff, 1973). Nevertheless, the above results always provide a basis for the analysis of the interfacial instabilities which eventually destroy the planar form.

3.3 Physico-mathematical analysis

In general, the effect of sensible heat may not be ignored, and the density difference between ice and water (or water vapour) must also be taken into account. A density difference introduces a fluid velocity relative to the ice, and since this motion acts to push (or pull) back neighbouring fluid it implies the expenditure of expansion work. Fig. 3.5 shows a given amount of planar ice Δ_i with an adjacent fluid phase Δ_f bounded initially by the arbitrary enclosure 1234, and by 12'3'4 a short time dt later. During the interval dt, the interface at $\hat{X}$ has moved to $\hat{X} + d\hat{X}$ causing a displacement $d\Delta_f$ in the fluid boundary as shown. Since the total mass initially enclosed by 1234 is fixed it follows that

$$\rho_i \, d\hat{X} = \rho_f \, d\hat{X} - \rho_f \, d\Delta_f$$

and hence

$$d\Delta_f = \left(\frac{\rho_f - \rho_i}{\rho_f}\right) d\hat{X}$$

The fluid velocity is therefore given, in general, by

$$V_f = \left(\frac{\rho_f - \rho_i}{\rho_f}\right) V_I \tag{3.25}$$

where $V_I = d\hat{X}/dt$ is the interface velocity. It is evident that when ice grows from water V_f is in the same sense as V_I, but in the opposite sense when ice grows from vapour ($V_f \approx -(\varrho_i/\varrho_f)V_I$ for vapour and $V_f \approx 0.08 V_I$ for water).

Referring to Fig. 3.5, the displacement work done by this material on the immediate surroundings during the interval dt is

$$đw = Pd\varDelta_f$$

per unit interface area. Hence the displacement work flux density j_W is

$$đw/dt = j_W = PV_f,$$

and using equation (3.25),

$$j_W = \frac{P(\rho_f - \rho_i)}{\rho_f} V_I \tag{3.26}$$

In the limit as the arbitrary thicknesses $\varDelta_i$ and $\varDelta_f$ tend to zero, the net heat flux, leaving the system is

$$(j_Q)_{net} = k_i(\partial T_i/\partial X)_I - k_f(\partial T_f/\partial X)_I \tag{3.27}$$

while the rate of increase of energy of the same system per unit interface area is given by

$$de/dt = V_I\rho_i(u_i - u_f) \tag{3.28}$$

where u is the specific internal energy (potential and kinetic energy are neglected). Now the energy principle, equation (2.2), requires that

$$de/dt = -(j_Q)_{net} - j_W$$

and hence

$$V_I\rho_i(u_i - u_f) = -k_i\left(\frac{\partial T_i}{\partial X}\right)_I + k_f\left(\frac{\partial T_f}{\partial X}\right)_I - \frac{P(\rho_f - \rho_i)}{\rho_f}V_I$$

Therefore

$$k_i\left(\frac{\partial T_i}{\partial X}\right)_I - k_f\left(\frac{\partial T_f}{\partial X}\right)_I = \rho_i V_I[(u_f - u_i) + P(v_f - v_i)]$$

where v is the specific volume. But

$$(u_f - u_i) + P(v_f - v_i) = (u_f + Pv_f) - (u_i + Pv_i) = h_f - h_i = \lambda_{if}$$

which defines the latent heat, λ_{if}, and hence

$$\rho_i\lambda_{if}V_I = k_i(\partial T_i/\partial X)_I - k_f(\partial T_f/\partial X)_I \tag{3.29}$$

which is an interface energy balance, and not merely a heat balance, linking the two phases.

Within either phase, the energy principle, given in point form by equation (2.9), requires that

$$\rho\,du/dt = -\nabla \cdot \mathbf{j}_E \tag{3.30}$$

again ignoring potential and kinetic energy. Expanding the energy flux density as

$$\mathbf{j}_E = \mathbf{j}_Q + P\mathbf{V} + \mathbf{j}_W'$$

where $\mathbf{j}_W'$ denotes work forms, other than displacement work, e.g., chemical or electrical work, and noting that (strictly this is only true if $\partial P/\partial t = 0$)

$$\nabla \cdot (P\mathbf{V}) = \rho\frac{d}{dt}(Pv)$$

equation (3.30) may be re-written as

$$\rho\frac{d}{dt}(u + Pv) = \rho\frac{dh}{dt} = -\nabla \cdot \mathbf{j}_Q - \nabla \cdot \mathbf{j}_W' \tag{3.31}$$

The second term on the right hand side is a heat generation or work dissipation term arising from the diffusion of solutes, electrons etc. In the absence of work dissipation, equation (3.31) becomes

$$\rho(\partial h/\partial t + \mathbf{V} \cdot \nabla h) = \nabla \cdot (k\nabla T) \tag{3.32}$$

after substituting Fourier's law.

Now if $h = h(P, T)$, it follows that

$$dh = \left(\frac{\partial h}{\partial T}\right)_P dT + \left(\frac{\partial h}{\partial P}\right)_T dP$$

so that if the pressure is fixed, or $(\partial h/\partial P)_T = 0$, or both,

$$dh = c_p\,dT$$

Therefore, when the material properties are constant, equation (3.32) reduces to

$$\partial T/\partial t + \mathbf{V} \cdot \nabla T = \varkappa\nabla^2 T \tag{3.33}$$

in which $\varkappa = k/\varrho c_p$ is the thermal diffusivity. When the frame of reference is embedded in the ice, it is clear that

$$\partial T_i/\partial t = \varkappa_i\nabla^2 T_i \tag{3.34}$$

governs heat transfer in the ice, whereas

$$\partial T_f/\partial t + \mathbf{V}_f \cdot \nabla T_f = \varkappa_f\nabla^2 T_f \tag{3.35}$$

applies to the fluid phase. In the absence of any imposed velocity field, $\mathbf{V}_f$ is determined solely from equation (3.25).

Before turning to the solutions of equations (3.34) and (3.35), together with equation (3.29), it is worthwhile making a brief examination of orders of magnitude and property dependence. The second of these is perhaps easier to deal with because, of the three most relevant thermal properties, ϱ, k and c_p, only the last needs further discussion: ϱ appears only as a multiplier in the general form (3.32); the treatment of k as a function of temperature or position was discussed in section 3.1, and it is only necessary to add that the transformations also apply under transient conditions. As seen above, the specific heat arises from the description of enthalpy as a function of temperature. Should the pressure dependency of enthalpy also be important, the approximation $dh = c_p dT$ is no longer adequate. It is often sufficient to incorporate c_p as a variable, usually as a function of temperature, though this renders equation (3.32) intrinsically non-linear in temperature. A particularly important example of this non-linearity is found in the absorption or release of latent heat over a range of temperatures: the enthalpy change must then be written as the sum of sensible and latent heat changes. For this situation,

$$dh = c_p(T) dT - \lambda dm_i$$

where dm_i is the mass fraction of ice[2] appearing in the temperature range dT. The apparent specific heat,

$$c_p' = (\partial h / \partial T)_P = c_p(T) - \lambda (\partial m_i / \partial T)_P \tag{3.36}$$

and the corresponding expression $dh = c_p'(T)dT$, may then be used in equation (3.32), thereby converting a Stefan problem into one of heat conduction with variable specific heat. The interface equation no longer contains the latent heat term and the temperature equation is generally more difficult to solve. One particular situation for which an exact solution is readily available occurs when the latent heat is released uniformly between two temperatures T_2 and T_1, in which case $\partial m_i / \partial T = 1/(T_2 - T_1)$. The solution to the corresponding Neumann problem proceeds along the lines developed in the previous section, with c_p' replacing c_p and the latent heat term in the interface equation being suppressed.

Orders of magnitude may be found by normalizing equations (3.29), (3.34) and (3.35) which, for simplicity, will be considered here in one-dimensional form. Using the normalized variables[3]

$$\phi = \frac{\theta}{\theta_C}, \tau = \frac{t}{t_C}, x = \frac{X}{X_C} \quad \text{and} \quad \xi = \frac{\hat{X}}{X_C}$$

the interface energy balance becomes

$$\frac{d\xi}{d\tau} = \left(\frac{\partial \phi_i}{\partial x_i}\right)_I - \left[\left(\frac{k_f}{k_i}\right)\left(\frac{\theta_{Cf}}{\theta_{Ci}}\right)\left(\frac{X_{Ci}}{X_{Cf}}\right)\right]\left(\frac{\partial \phi_f}{\partial x_f}\right)_I \qquad (3.37)$$

if the time scale is defined by

$$t_C = \rho_i \lambda_{if} X_{Ci}^2 / k_i \theta_{Ci} \qquad (3.38)$$

It is immediately evident from equation (3.37) that the *relative* importance of heat conduction in the fluid phase is given by the magnitude of the square bracketed parameter in relation to unity. The intrinsic scales for temperature, length and conductivity are not generally equal in the ice and fluid systems; their ratios become important parameters whose role is now apparent. It is useful to define $\theta_{Cf}/\theta_{Ci} = (T_\infty - T_1)/(T_1 - T_0)$ as the superheat ratio. Although more difficult to define, the length ratio X_{Ci}/X_{Cf} helps distinguish[4] between 'thin' ice systems, in which $X_{Ci} \ll X_{Cf}$ and 'thick' ice systems for which $X_{Ci} \simeq X_{Cf}$.

The same normalized variables transform the ice energy equation (3.34) into

$$\partial^2 \phi_i / \partial x_i^2 = Ste_i (\partial \phi_i / \partial \tau) \qquad (3.39)$$

revealing that the relative importance of the sensible heat in a Stefan problem is given by the magnitude of the Stefan number, as noted earlier. (Note that $c_p \theta_C / \lambda$ may be different in each phase.) Similarly the fluid energy equation (3.35) becomes

$$\frac{\partial^2 \phi_f}{\partial x_f^2} = \left(\frac{X_{Cf}}{X_{Ci}}\right)^2 \left(\frac{\varkappa_i}{\varkappa_f}\right) Ste_i \left\{\frac{\partial \phi_f}{\partial \tau} + \left[\left(\frac{X_{Ci}}{X_{Cf}}\right)\left(\frac{\rho_f - \rho_i}{\rho_f}\right)\right]\frac{d\xi}{d\tau}\frac{\partial \phi_f}{\partial x_f}\right\} (3.40)$$

in the absence of an imposed velocity field. This shows that the relative importance of a change in density depends not only upon the magnitude of the change itself but upon the relative thickness of the ice. It is apparent that the effect of a change in density is likely to be small, especially in 'thin' ice.

3.4 Analytic techniques

Stefan problems have long intrigued mathematicians, some of whom have proved to be very ingenious in devising solution techniques capable of handling the intrinsic non-linearity caused by the moving interface. Conceptually, phase change problems may be regarded simply as a special class of moving heat source problems, but the coupling between the interface motion and the temperature distribution makes them rather unusual and particularly challenging. The ability to use linear solutions within the framework of a non-linear problem is especially intriguing. However, given the anticipated readership of this book no attempt will be made here to deal with purely mathematical problems, or with analytic solutions in general. Instead, a number of selected techniques will be used to uncover the structure and tractability of the Stefan problem, and to offer some insight into the roles played by various terms. Each technique will be applied to the essential Stefan problem (i.e. freezing at the edge of a semi-infinite domain: $\phi(x,\tau) \leq 0$.) posed as a search for solutions to the equations

$$\left. \begin{array}{l} \partial^2\phi/\partial x^2 = \sigma(\partial\phi/\partial\tau) \\ d\xi/d\tau = (\partial\phi/\partial x)_\xi \end{array} \right\} \tag{3.41}$$

subject to suitable boundary and initial conditions: it will be assumed that $\phi = 0$ if $x \geq \xi$ (for all τ) and when $\tau = 0$ (for all x). The Stefan number has been represented by σ.

3.4.1 Series solutions

Evans, Isaacson and Macdonald (1950) have demonstrated that

$$\phi(x,\tau) = \sum_{i,j=0}^{\infty} a_{ij} x^i \tau^j$$

is a solution of the first of equations (3.41). This may be combined with the form

$$\xi(\tau) = \sum_{i=0}^{\infty} b_i \tau^i / i!$$

to yield a complete solution in which the coefficients a_{ij} and b_i may be determined through substitution into the interface equation and the boundary condition $\phi(\xi, \tau) = 0$. For a system consisting initially of one phase, and a prescribed rate of heat removal at $x = 0$, $\xi(0) = 0$ and $f(\tau) = (\partial\phi/\partial x)_0$. The corresponding coefficients describing interface growth are then

$$b_0 = 0$$
$$b_1 = f(0)$$
$$b_2 = -\sigma f^3(0) + \dot{f}(0)$$
$$b_3 = 5\sigma^2 f^5(0) - 6\sigma \dot{f}(0) f^2(0) + \ddot{f}(0)$$
etc.

where the dot indicates differentiation with respect to time. Thus, when the heat flux density is held constant at f_0, interface growth is described by

$$\xi = \xi^0 \left[1 - \sigma\tau\frac{f_0^2}{2!} + 5(\sigma\tau)^2\frac{f_0^4}{3!} - \cdots \right] \tag{3.42}$$

where $\xi^0 = \tau f_0$ is the solution when sensible heat is neglected.

The use of the Stefan number as a regular perturbation parameter has been suggested (Lock, 1971; Lock, Gunderson, Quon and Donnelly, 1969). Taking

$$\phi(x,\tau,\sigma) = \sum_{i=0}^{\infty} \sigma^i \phi_i(x,\tau)$$

and

$$\xi(\tau,\sigma) = \sum_{i=0}^{\infty} \sigma^i \xi_i(\tau),$$

generates a set of functions ϕ_i and ξ_i satisfying equations (3.41). When the temperature at $x = 0$ is a prescribed function $\phi_0(\tau)$, interface growth is given by

$$\xi = \xi^0 \left[1 + \frac{\sigma}{6}\phi_0(\tau) + O(\sigma^2) \right] \tag{3.43}$$

where

$$\xi^0 = \left[-2\int_0^\tau \phi_0(\zeta)\,\mathrm{d}\,\zeta \right]^{\frac{1}{2}}$$

again represents the solution when $\sigma = 0$. It is interesting to note that if $\phi_0(\tau)$ is periodic, the interface location is given by ξ^0 every half cycle, at least to an accuracy of σ^2. This helps explain why degree–day estimates of soil freezing (and active zone) depths based on half cycles are so insensitive to the effect of sensible heat.

In many natural examples of ice forming from water, $\sigma \ll 1$, in which case series solutions such as the above are simple, rapid and easily tailored to any desired accuracy; this can be said with less confidence when $\sigma \to 1$. Should $\sigma > 1$, such series solutions may not converge unless $\tau \ll 1$. In general, therefore, alternative formulations are necessary.

3.4.2 *Transformation and similarity*

The difficulty associated with motion of the interface may often be reduced by means of a transformation suggested in the work of Landau (1950). If displacement is defined in relation to the interface by taking

$$\eta = x/\xi(\tau) \tag{3.44}$$

equations (3.41) transform to

$$\frac{\partial^2 \Phi}{\partial \eta^2} + \sigma \eta \xi \dot{\xi} \frac{\partial \Phi}{\partial \eta} = \sigma \xi^2 \frac{\partial \Phi}{\partial \tau} \left.\begin{array}{c} \\ \\ \\ \\ \\ \\ \end{array}\right\}$$

and

$$\frac{d\xi}{d\tau} = \frac{1}{\xi}\left(\frac{\partial \Phi}{\partial \eta}\right)_1$$

(3.45)

where $\Phi(\eta, \tau) = \phi(x, \tau)$. In this form, the equations have been the subject of mathematical studies (Portnov, 1962; Jackson, 1964) in which solutions were sought through the use of the Laplace transformation. Here they will be used to explore the prospect of similarity solutions.

Similarity techniques are common in boundary layer theory which suggests a search for solutions in the form

$$\Phi(\eta, \tau) = g(\tau)f(\eta)$$

Substituting into equations (3.45) gives

$$f'' + \sigma(\xi \dot{\xi})\eta f' = \sigma(\dot{g}\xi^2/g)f \left.\begin{array}{c} \\ \\ \end{array}\right\}$$

and

$$d\xi/d\tau = g(f'(1)/\xi)$$

(3.46)

where the prime denotes differentiation with respect to η. The requirement of similarity is met when the first of these equations is reduced to an ordinary differential equation for $f(\eta)$. This is possible only if the bracketed terms are constants. From the left hand side of the equation, it is evident that such a constraint requires

$$\xi(d\xi/d\tau) = \text{constant}$$

leading to the familiar interfacial result $\xi \propto \tau^{\frac{1}{2}}$ which, when used with the constraint on the right hand side, demands that

$$\frac{\tau}{g}\frac{dg}{d\tau} = \text{constant} = m, \text{ say}$$

i.e. $g \propto \tau^m$. But the interface equation requires that g is independent of time when $\xi \propto \tau^{\frac{1}{2}}$, and hence $m = 0$ provides the only prospect for similarity. The single similarity solution may thus be found from the integration of

$$f'' + \sigma \eta f' = 0$$

which yields the error function form found by Neumann for a step change in surface temperature: i.e., when g is a constant.

Similarity analysis is not limited to materials with constant properties or to materials which do not undergo a density change upon freezing (Hamill and Bankoff, 1964). The analysis may also be extended to non-similar situations by setting $\eta = x/\tau^{\frac{1}{2}}$ so that equations (3.41) transform to

$$\left.\begin{aligned} \frac{\partial^2 \Phi}{\partial \eta^2} + \frac{\sigma \eta}{2} \frac{\partial \Phi}{\partial \eta} &= \sigma \tau \frac{\partial \Phi}{\partial \tau} \\[2em] \frac{\mathrm{d}\xi}{\mathrm{d}\tau} &= \frac{1}{\tau^{\frac{1}{2}}} \left(\frac{\partial \Phi}{\partial \eta}\right)_{x=\xi} \end{aligned}\right\} \tag{3.47}$$

and

Solutions may then be sought by taking

$$\Phi(\eta, \tau) = \sum_{i=0}^{\infty} \Phi_i(\eta) \tau^i$$

and

$$\xi(\tau) = \sum_{i=0}^{\infty} \beta_i \tau^{(i+1)/2}$$

in which $\Phi_i(\eta)$, $\beta_i(\tau)$ are determined through substitution into equations (3.47). Convergence is the principal limitation of the method (Poots, 1962).

3.4.3   Integral solutions

The Von Karman–Pohlhausen integral technique (Von Karman, 1921; Pohlhausen, 1921) is well developed in boundary layer theory where the form of the governing equations, the heat conduction equation in particular, has certain similarities with the equations describing freezing. It is not surprising that the technique re-appears in the solution of Stefan problems cast in like form. Boundary layer usage is frequently limited to similarity forms and therefore the application of the technique to Stefan problems, if it is not to be restricted to the Neumann solution, must proceed along slightly different lines.

The technique is well illustrated by the work of Goodman (1964) using the example of an imposed boundary heat flux applied to material initially at the equilibrium freezing temperature. Integrating the first of the equations (3.41) with respect to x over $0 \leqslant x \leqslant \xi$ gives

$$\left(\frac{\partial \phi}{\partial x}\right)_\xi - \left(\frac{\partial \phi}{\partial x}\right)_0 = \sigma \int_0^\xi \frac{\partial \phi}{\partial \tau} \mathrm{d}x = \sigma \frac{\mathrm{d}}{\mathrm{d}\tau} \int_0^\xi \phi \, \mathrm{d}x$$

using Reynolds' transport theorem. But in view of the initial condition, $(\partial\phi/\partial x)_\xi = d\xi/d\tau$, while $(\partial\phi/\partial x)_0 = f(\tau)$, the surface heat flux density, and hence

$$\frac{d\xi}{d\tau} = f(\tau) + \sigma\frac{d}{d\tau}\int_0^\xi \phi\, dx$$

Integrating,

$$\xi(\tau) = \xi^0(\tau) + \sigma\int_0^\xi \phi\, dx \qquad (3.48)$$

where $\xi^0(\tau) = \int_0^\tau f(\zeta)\,d\zeta$. The variation of ϕ with x and τ may now be introduced as a guess, consistent with the boundary conditions. Obviously such a guess can benefit from insight and experience. Furthermore, since the term in question reflects the effect of sensible heat, the magnitude of any error is limited by the Stefan number which, as noted above, is commonly less than 1.0 in ice formation problems.

Taking a solution in polynomial[5] form

$$\phi(x,\tau) = a_1(\tau) + a_2(\tau)x + a_3(\tau)x^2$$

requires three conditions to be satisfied in order to determine the coefficients $a(\tau)$. If a constant heat flux of density f_0 is being removed, then

$$f_0 = (\partial\phi/\partial x)_0 = a_2(\tau)$$

and since $\phi(\xi, \tau) = 0$,

$$a_1(\tau) = -a_2(\tau)\xi - a_3(\tau)\xi^2$$

The third condition arises from the fact that the rate of change of temperature on the interface

$$\frac{d\Phi_\xi}{d\tau} = \frac{\partial\phi_\xi}{\partial\tau} + \frac{d\xi}{d\tau}\left(\frac{\partial\phi_\xi}{\partial x}\right)_\xi$$

is zero. Hence

$$(\partial\phi/\partial\tau)_\xi = -(\partial\phi/\partial x)_\xi^2$$

using the second of equations (3.41), and therefore

$$(\partial^2\phi/\partial x^2)_\xi = -\sigma(\partial\phi/\partial x)_\xi^2$$

using the first equation. The third condition is thus provided by the implicit expression

$$a_3(\tau) = -\frac{\sigma}{2}[a_2^2(\tau) + 4a_2(\tau)a_3(\tau)\xi + 4a_3^2(\tau)\xi^2]$$

Using these relations to find $\phi(x, \tau)$, and hence the second term on the right hand side of equation (3.48), Goodman (1958, 1964) has obtained a series expression for interface growth which agrees with the first four terms of the exact result presented earlier in equation (3.42).

In general terms, integral methods work better with smooth, slowly-varying boundary functions. For phase change problems this corresponds to the surface temperature or heat flux varying monotonically with time. Periodic or step functions, for example, are difficult to treat, largely because the ice temperature profile cannot then be represented adequately by a simple polynomial. When the Stefan number is much less than 1, the method may be simplified further without much loss in overall accuracy by substituting the steady state temperature profile, which eliminates the transient effect of sensible heat but retains the quasi-steady effect. This removes the need to satisfy the second of the two interface conditions used above. For example, for freezing caused by the withdrawal of a constant heat flux of density f_0,

$$\int_0^\xi \phi\,dx = -\int_0^\xi f_0(\xi - x)\,dx = -\frac{f_0\xi^2}{2}$$

and therefore equation (3.48) becomes

$$\xi(\tau) = f_0\tau - \sigma\frac{f_0}{2}\xi^2$$

Using the perturbation expansion

$$\xi(\tau) = \xi_0(\tau) + \sigma\xi_1(\tau) + \sigma^2\xi_2(\tau) + \dots$$

it may be shown that

$$\xi_0(\tau) = f_0\tau, \; \xi_1(\tau) = -\frac{f_0^3}{2}\tau^2, \; \xi_2(\tau) = \frac{f_0^5\tau^3}{2} \text{ etc,}$$

and hence

$$\xi(\tau) = \xi^0\left[1 - \sigma\tau\frac{f_0^2}{2} + O(\sigma\tau)^2\right]$$

which agrees with the exact solution, equation (3.42), to an accuracy of the Stefan number squared.

Analytic techniques constitute a powerful group of tools. Taken collectively, they provide the means whereby Stefan problems may be properly phrased and solved in forms which are computationally appropriate. In principle, analytic techniques are capable of handling multi-dimensional systems within which the properties are not constant and around which the boundary conditions are time dependent, but this is seldom achieved in practice. However, the few selected techniques discussed above, along with others beyond the scope of this text, must not be disregarded on that account. Apart from the insight and structure they bring, they apply directly to a great many practical problems at the Earth's surface and in the near-surface regions of three-dimensional bodies. Typically, they offer a simple solution, often in closed form (see, e.g. Zarling, 1987; Lunardini, 1987) which is not only convenient but also provides a comparison for a more comprehensive new technique, or a starting solution for more elaborate procedures such as those used in numerical analysis.

3.5   Numerical techniques

The high-speed digital computer has dramatically altered procedures for the calculation of temperature fields and interface motion. As computing speeds have increased, a variety of numerical formulations and techniques have become increasingly practical, and their scope has gradually been enlarged. This process, which has been accompanied by a commensurate growth in storage capacity, has stimulated the development of new and refined methods. In this continuing, successful expansion it is important to recognize that numerical methods are no better than the physico-mathematical model on which they are based, and that the full cost of numerical results must always include the effort expended on the formulation, the programming and the testing of the scheme.

In the absence of clear cut criteria for stability and convergence, a set of numerical results must be treated with a healthy degree of suspicion until the data have been compared with another independent set, an experiment or, as often happens, with a set of analytic results. A typical text on advanced numerical methods will demonstrate how given problems may be, and have been, solved; less frequently will it be able to assist the reader who wishes to solve a different problem. Numerical methods in general, and those applicable to ice formation in particular, are in a period of rapid development, and for this reason it would be unwise to present them here in detail; indeed space limitations alone preclude a treatment that would do them justice. However, since they are in widespread and growing use a brief introduction would appear to be worthwhile for student and practitioner alike. Broadly, they are divided into two categories: *finite difference* methods and *finite element* methods.

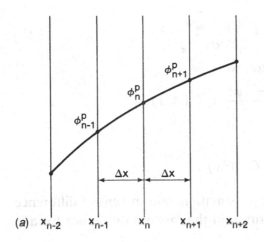

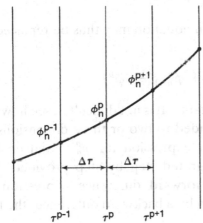

Fig. 3.6. Discretization of temperature: (*a*) spatial variation at a given time, $p\,\Delta\tau$; (*b*) temporal variation at a given location, $n\,\Delta x$.

3.5.1 The finite difference method

Consider a Taylor expansion of the temperature about the spatial point n in Fig. 3.6:

$$\phi^p_{n+1} = \phi^p_n + \left(\frac{\partial \Phi}{\partial x}\right)^p_n \Delta x + \frac{1}{2!}\left(\frac{\partial^2 \phi}{\partial x^2}\right)^p_n \Delta^2 x + \dots$$

and

$$\phi^p_{n-1} = \phi^p_n - \left(\frac{\partial \phi}{\partial x}\right)^p_n \Delta x + \frac{1}{2!}\left(\frac{\partial^2 \phi}{\partial x^2}\right)^p_n \Delta^2 x + \dots$$

$$(3.49)$$

where p denotes a particular time. Similarly, an expansion about the temporal point p is given by

$$\phi_n^{p+1} = \phi_n^p + \left(\frac{\partial \phi}{\partial \tau}\right)_n^p \Delta\tau + \frac{1}{2!}\left(\frac{\partial^2 \phi}{\partial \tau^2}\right)_n^p \Delta^2\tau + \dots \tag{3.50}$$

Equations (3.49) may be added to give

$$\left(\frac{\partial^2 \phi}{\partial x^2}\right)_n^p = \frac{\phi_{n+1}^p + \phi_{n-1}^p - 2\phi_n^p}{\Delta^2 x} + O(\Delta^2 x)$$

and subtracted to yield

$$\left(\frac{\partial \phi}{\partial x}\right)_n^p = \frac{\phi_{n+1}^p - \phi_{n-1}^p}{2\Delta x} + O(\Delta^2 x)$$

Both of these finite difference approximations are in central difference form. Equation (3.50) may be written in the forward difference form

$$\left(\frac{\partial \phi}{\partial \tau}\right)_n^p = \frac{\phi_n^{p+1} - \phi_n^p}{\Delta\tau} + O(\Delta\tau)$$

The one-dimensional heat conduction equation may thus be replaced by the different equation[6]

$$\sigma(\phi_n^{p+1} - \phi_n^p) = \frac{\Delta\tau}{\Delta^2 x}(\phi_{n+1}^p + \phi_{n-1}^p - 2\phi_n^p) \tag{3.51}$$

in which $\Delta\tau$ and Δx are taken as constants. It is not difficult to see how this form of representation may be extended to two or three dimensions.

Equation (3.51) is a fully explicit expression for ϕ_n^{p+1} and enables the differential equation to be integrated step by step provided[7] that $\Delta\tau/\sigma\Delta^2 x \leqslant \frac{1}{2}$. Alternatively, if the forward difference expression replacing the time derivative is replaced by a backward difference, the heat conduction equation becomes

$$\sigma(\phi_n^{p+1} - \phi_n^p) = \frac{\Delta\tau}{\Delta^2 x}(\phi_{n+1}^{p+1} + \phi_{n-1}^{p+1} - 2\phi_n^{p+1}) \tag{3.52}$$

which is an implicit expression carrying with it the requirement of matrix inversion, and therefore increased computational time, but having no restriction imposed by stability.[8] The solution of a set of linear equations such as those expressed in (3.51) or (3.52) presents no special difficulty from a computational standpoint (Jennings, 1977). Accuracy, speed and total cost will dictate which formulation is preferable in given circumstances (Murray and Landis, 1959; Lockwood, 1966; Goodrich, 1982b).

Solutions to the nodal equations (3.51) or (3.52) must satisfy appropriate initial and boundary conditions, the former being introduced simply by prescribing $\phi(x, 0)$. Given that there are N elements, with $n = 1, 2, \dots, N, N + 1$, the boundary condition at x_1 may be satisfied in the following manner:

(1) First kind (Dirichlet): $\phi_1(\tau)$ is prescribed at x_1, the first nodal equation being satisfied at x_2.

(2) Second kind (Neumann): $(\partial\phi/\partial x)_1$ is prescribed at x_1, with the unknown ϕ_1, being determined from an additional finite difference expression. For example,[9] $(\partial\phi/\partial x)_1 \approx (\phi_2 - \phi_1)/\Delta x$.

(3) Third kind (linear convection): a relation of the form $(\partial\phi/\partial x)_1 = \alpha(\phi_1 - \phi_\infty)$, where ϕ_∞ represents the distant fluid temperature, is given. The unknown ϕ_1 is thereby incorporated by using a suitable approximation for the gradient, as above.

(4) Third kind (nonlinear radiation): a relation such as $(\partial\phi/\partial x)_1 = \beta T_1^4$ where T_1 is the absolute temperature at x_1, is given: this again permits ϕ_1 to be expressed in terms of prescribed quantities and other (unknown) nodal temperatures.

In general, the boundary condition at x_1 (or x_{N+1}) may be satisfied either by prescribing the nodal temperature or by specifying a suitable heat flux relationship at the node.

The principal computational difficulty arises from the interface equation represented by

$$\frac{\mathrm{d}\xi}{\mathrm{d}\tau} = \left(\frac{\partial\phi_i}{\partial x}\right)_I - \beta\left(\frac{\partial\phi_f}{\partial x}\right)_I$$

where β is a known coefficient (see equation (3.37)). From earlier discussion it is apparent that the temperature gradient at the interface may be written in a form such as

$$\left(\frac{\partial\phi}{\partial x}\right)_n = \frac{\phi_{n+1} - \phi_n}{\Delta x} + O(\Delta x)$$

If this error is unacceptable when calculating the interface motion, the linear approximation must be replaced by second or higher order polynomials extending further from the interface. Although this does not create any difficulty in principle, it does assume that the interface position coincides with a node or, at the very least, that the interface can be located precisely in relation to the nodal network. When the network is fixed in space the interface generally moves through it non-uniformly and is therefore unlikely to coincide with a node at many, or any, of the prescribed moments $p\Delta\tau$ in time. Tracking of the interface in relation to the fixed network thus becomes crucial. On the other hand, the use of the Landau transformation, equation (3.44), leading to equation (3.45), locks the network to the interface. This avoids the nodal mismatch but denies the use of the nodal equations (3.51) or equation (3.52) in representing the

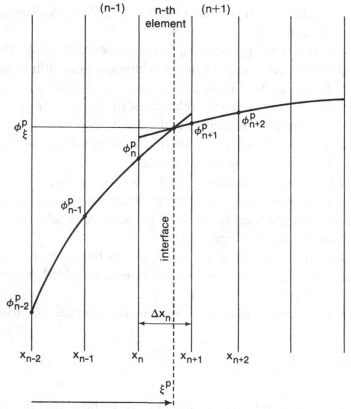

Fig. 3.7. Interface location and temperature distribution at given time $p \, \Delta \tau$.

temperature field fully because the temporal rate of change of temperature at a moving network node must be written

$$\frac{d\phi}{d\tau} = \left(\frac{\partial \phi}{\partial x}\right)_\tau \frac{dx}{d\tau} + \left(\frac{\partial \phi}{\partial \tau}\right)_x$$

in which $dx/d\tau$ is the local network (relative) velocity, determined from its proximity to the interface where $dx/d\tau = d\xi/d\tau$. Transformed equations such as (3.45) introduce the intrinsic non-linearity into the conduction equation and require special handling whenever an ice layer appears or disappears, carrying the embedded nodal network with it (Lock, Gunderson, Quon and Donnelly, 1969). In two or three dimensions the Landau transformation is effectively limited to situations in which the interface coincides with the coordinate system: e.g., with cylindrical or spherical symmetry.

For a fixed network, the position and effect of the interface may be described in a number of ways. The simplest, and crudest, treatment assumes that latent heat exchange takes place uniformly and isothermally in the element which contains the interface. Calculation of the temperature distribution on either side of this element is then reduced to the solution of a transient heat conduction problem with both of the element nodal temperatures being fixed at the equilibrium freezing temperature until the interface leaves the element. The temperature gradients on either side of the element are then only rough approximations to the actual interfacial gradients but the error can obviously be controlled by choosing an acceptable element size. Care must be exercised to ensure that the interface moves smoothly from one element to the next without missing any.

More accurately, the interface location may be described according to Fig. 3.7. At time $p \triangle \tau$ the location is at ξ^p where the temperature is ϕ_ξ^p. The instantaneous temperature gradients on either side are given approximately by

$$
\left.\left(\frac{\partial \phi}{\partial x}\right)_{\xi,p} = \frac{\phi_\xi^p - \phi_n^p}{\xi^p - x_n}\right\}
$$

for the left hand[10] (ice) side, and $\quad\quad\quad\quad\quad\quad\quad\quad$ (3.53)

$$
\left.\left(\frac{\partial \phi}{\partial x}\right)_{\xi,p} = \frac{\phi_{n+1}^p - \phi_\xi^p}{x_{n+1} - \xi^p}\right\}
$$

for the right hand (fluid) side. These expressions ignore curvature in the temperature profile, consistent with the low Stefan number of an individual element. Depending upon the numerical scheme employed, the temperatures at the fixed nodes on either side of the interface may not be generated automatically, in which case they must be found by interpolation/extrapolation using a sufficient number of neighbouring nodal values to give the required accuracy.

Returning to the interface equation, it is now evident that it may be put in the discrete form

$$
\frac{\Delta \xi}{\Delta \tau} = \left(\frac{\phi_\xi - \phi_n}{\xi - x_n}\right) - \beta \left(\frac{\phi_{n+1} - \phi_\xi}{x_{n+1} - \xi}\right)
$$

which facilitates step by step integration. This makes the interrelatedness of the interface velocity and the temperature distribution abundantly clear. To determine $\triangle \xi$, the interface location and flanking nodal temperatures [11] must be known. Once the interface has stepped forward, a new interface location is prescribed but a new temperature distribution must also be calculated before the next step may be taken. For an explicit scheme, the difference quotients given in equation (3.53) may be used, so that

$$\frac{\xi^{p+1} - \xi^p}{\Delta \tau} = \left(\frac{\phi_{\xi}^p - \phi_n^p}{\xi^p - x_n} \right) - \beta \left(\frac{\phi_{n+1}^p - \phi_{\xi}^p}{x_{n+1} - \xi^p} \right)$$

whereas for a typical (Crank–Nicolson) explicit/implicit scheme, an averaged representation, such as

$$\frac{\xi^{p+1} - \xi^p}{\Delta \tau} = \frac{1}{2} \sum_{q=p,p+1} \left[\left(\frac{\phi_{\xi}^q - \phi_n^q}{\bar{\xi} - x_n} \right) - \beta \left(\frac{\phi_{n+1}^q - \phi_{\xi}^q}{x_{n+1} - \bar{\xi}} \right) \right]$$

in which $\bar{\xi} = (\xi^{p+1} + \xi^p)/2$, would be more appropriate (Goodrich, 1978).

An alternative approach to the interface equation is to eliminate it entirely by permitting phase change in a small zone within which the problem is treated as one of transient heat conduction with highly variable thermal properties: in particular, to use within this zone the effective specific heat (equation (3.36))

$$c_p' = c_p - \Lambda(T),$$

where $\Lambda(T)$ is a distributed latent heat capacity (the specific heat and latent heat are taken here as volumetric, not specific, quantities). This enables the one-dimensional heat conduction equation to be taken in the form

$$c_p' \frac{\partial T}{\partial t} = \frac{\partial}{\partial X} \left(k \frac{\partial T}{\partial X} \right)$$

and integrated over the region ($\pm \Delta$) in which the latent heat is released: thus

$$\int_{\dot{X} - \Delta}^{\dot{X} + \Delta} c_p' \frac{\partial T}{\partial t} \, dX = \int_{\theta_1}^{\theta_2} c_p' \frac{dX}{dt} \, d\theta = \int_{\dot{X} - \Delta}^{\dot{X} + \Delta} \frac{\partial}{\partial X} \left(k \frac{\partial \theta}{\partial X} \right) dX$$

in which $\theta = T - T^i$, T^i is the mean freezing temperature within the zone, and $\theta_2 - \theta_1$ is defined by the latent heat characteristics of the freezing material. When the latent heat is released only at an interfacial discontinuity, the effective specific heat may be written as

$$c_p' = c_p - \lambda \delta(\theta)$$

where λ is the latent heat and $\delta(\theta)$ is the Dirac delta[12] (Oleinik, 1960; Samarskii and Moiseynko, 1965; Bonacina, Comini, Fasono and Primicerio, 1973). Hence, as $\theta_2 - \theta_1 \to 0$,

$$\int\limits_{\theta_1}^{\theta_2} c_p' \frac{\mathrm{d}X}{\mathrm{d}t}\,\mathrm{d}\theta = \int\limits_{\theta_1}^{\theta_2} \frac{\mathrm{d}X}{\mathrm{d}t}\,[c_p - \lambda\,\delta(\theta)]\,\mathrm{d}\theta \rightarrow$$
$$-\lambda \frac{\mathrm{d}\hat{X}}{\mathrm{d}t} \int\limits_{\theta_1 \to \theta_2} \delta(\theta)\,\mathrm{d}\theta \rightarrow -\lambda \frac{\mathrm{d}\hat{X}}{\mathrm{d}t},$$

while

$$\int\limits_{\hat{X}-\varDelta}^{\hat{X}+\varDelta} \frac{\partial}{\partial X}\left(k\frac{\partial\theta}{\partial X}\right)\mathrm{d}X \rightarrow k_\mathrm{f}\left(\frac{\partial\theta_\mathrm{f}}{\partial X}\right)_\mathrm{I} - k_\mathrm{i}\left(\frac{\partial\theta_\mathrm{i}}{\partial X}\right)_\mathrm{I}$$

thus recovering the classical interface equation.

More generally, the discontinuity at the interface must be replaced by a two-phase zone in which thermal properties may vary rapidly. The specific heat and thermal conductivity, in particular, may then be represented by continuous functions whose *gradients* may be taken as discontinuous at the extremities of the two-phase zone. When such a zone extends over the entire temperature field the problem reverts to one of transient heat conduction; as the zone becomes vanishingly small, we recover the classical Stefan problem. In practice, the extent of the two-phase zone may be almost anywhere between the two extremes, but a small zone is commonly observed in the growth or decay of ice (porous media may provide exceptions). In such situations, the thermal conductivity may be represented by a linear function, but the effective specific heat is perhaps best treated as the derivative of a smoothed enthalpy distribution (Bonacina *et al.*, 1973; Comini, Del Guidice, Lewis and Zienkiewicz, 1974). The enthalpy may, in fact, be used instead of the temperature in an alternative general formulation which automatically includes a distributed latent heat (Shamsundar and Sparrow, 1975; Voller and Cross, 1983). However, such formulations do not appear to be well adapted to situations where, for all practical purposes, there exists an interfacial discontinuity with a precise location: for example, when polycrystalline ice is growing or ablating in the presence of forced or natural convection (Kushner and Walston, 1977).

3.5.2 The finite element method

Finite element techniques are not substantially different from finite difference techniques. From a formulational standpoint they tend to be more complex but they possess the significant advantage of being more adaptable to the curved surfaces of three-dimensional spaces and the interfaces which move through them. The scope of this book precludes a full development of the finite element technique which is thoroughly covered elsewhere (Allaire, 1985; Norrie and DeVries, 1973; Zienkiewicz,

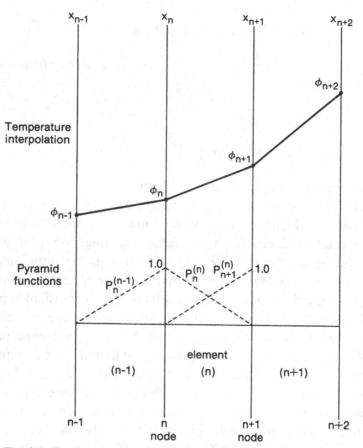

Fig. 3.8. Temperature interpolation and pyramid functions in one dimension.

1977). Instead, a one-dimensional treatment will be used to illustrate the basic approach to problems of freezing and melting.

In essence, the finite element technique replaces a continuum with a set of contiguous elements within each of which the form of the temperature [13] distribution is prescribed by an interpolation function. The simplest geometry for these elements is linear, triangular and tetrahedral for one, two and three dimensions, respectively. The simplest form of representation for the temperature distribution $\phi^{(e)}$ within a three-dimensional element (e) is linear and may, in general, be written

$$\phi^{(e)} = \alpha + \beta x + \gamma y + \delta z$$

or, more succinctly, as

$$\phi^{(e)} = P_i^{(e)} \phi_i + P_j^{(e)} \phi_j + P_k^{(e)} \phi_k + P_l^{(e)} \phi_l \tag{3.54}$$

where ϕ_n is the temperature at node n, and $P_n^{(e)}$ is the pyramid function for the element at the same node. In this relation, the bracketed superscript identifies the element while the subscript identifies the node (see Fig. 3.8). $P_n^{(e)}(x, y, z)$ is defined as 1.0 at the node n and decreases linearly to zero at the other three nodes of the element; *outside of the element*, $P_n = 0$. Fig. 3.8 illustrates the assumed temperature distribution and the pyramid functions P for plane (simplex) elements. It is apparent that for such one-dimensional elements

$$
\left.
\begin{aligned}
P_n^{(n-1)} &= \frac{x - x_{n-1}}{x_n - x_{n-1}} \\[2mm]
P_n^{(n)} &= 1 - \frac{x - x_n}{x_{n+1} - x_n} \\[2mm]
P_{n+1}^{(n)} &= \frac{x - x_n}{x_{n+1} - x_n}
\end{aligned}
\right\} \tag{3.55}
$$

Given such interpolation functions as those prescribed in equations (3.55), it is now necessary to find a strategy by which it is possible to determine the nodal temperatures ϕ_n satisfying the differential equation within each element, at least to some specified accuracy. Clearly, if this is done for every element the differential equation will then be satisfied approximately throughout the entire region of interest, assuming the boundary and initial conditions have also been incorporated in some manner. Of the two main avenues for attacking this problem, only the method of weighted residuals will be considered here (variational methods form the other avenue). In essence, the method of weighted residuals consists of first guessing the nodal temperatures and multiplying the error thus generated around each node by a weighting function; a residual for the node is then defined by the integral of this product over the entire solution domain. Nodal values of temperature which force the residuals to zero are taken as the solution.

For the differential equation

$$
\sigma(\partial\phi/\partial\tau) = \partial^2\phi/\partial x^2
$$

the error may be defined by

$$
r(x,\tau) = \partial^2\phi^A/\partial x^2 - \sigma(\partial\phi^A/\partial\tau)
$$

where $\phi^A(x, \tau)$ is a guessed approximation provided through the linear interpolations. If the weighting functions are now taken as the pyramid functions defined earlier by equations (3.55) (this is Galerkin's method), the nodal residual may be written

$$
R_n = \int_x P_n(x)\, r(x,\tau)\, dx \tag{3.56}
$$

and is equated to zero. Although the integration is carried out over the full range of x, the functions P_n are zero outside of the range $x_{n-1} \leqslant x \leqslant x_{n+1}$; as Fig. 3.8 indicates, $P_n^{(n-1)}$ is non-zero within $x_{n-1} \leqslant x \leqslant x_n$ and $P_n^{(n)}$ is non-zero within $x_n \leqslant x \leqslant x_{n+1}$. Hence

$$
R_n = \int_{x_{n-1}}^{x_{n+1}} P_n(x)\, r(x,\tau)\, dx
$$

$$
= \int_{x_{n-1}}^{x_n} P_n^{(n-1)}(x)\, r(x,\tau)\, dx + \int_{x_n}^{x_{n+1}} P_n^{(n)}(x)\, r(x,\tau)\, dx
$$

or

$$
R_n = R_n^{(n-1)} + R_n^{(n)} = 0
$$

in which a bracketed superscript again indicates an element and an unbracketed subscript indicates a node.

Since ϕ is a function of both x and τ, the residuals will contain contributions from both spatial and temporal errors. Thus

$$
R_n = R_{nx} + R_{n\tau}
$$

in which R_{nx} and $R_{n\tau}$ are separated through the differential equation. That is,

$$
R_n = \int_x P_n(x) \left[\frac{\partial^2 \phi^A}{\partial x^2} - \sigma \frac{\partial \phi^A}{\partial \tau} \right] dx
$$

$$
= \int_x P_n(x) \frac{\partial^2 \phi^A}{\partial x^2}\, dx - \sigma \int_x P_n(x) \frac{\partial \phi^A}{\partial \tau}\, dx
$$

It therefore follows that

$$
R_{nx}^{(n-1)} = \int_{x_{n-1}}^{x_n} P_n^{(n-1)} \frac{\partial^2 \phi^A}{\partial x^2}\, dx
$$

and

$$
R_{nx}^{(n)} = \int_{x_n}^{x_{n+1}} P_n^{(n)} \frac{\partial^2 \phi^A}{\partial x^2}\, dx
$$

while

$$
R_{n\tau}^{(n-1)} = -\sigma \int_{x_{n-1}}^{x_n} P_n^{(n-1)} \frac{\partial \phi^A}{\partial \tau}\, dx
$$

and

$$R_{n\tau}^{(n)} = -\sigma \int_{x_n}^{x_{n+1}} P_n^{(n)} \frac{\partial \phi^A}{\partial \tau} \, dx$$

Thus

$$R_{nx}^{(n-1)} + R_{nx}^{(n)} + R_{n\tau}^{(n-1)} + R_{n\tau}^{(n)} = 0$$

To calculate the spatial residuals it is necessary to recall that the integration extends over the entire domain. Therefore

$$R_{nx} = \int_x P_n \frac{\partial^2 \phi^A}{\partial x^2} \, dx$$

$$= \int_x \frac{\partial}{\partial x} \left(P_n \frac{\partial \phi^A}{\partial x} \right) dx - \int_x \frac{\partial P_n}{\partial x} \frac{\partial \phi^A}{\partial x} \, dx$$

But

$$\int_x \frac{\partial}{\partial x} \left(P_n \frac{\partial \phi^A}{\partial x} \right) dx = \left[P_n \frac{\partial \phi^A}{\partial x} \right]_1^{N+1}$$

where $x = 1$, $N + 1$ represent the boundaries of the system where $n = 1$ or $N + 1$. Thus for *interior* nodes,

$$R_{nx} = -\int_{x_{n-1}}^{x_{n+1}} \frac{\partial P_n}{\partial x} \frac{\partial \phi^A}{\partial x} \, dx$$

and hence

$$R_{nx}^{(n-1)} = -\int_{x_{n-1}}^{x_{n+1}} \frac{\partial P_n^{(n-1)}}{\partial x} \frac{\partial \phi^A}{\partial x} \, dx = \frac{\phi_{n-1}^A - \phi_n^A}{L^{(n-1)}}$$

after substituting the pyramid and interpolation functions, and setting $L^{(n-1)} = x_n - x_{n-1}$. Similarly

$$R_{nx}^{(n)} = -\int_{x_n}^{x_{n+1}} \frac{\partial P_n^{(n)}}{\partial x} \frac{\partial \phi^A}{\partial x} \, dx = \frac{\phi_{n+1}^A - \phi_n^A}{L^{(n)}}$$

Therefore, for a uniform grid,

$$R_{nx} = \frac{1}{L} (\phi_{n+1}^A + \phi_{n-1}^A - 2\phi_n^A) \tag{3.57}$$

which is precisely the same as the interior node form developed with the finite difference technique.

The temporal residuals are evaluated after first introducing the forward difference approximation

$$\partial \phi^A / \partial \tau \approx (\phi^{p+1} - \phi^p)/\Delta \tau$$

where the unbracketed superscripts p, $p+1$, etc. indicate moments in time. Thus

$$R_{n\tau}^{(n-1)} = -\sigma \int_{x_{n-1}}^{x_n} P_n^{(n-1)} \left(\frac{\phi_n^{(p+1)} - \phi_n^p}{\Delta \tau} \right) dx$$

and

$$R_{n\tau}^{n)} = -\sigma \int_{x_n}^{x_{n+1}} P_n^{(n)} \left(\frac{\phi_n^{(p+1)} - \phi_n^p}{\Delta \tau} \right) dx$$

Now if it is assumed that the temporal rate of change does not vary within an element,

$$R_{n\tau}^{(n-1)} = -\sigma \left(\frac{\phi_n^{p+1} - \phi_n^p}{\Delta \tau} \right) \frac{L^{(n-1)}}{2}$$

and

$$R_{n\tau}^{(n)} = -\sigma \left(\frac{\phi_n^{p+1} - \phi_n^p}{\Delta \tau} \right) \frac{L^{(n)}}{2}$$

Hence, for a uniform grid,

$$R_{n\tau} = -\sigma \left(\frac{\phi_n^{p+1} - \phi_n^p}{\Delta \tau} \right) L \tag{3.58}$$

which, combined[14] with equation (3.57), leads to

$$R_n = \frac{1}{L} (\phi_{n+1}^p + \phi_{n-1}^p - 2\phi_n^p) - \sigma \left(\frac{\phi_n^{p+1} - \phi_n^p}{\Delta \tau} \right) L = 0$$

so that

$$\sigma(\phi_n^{p+1} - \phi_n^p) = \frac{\Delta \tau}{L^2} (\phi_{n+1}^p + \phi_{n-1}^p - 2\phi_n^p)$$

the same explicit result for interior nodes previously found using finite differences (see equation (3.51)).

If the temporal rate of change is not assumed to be constant within an element,

$$R_{n\tau}^{(n-1)} = -\frac{\sigma}{\Delta \tau} \int_{x_{n-1}}^{x_n} P_n^{(n-1)} \phi_n^{p+1} dx - \frac{\sigma}{\Delta \tau} \int_{x_{n-1}}^{x_n} P_n^{(n-1)} \phi_n^p dx$$

Substituting the interpolation function for the $(n-1)$-th element,

$$\phi^{(n-1)} = P_{n-1}^{(n-1)} \phi_{n-1} + P_n^{(n-1)} \phi_n$$

it is found that

$$R_{n\tau}^{(n-1)} = -\frac{\sigma L^{(n-1)}}{6\Delta\tau} (\phi_{n-1}^{p+1} + 2\phi_n^{p+1}) + \frac{\sigma L^{(n-1)}}{6\Delta\tau} (\phi_{n-1}^p + 2\phi_n^p)$$

Similarly,

$$R_{n\tau}^{(n)} = \frac{-\sigma L^{(n)}}{6\Delta\tau} (\phi_{n+1}^{p+1} + 2\phi_n^{p+1}) + \frac{\sigma L^{(n)}}{6\Delta\tau} (\phi_{n+1}^p + 2\phi_n^p)$$

and hence, for a uniform grid,

$$R_{n\tau} = \frac{\sigma L}{6\Delta\tau} [(\phi_{n-1}^p + \phi_{n+1}^p + 4\phi_n^p) - (\phi_{n-1}^{p+1} + \phi_{n+1}^{p+1} + 4\phi_n^{p+1})] \quad (3.59)$$

Combining this with equation (3.57) evaluated at $p\Delta\tau$,

$$R_n = \frac{1}{L} (\phi_{n+1}^p + \phi_{n-1}^p - 2\phi_n^p) + \frac{\sigma L}{6\Delta\tau} [(\phi_{n-1}^p + \phi_{n+1}^p + 4\phi_n^p) - (\phi_{n-1}^{p+1} + \phi_{n+1}^{p+1} + 4\phi_n^{p+1})]$$

and hence

$$\sigma(\phi_{n-1}^{p+1} + 4\phi_n^{p+1} + \phi_{n+1}^{p+1}) = \frac{6\Delta\tau}{L^2} (\phi_{n-1}^p - 2\phi_n^p + \phi_{n+1}^p) - \sigma(\phi_{n-1}^p + 4\phi_n^p + \phi_{n+1}^p) \quad (3.60)$$

which is an implicit result requiring that the ϕ^{p+1} must be found by matrix inversion.

For purposes of comparison with the finite difference technique, the above result was obtained by focussing on the n-th node. Alternatively, and in general, the residuals are more conveniently expressed in terms of elements and element matrices. Thus for a pair of nodes i,j flanking an element (e), the corresponding pair of residuals appropriate to the differential equation considered above is written[15]

$$R_i^{(e)} = \frac{1}{L} (\phi_i^p - \phi_j^p) + \frac{\sigma L}{6\Delta\tau} (2\phi_i^{p+1} + \phi_j^{p+1} - 2\phi_i^p - \phi_j^p)$$

and

$$R_j^{(e)} = \frac{1}{L} (\phi_j^p - \phi_i^p) + \frac{\sigma L}{6\Delta\tau} (\phi_i^{p+1} + 2\phi_j^{p+1} - \phi_i^p - 2\phi_j^p)$$

$\left.\right\} \quad (3.61)$

again assuming a uniform grid. These residuals are not to be added together, because they do not refer to the same node, but may be expressed in the matrix form

$$\begin{Bmatrix} R_i \\ R_j \end{Bmatrix}^{(e)} = \{B\}^{(e)} \begin{Bmatrix} \phi_i^{p+1} \\ \phi_j^{p+1} \end{Bmatrix} + \{C\}^{(e)}$$

in which, from equations (3.61),

$$\{B\}^{(e)} = \frac{\sigma L}{6\Delta\tau} \begin{Bmatrix} 2 & 1 \\ 1 & 2 \end{Bmatrix}$$

and

$$\{C\}^{(e)} = \frac{1}{L} \begin{Bmatrix} 1 & -1 \\ -1 & 1 \end{Bmatrix} \begin{Bmatrix} \phi_i^p \\ \phi_j^p \end{Bmatrix} - \frac{\sigma L}{6\Delta\tau} \begin{Bmatrix} 2 & 1 \\ 1 & 2 \end{Bmatrix} \begin{Bmatrix} \phi_i^p \\ \phi_j^p \end{Bmatrix}$$

The nodal residuals may then be compiled by forming

$$R_j^{(e-1)} + R_i^{(e)} = 0$$
$$R_j^{(e)} + R_i^{(e+1)} = 0$$

etc. The element matrix formulation is especially useful when working with dense grids for which assembly of the global matrix demands careful and systematic treatment.

The boundary conditions are also stated conveniently in terms of the element residuals, the spatial residuals in particular. It will be recalled that the R_{nx} were given earlier for interior nodes only because the full integration process generated an extra pair of terms $[P_n(\partial\phi/\partial x)]_1^{N+1}$ to be evaluated only at the system boundaries where $n = 1$ or $N+1$. If ϕ_1 and ϕ_{N+1} are not prescribed, then

$$-\left[P_n\frac{\partial\phi}{\partial x}\right]_1^{N+1} = -\left(\frac{\partial\phi}{\partial x}\right)_{N+1} + \left(\frac{\partial\phi}{\partial x}\right)_1$$

must be added to the expressions for R_{nx} in the boundary elements. This is equivalent to adding $-(\partial\phi/\partial x)_{N+1}$ to $R_{N+1}^{(N)}$ and $(\partial\phi/\partial x)_1$ to $R_1^{(1)}$ in equations (3.61) applied to the full set of elements. When $(\partial\phi/\partial x)$ is prescribed at the boundaries in terms of other effects, such as convection or radiation, the equations for $R_1^{(1)}$ and $R_{N+1}^{(N)}$ must be modified accordingly. On the other hand, if either ϕ_1 or ϕ_{N+1} is prescribed, and therefore no longer an unknown quantity, the corresponding weighting function is not defined and the element residual then vanishes. Thus, for example, if temperature is prescribed at x_1 but not at x_{N+1}, $R_1^{(1)}$ is excluded but

$$R_{N+1}^{(N)} = \frac{1}{L}(\phi_{N+1}^p - \phi_N^p) + \frac{\sigma L}{6\Delta\tau}(\phi_N^{p+1} + 2\phi_{N+1}^{p+1}$$

$$- \phi_N^p - 2\phi_{N+1}^p) - \left(\frac{\partial\phi}{\partial x}\right)_{N+1}$$

which is set equal to zero because the weighting function consists of $P_{N+1}^{(N)}$ only.

Beyond their formulational differences, the finite element technique and the finite difference technique are very similar. Each may be used to describe transient heat conduction in a manner which leads to a system of matrix equations to be solved using similar, and frequently standard, methods: for example, Gaussian elimination. Each may be used in the fully explicit form, the fully implicit form, or in some compromise between them. The incorporation of latent heat presents similar difficulties for the two techniques: as a distributed function or a discontinuous function; in a fixed coordinate system or a moving coordinate system. All of these matters were discussed earlier in Section 3.5.1. Both techniques have merits and demerits, and therefore the choice between them on a particular occasion will continue to be influenced strongly by the particular application, its scope and the attendant economics.

4

Ice and water

Of the two principal ways in which ice may be formed, perhaps the freezing of water comes most readily to mind. For many people, the evidence is spread before them every winter: lakes freeze, rivers freeze, pipes freeze; and in the summer, the tinkling in a cocktail glass serves as a friendly reminder. The range of situations in which water may be transformed into ice is enormous, and in this chapter it will only be possible to give the briefest of descriptions of the most common occurrences. Each represents an important aspect of *hydroglaciology*.

To begin with, the processes and events which occur in the freezing of aqueous solutions and suspensions will be considered, largely from a thermophysical point of view. This complements the foundation laid down in Chapter 2 and leads quite naturally to the discussion of ice formation on the surface of large bodies of water, notably lakes, rivers and seas. The existence of a thin ice cover then paves the way for the treatment of its subsequent growth, again on the typically large scale of natural ice.

The chapter then continues with a more detailed discussion of the ice–water interface under specific conditions: when heat transfer from the water is by forced convection, and when by natural convection. The discussion is limited to common situations in which the interface is nominally planar, although some scope is given for departures from this strictly one-dimensional situation. Curved interfaces are considered more fully in the last two sections which explore ice formation on the inside and outside of three-dimensional surfaces.

4.1   Freezing of solutions and suspensions

Seldom is ice formed from pure water, although it is not uncommon for the water to have impurity traces small enough to render their neglect safe and practical, at least from the point of view of human use.

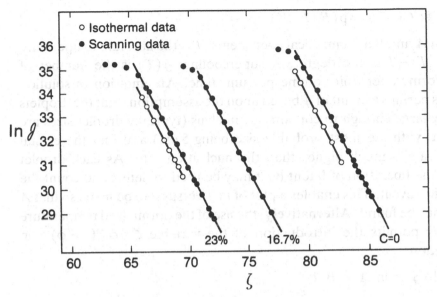

Fig. 4.1. Temperature and concentration dependence of ice nucleation rate in aqueous solutions of polyethylene glycol (following Michelmore and Franks (1982)).

More generally, the presence of impurities, whether in solution or suspension, may modify the freezing process and the type of ice formed, often to a substantial degree. The details of the process depend upon the concentration, distribution and type of impurity; they also depend upon whether the impurity is dissolved or particulate, and upon the rate at which the freezing takes place.

Tap water is certainly not pure despite the fact that it is often drinkable. Apart from microscopic organisms, usually various types of bacteria, it contains many dissolved substances; depending on geographical location, these commonly include mineral salts and atmospheric gases. Pond, lake, river and sea water contain similar impurities, often in greater concentrations; and they will likely contain organic and inorganic debris in varying sizes and amounts.

Despite the extensive study of ice formation in aqueous solutions under equilibrium or quasi-equilibrium conditions, it is only recently that attempts have been made to uncover the physics of nucleation. The work of Franks and his coworkers is noteworthy (Michelmore and Franks, 1982; Franks, Mathias, Parsonage and Tong, 1983; Franks, Mathias and Trafford, 1984). This rests upon the theory of homogeneous nucleation outlined in Section 2.8. The nucleation rate may be expressed (Wood and Walton, 1970) in the form

$$\mathcal{J}(T) = A \exp(B/T^3\theta^2)$$

in which A and B are empirical coefficients, T is the absolute temperature and $\theta = T^i - T$ is the degree of supercooling: $\mathcal{J}(T)$ is the number of nuclei formed per unit volume per unit time. An emulsion of solution droplets permits a count of $\mathcal{J}$ based upon the assumptions that the droplets are only large enough to contain one nucleus (typically droplet sizes are 2–15 μm, with the most probable size being 5 μm) and that the crystal growth rate is much higher than the nucleation rate. As each droplet freezes the liberation of latent heat may be used to detect and count the nucleation event. This enables a plot of $\ln \mathcal{J}$ versus θ to be made so that A and B may be found. Alternatively, the use of the normalized temperature $\phi = T/T^i$ permits the introduction of the variable $\zeta = \phi^{-3}(1 - \phi)^{-2}$ in the relation

$$\ln \mathcal{J} = \ln A + B'\zeta$$

which fits representative data over several orders of magnitude in $\mathcal{J}$.

Fig. 4.1 shows experimental data for water and an aqueous solution of polyethylene glycol (Michelmore and Franks, 1982). Data for concentrations other than those shown are omitted but follow the same trend. The theoretical model incorporates resistance to diffusion and accommodation of the water molecules in the coefficient A, as noted in Section 2.8. The figure suggests that the barrier to nucleation depends upon the concentration C in a simple manner. In fact the data fit the correlation

$$\mathcal{J} = 10^{50} A_0 \exp(-0.51C) \exp(1.02\zeta)$$

in the range 230 K $< T <$ 273 K where $A_0 = O(1)$. However, the effect of solutes is generally complex: they reduce the volume fraction of the water; increase the critical cluster radius; and limit diffusion through the slowest species present. It should also be mentioned that the properties of supercooled water itself were not known accurately until quite recently, and have revealed unexpected behaviour which has, at the very least, invalidated extrapolations far from the vicinity of 273 K.

The effect of dissolved salts on freezing behaviour may be seen by reference to a simple binary solution, such as described in Fig. 2.5. At very low freezing rates (i.e., interface velocities of the order of 10^{-7} m s^{-1} or 1 m in 4 months) the equilibrium phase diagram applies; salt is rejected into the solution as ice forms. This constitutes the basis of the freeze-desalination process. At high freezing rates, the salt cannot diffuse away from the interface fast enough to avoid being partially trapped in the advancing ice which, in molecular terms, is not very well designed to

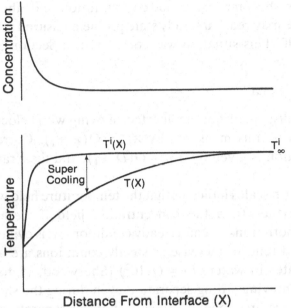

Fig. 4.2. The origin of constitutional supercooling.

accommodate it. With the exception of a few substitutional molecules (e.g., HF, NH$_3$) ice cannot accept impurities without substantial alterations to its crystalline structure. These alterations take the form of defects which, at least in principle, permit diffusion of the impurities by mechanisms similar to those of self-diffusion (see Section 2.6). In practice, however, molecular diffusion rates are often low enough to be ignored, thus enabling the impregnated ice to be treated as a pure substance. For the most part, solute molecules are not incorporated in the ice lattice which, in growing, further increases solute concentration in the neighbouring solution.

Freeze concentration has its principal effect near the interface where temperature and concentration gradients create simultaneous diffusive fluxes. If the coupling between these fluxes is ignored (see Section 2.4), their respective penetrating abilities may be determined independently from the temperature and concentration fields which, in general, are transient. For one-dimensional heat conduction in a quiescent aqueous solution, the thermal penetration distance is given by $X_{CT} = O(\varkappa t_C)^{\frac{1}{2}}$ where t_C is the time scale. Similarly, the mass penetration distance is given by $X_{CM} = O(Dt_C)^{\frac{1}{2}}$. The penetrational ability of heat relative to that of mass is therefore given by $(\varkappa/D)^{\frac{1}{2}}$, where $\varkappa/D = Le$, known as the Lewis number.

If sufficient time elapses, the temperature and concentration fields ahead of the advancing interface may reach a steady state profile measured relative to the interface itself. This situation was considered in Section 3.2 where the solution

$$\theta(X') = \theta_1 \exp(-V_1 X'/\varkappa)$$

was presented: X' is the distance from the interface moving with velocity V_I. Thermal penetration is thus measured by $X_{CT} = O(\varkappa/V_I)$. Correspondingly, mass penetration is given by $X_{CM} = O(D/V_1)$, and their ratio is therefore $Le = \varkappa/D$.

This comparison of length scales indicates that the temperature field will propagate faster and further than the concentration field if $Le > 1$. Although this is true for both transient and steady conditions, it is evident that the separation would tend to increase as steady conditions are approached. For most solutes in water, $Le = O(10^2)$ (Sherwood, Pigford and Wilke, 1975), thus implying strong separation even during the short period immediately after ice first appears. The effect of this separation is illustrated in Fig. 4.2 which shows two steady temperature profiles in a solution initially at its equilibrium freezing temperature T_∞^i. Should a substrate temperature less than T_∞^i cause ice to appear, the resulting concentration profile creates a liquidus temperature profile beneath the initial value. However, this effect propagates more slowly than heat conduction and therefore the solution temperature at any location is always beneath the local equilibrium freezing temperature. That is, the solution is supercooled, a state known as *constitutional* supercooling.

Under such conditions, the ice–solution interface is unstable because the degree of supercooling increases with distance near the interface. A small perturbation ahead of the nominal interface would be surrounded by liquid which is slightly more supercooled and thereby better able to accelerate the perturbation further. This situation favours the faster a-axis growth in the form of dendrites, and is illustrated in Fig. 4.3(a) which depicts the effect of lowering the temperature of an ice substrate T_S below the initial freezing temperature T^i of a solution when the bulk temperature and concentration are T_∞ and C_∞, respectively. Constitutional supercooling induces dendrite growth in the region defined by $T^i < T < T_S$, but the magnitude of the Lewis number ($Le \gg 1$) dictates that the temperature profile is essentially independent of concentration. On the other hand, if stable equilibrium exists at any height in the freezing zone, the phase diagram demands that the local concentration is uniquely determined by the temperature, providing the substrate temperature remains

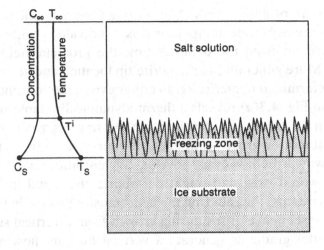

(a)

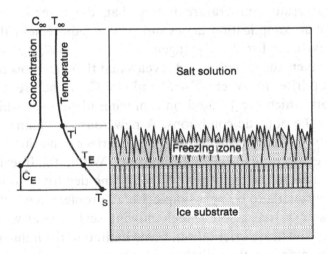

(b)

Fig. 4.3. Dendritic ice growth in a dilute salt solution (a) $T_S > T_E$ (b) $T_S < T_E$.

above the eutectic temperature T_E. This quasi-equilibrium freezing zone model provides a basis for analysis of growth in aqueous solutions (Fang, Cheung, Linehan and Pederson, 1984; O'Callaghan, Gravalho and Huggins, 1982). The experimental work of Fang *et al.* (1984) indicates that the number of dendrites growing on the substrate does not change with time, but may be made to increase with the initial temperature deficit $T^i - T_s$.

Despite the presence of growing dendrites, this problem is essentially a one-dimensional Stefan problem in which a freezing zone is bounded by two planes. When the substrate temperature is fixed, along with the solution temperature far from the dendrite tips, the problem yields a similarity solution. More generally, the dendrite tip location and velocity would have to be determined numerically. In either event, the concentration profile shown in Fig. 4.3(a) reveals a thermodynamically stable situation: concentration increases with depth within the freezing zone. The attendant absence of motion greatly simplifies the energy equation. If the dendrites grew downwards, however, adverse concentration and temperature gradients would exist and might destabilize the liquid in the vicinity of the dendrite tips, thus altering the heat transfer process in that region. A similar consequence arises during growth from a vertical substrate when horizontal gradients generate a vertical buoyant flow and thus introduce advective effects.

Whenever the substrate temperature is less than the eutectic temperature an additional complication arises out of the requirement that the concentration is fixed for $T < T_E$ (again assuming stable equilibrium). Fig. 4.3(b) depicts the situation and reveals that there are now two interfaces: at the dendrite tips where $T = T^i$ and $C = C_\infty$, as before, and at the dendrite roots which are located on a moving plane over which $T = T_E$ and $C = C_E$, the eutectic conditions. A new zone containing the eutectic thus appears and grows according to the interface condition that couples it to the freezing zone at the dendrite roots. Within the freezing zone, $C_\infty < C < C_E$, and in the event that $C_\infty = C_E$, the dendrites do not appear; the problem reverts to being a simple Stefan problem in which a pure (eutectic) substance freezes on a single moving surface over which $T = T_E$. The effect of substrate orientation, being limited to the influence of thermal buoyancy alone, is then likely to be reduced.

The freezing of particulate suspensions in aqueous solutions is a complex interaction between the particles and the solution. This behaviour is not well understood. Particulates in pure water have received enough study to reveal that, in general terms, they may not be treated as passive modifiers of the bulk water properties. It has been found that if the particles are large enough, and the freezing rate high enough, an advancing ice–water interface may engulf them: interaction between particle and interface may then be neglected. For smaller particles and lower freezing rates, however, engulfment does not take place, and the particles are swept ahead of the interface. The conditions under which particles will or will not be incorporated into the ice are important in processes which seek to produce either purified or strengthened ice, and

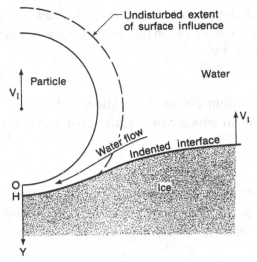

Fig. 4.4. Steady sweeping of a spherical particle by an interface moving with velocity V_I.

should the particle be a living organism the question may become a matter of survival.

In recent years, particulate behaviour in the presence of an ice surface has been explained by assuming the presence of a thin enveloping film of unfrozen water whose existence may be attributed to hydrophilic forces at the particle surface (Fletcher, 1973; Gilpin, 1979a). Such a film, whose thickness is estimated to be of the order of 10–100 nm, is defined by the extent of surface-induced alterations to the chemical potential and by concomitant alterations in the pressure field. For undisturbed, equilibrium conditions, the pressure field simply balances the hydrophilic 'body' force fields, but the encroachment of an ice surface is capable of distorting the pressure field, thus providing a driving force which squeezes water between the particle and the ice surface and thereby prevents engulfment. Fig. 4.4 describes the situation.

Following Gilpin (1979b) the chemical potential of the water may be written

$$\mu = \mu_\infty - \mu'(Y) \tag{4.1}$$

where μ_∞ is the bulk value and μ' is the surface-induced change at a distance Y from the surface. In the bulk,

$$d\mu_\infty = v_w dP - s_w dT$$

which may be integrated approximately to give

$$\mu_\infty - \mu_0 = v_w(P - P_0) - s_w(T - T_0) \tag{4.2}$$

where the subscript 0 refers to a datum, here taken to be the bulk equilibrium freezing condition. The form of $\mu'(Y)$, on the other hand, is unknown *a priori* but may be represented[1] by

$$\mu'(Y) = aY^{-\alpha} \tag{4.3}$$

When the bulk is at the equilibrium freezing condition and stable conditions exist, $\mu = \mu_0$ throughout, in which case equations (4.1), (4.2) and (4.3) may be combined to give

$$\mu'(Y) = v_w(P_w - P_0)$$

if temperature variations across the film are negligible. That is, the variation of pressure and chemical potential across the film have essentially the same form. In particular, the water pressure is given by

$$P_w(Y) = P_w(H) + \frac{1}{v_w}[\mu'(Y) - \mu'(H)] \tag{4.4}$$

where H is the film thickness.

It is the variation in pressure parallel to the ice and particle surface which gives rise to the ingress of water preventing engulfment. At first glance, it appears that the pressure gradient may act to expel water but closer consideration reveals the opposite. Again appealing to stable equilibrium, the chemical potential of the ice and the water must be equal at the ice–water interface, and hence

$$v_i(P_i - P_0) - s_i(T_i - T_0) = v_w(P_w - P_0) - s_w(T_w - T_0) - aH^{-\alpha}$$

in which the pressures and temperatures are evaluated at the interface where $T_i = T_w = T_H$ but $P_i \neq P_w$ because, in general, the interface has a curvature K. Thus

$$(v_w - v_i)(P_w - P_0) + v_i \sigma_{iw} K + \lambda_{iw} \theta_I / T_I = aH^{-\alpha}$$

in which σ_{iw} and λ_{iw} are the surface tension and latent heat, respectively, and $\theta_I = T_0 - T_I$ is the freezing point depression. For a flat surface at atmospheric pressure this expression provides the freezing point depression directly in terms of the film thickness. Gilpin (1979a) recommends $\alpha = 2$ and $H(\theta) = H_1/\theta^{\frac{1}{3}}$ where $H_1 = H(1) = O(1)$.

The above results demonstrate that the local water pressure in the film is determined not only by the interface pressure $P_w(H)$ but by the local curvature and temperature of the interface. The results may be used with equation (4.4) in satisfying the requirements of continuity and a force balance, based on the assumption of a Newtonian–Poiseuille type of flow in which water is drawn in under the particle at a rate just sufficient to

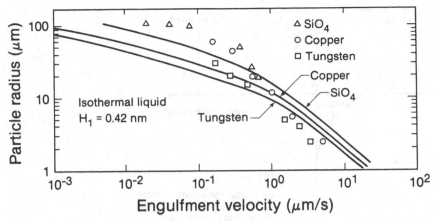

Fig. 4.5. Engulfment velocities of small particles (following Gilpin (1979b)).

maintain the gap while both particle and ice advance at the same rate V_I. Substitution of an appropriate temperature distribution (Nye, 1967) then yields an expression for the gap as a function of location on the particle's lower surface. The analysis is completed by balancing the pressure distribution over the particle with the force (e.g., its weight) steadily applied.

The details of this quasi-steady analysis produce, among other things, the shape of the interface indentation and the corresponding pressure distribution acting on the particle. The criterion for engulfment is simply that the indentation–pressure relationship should generate a net upward force unable to balance the applied downward force, typically the particle's weight. This critical balance leads to a relationship between the engulfment velocity and such variables as the particle radius, density, thermal conductivity, etc. Fig. 4.5 compares Gilpin's theoretical curves with selected experiments and indicates good agreement, bearing in mind the difficulties in measurement. It must be noted, however, that the empirical constants, notably those in $\mu'(Y)$, have been chosen to ensure the fit. It is significant that this theory tends to support the experimental discovery of two regimes separated by an engulfment velocity of the order of $1\,\mu\mathrm{m\,s^{-1}}$. For velocities higher than this, engulfment is believed to be a function of particle radius only; for lower values, particle density and thermal conductivity become important.

The presence of solutes in the water may be expected to add further complexity. Freezing point depression of the bulk solution merely constitutes a shift in datum but freeze concentration ahead of the interface alters the chemical potential field around the particles in a way which

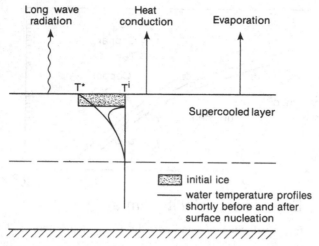

Fig. 4.6. Incipient formation of ice in a shallow, quiescent pool.

will be dependent on constitutional supercooling and the dendrites which it creates. The behaviour of the particles in the face of this 'mushy' advance is currently unknown, despite the fact that it may play an important role in the capture of microscopic sea biota in bottom ice (Horner, 1985). Equally, engulfment may be a contributing factor in the ability of frazil particles (discussed below) to collect algal cells while rising, and perhaps growing, in supercooled sea water beneath the ice–water interface (Ackley, 1982).

4.2   Formation of an ice cover

The appearance of an ice cover on the surface of a body of water is a familiar occurrence in cold climates. The covers of shallow pools and puddles appear to have no particular human significance, except perhaps to the children who like to hear them snap and squeak, but they are significant to the organisms living beneath. To a certain extent, this comment may be applied to many rivers, lakes and oceans, although it is obvious that a thick enough ice cover does lend itself to recreation. However, where mankind has established demands for a supply of fresh water, electrical power generation and various types of transportation, the formation of an ice cover may assume the proportions of a crucial event. The circumstances leading to, and during, such an event vary widely and it is therefore to be expected that the corresponding details of formation would be equally variable and complex.

The simplest formation event occurs in a shallow pool of quiescent water, as depicted in Fig. 4.6. The mean bulk temperature of the pool may be determined from a complete heat balance, and if ambient conditions are close to freezing this balance is easily tilted in favour of a net heat loss by conduction, evaporation or long wave radiation; any of these, or any combination of them, may be sufficient to lower the surface temperature beneath the equilibrium freezing temperature. For broad, shallow pools, a supercooled layer may form just beneath the free surface and begin to deepen at a rate dependent on the rate of heat loss from the surface. The deepening of the supercooled layer continues until its coolest point, the free surface, has reached the temperature for spontaneous nucleation T^*. The maximum degree of supercooling which may occur at the free surface determines the rate at which nuclei are formed per unit surface area and subsequently controls the rate at which these nuclei grow. It will be recalled from Sections 2.8 and 2.9 that above $-40\,°C$ the greater the supercooling the greater the nucleation rate and the greater the crystal growth velocity. In most natural bodies of water the amount of supercooling is limited by the presence of distributed particles and bounding surfaces. These tend to promote heterogeneous nucleation to an extent dependent on their geometry and surface chemistry. Should supercooling be completely suppressed, freezing begins at the moment the free water surface passes through the equilibrium freezing temperature T^i. Subsequent growth of the ice requires efficient removal of the latent heat released at the interface which therefore tends to propagate along the crystallographic axis with the greater thermal conductivity. Growth then occurs with the c-axis parallel to the direction of the heat flux (or temperature gradient).

Supercooling of only a few degrees creates a radical change in the initial growth pattern in the same quiescent pool. When the rate of surface heat loss is low enough to correspond to supercoolings of about $0.3\,°C$ or less, crystal growth begins essentially in the plane of the free surface. Faster growth along the a-axes into the surrounding supercooled water layer again tends to initiate crystals with vertical c-axes, a feature which also lends itself to the more rapid removal of latent heat through the exposed upper surface of the crystal. This process may start around the perimeter of the pool, over the free surface, or both simultaneously. It will continue on calm water (Knight, 1962; Lyons and Stoiber, 1959a) until the crystals are large enough to contact each other and interlock, thus forming a continuous sheet of polycrystalline ice, often called *skim*. This thin type of cover, which is usually transparent, may be seen on pools, reservoirs and, depending upon local conditions, on the surfaces of lakes[2] and sluggish (typically with velocities $<0.5\,\mathrm{m\,s^{-1}}$) streams and rivers.

For higher rates of surface heat loss, the degree of supercooling and the depth of the supercooled layer are both likely to be greater, thus increasing the crystal growth velocity and relaxing the planar 'constraint'. Under these conditions, the limited room for horizontal growth quickly leads to *a*-axis growth inclined into the deeper, supercooled layer. The *c*-axes no longer lie parallel to the overall temperature gradient but are inclined to it in amounts which are limited by the initial depth of the supercooled layer i.e., with the rate of surface heat loss. Downward *a*-axis growth is maintained as long as the crystal front continues to penetrate supercooled water. Typically, some latent heat will be released into the supercooled water thus reducing both the local degree of supercooling and the corresponding crystal front velocity. The release of latent heat tends to warm the water through which the *a*-axis front has already passed and thereby reduces the lateral *c*-axis growth velocity even further.

The *c*-axis growth of a planar interface into superheated water is a stable situation whereas *a*-axis growth into supercooled water is unstable. Under the latter conditions, perturbations of the interface tend to be amplified because any point slightly ahead of its nominal position is surrounded by cooler water (see Fig. 4.6) and therefore enjoys a greater ability to release latent heat, enabling it to move further ahead; similarly, trailing points tend to fall further behind. Growth in a deep supercooled layer of pure water is therefore characterized by dendritic needles. These needles spread laterally and interlock to form a fairly random matrix whose 'fabric' reflects the location of the initial nucleation sites distributed over the surface and around its perimeter. The final ice cover tends to be thicker and less transparent than that produced by smaller supercoolings, especially if dissolved gases are rejected in bubble form. Subsequent growth of either of these two types of surface ice depends upon the prevailing conditions, as will be discussed later. In the simplest circumstances, when the heat removed upwards is the direct cause of downward interface penetration into water no longer supercooled, the problem reverts to the classical Stefan form, and interface growth once more favours the axis along which the thermal conductivity is greater. Such growth is merely a crystallographic continuation for skim produced from low supercooling, but will require considerable re-orientation for surface ice produced from deep supercooling.

The presence of solutes may also influence crystallographic orientation. For small supercoolings, the situation is almost identical to that for pure water, except that freeze concentration is produced. Weeks and Ackley (1982) have observed crystals with their *c*-axis vertical growing on calm sea

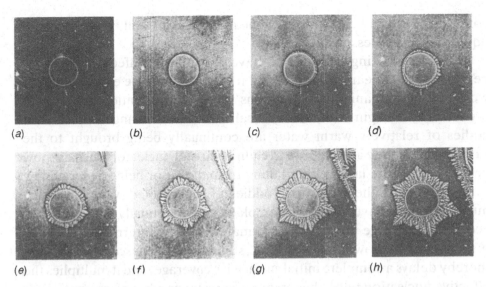

Fig. 4.7. Growth of a frazil disc: (*a*) and (*b*) show stable interfacial growth; (*c*) and (*d*) reveal the emergence of protruberances caused by interfacial instability; and (*e*)–(*h*) display the gradual development of a feathered stellar structure (following Arakawa (1955)).

water. With larger supercoolings, the tendency for vertical *a*-axis growth is again observed. In common with pure water, deeply supercooled aqueous solutions create unstable conditions for ice growth beneath the free surface, and thus generate downward growing dendrites which are essentially solute free. These dendrites are surrounded by an enriched solution which, together with latent heat released at the ice–solution interface, tends to limit growth. However, as noted in the previous section, a characteristically large Lewis number causes constitutional supercooling which continues to promote dendritic growth regardless of the depth of the initial supercooling. In fact, constitutional supercooling would occur in the absence of initial supercooling. Once begun, this *a*-axis growth evidently continues indefinitely.

A more dramatic alteration in the sequence of formative events occurs if the water is not calm, and it is well known that most natural bodies of water are seldom quiescent for very long. River currents in excess of about $0.5\,\mathrm{m\,s^{-1}}$ invariably imply that the water is turbulent and, even in the absence of significant currents, large stretches of water are frequently covered in wind-generated waves below which there is turbulent mixing. At certain times of the year, such mixing may be augmented by, or even caused by, the phenomenon of turnover if vertical temperature and concentration gradients induce a near-surface instability when dense liquid lies above less dense liquid.

Subsurface mixing is bound to have a significant effect on the temperature regime near the surface and it is therefore to be expected that it would be accompanied by alterations in the ice nucleation and growth processes. To begin with, the vertical turbulent exchange implies that eddies of relatively warm water are continually being brought to the surface where they are exposed to chilling air and nucleators such as snow or ice crystals, the latter either falling from the air or being generated by wave action. At the same time, eddies of supercooled water, in which nucleation may have already taken place, are continually being submerged. This exchange process has two main effects: it interrupts the undisturbed crystal growth observed in supercooled quiescent water and thereby delays a complete initial surface ice coverage; and it multiplies the effective nucleation rate[3], thus creating a greater number of ice crystals per unit surface area than would otherwise appear. The crystals which grow under these conditions have a characteristic form and are collectively known as *frazil*.

Each frazil crystal appears initially as a clear transparent disc such as that shown in Fig. 4.7(*a*). The *c*-axis coincides with the axis of cylindrical symmetry. The diameters of discs commonly observed in nature range over a spectrum from 0.5 mm to 5 mm, while their thicknesses are roughly one-tenth of their diameter. Frazil crystals grow to sizes which depend principally on the amount of supercooling and the time they are exposed to it, but even slight supercooling, of the order of 0.01 °C, may be accompanied by large amounts of frazil (Osterkamp, 1978; Martin, 1981; Ashton, 1983). Turbulent mixing and the release of latent heat both tend to promote temperatures close to 0 °C. This is especially true of a fast moving stretch of river where bulk motion and wind-generated surface motion may combine to facilitate the high advective heat fluxes which are necessary to accommodate the large latent heat flux that accompanies substantial frazil generation rates in nearly isothermal water.

Frazil crystals in discoid or needle[4] form are subject to two opposing tendencies: fracture and, to coin a word, geli-flocculation. Both of these result from the relative motion of crystals having natural differential rise velocities[5] superimposed on the random fluctuations characteristic of turbulent eddies. Such motion inevitably leads to brief contacts between crystals, the nature and duration of the contact determining the outcome. Violent contacts between large crystals, as when the edge of one impacts suddenly on the face of another, are capable of fracturing either or both of

Fig. 4.8. Pancake ice near Antarctic ice edge (courtesy R. Masson).

the crystals, with the resulting fragments continuing to grow and perhaps fracturing again. This splinter mechanism, sometimes called collision breeding, is most effective when the edges of the discoids have undergone a stability transition and grown the dendritic feathers evident in Fig. 4.7(*h*). On the other hand, a glancing or rolling contact, in which the relative velocity between the crystals remains very close to zero for a brief period, may permit the crystals to become stuck together by the freezing of supercooled water near their point of contact.

This 'sintering' mechanism, which is only effective in supercooled water, is the basis of the geli-flocculation process which generates first clusters then larger aggregates of frazil crystals having a natural tendency to rise to the nearest free surface. While the crystals and the flocs they produce are distributed throughout supercooled water they are capable of sticking to almost any submerged solid surface they contact. The accretion thus formed is known as *anchor* ice which, as its name suggests, will remain attached so long as erosive or warm currents permit. Anchor ice up to 1.0 m thick (Altberg, 1936) has been observed on river bottoms from which, upon eventual release, it may lift rocks and plants to the surface. However, it is the unattached frazil floes which usually have the greatest effect. As they accumulate on the surface they have a tendency to coagulate further. Very turbulent water may generate slush balls from

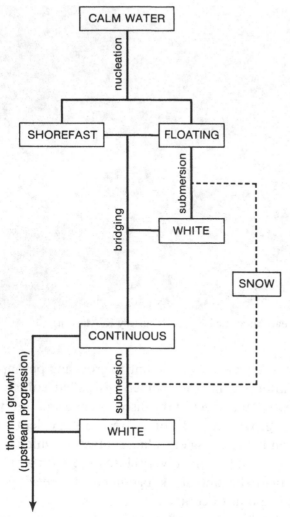

Fig. 4.9. Ice cover formation on calm water.

them while on less turbulent water they form slush patches which eventually become slush pans. Frazil slush tends to dampen free surface motion thus allowing subzero air temperatures to freeze intercrystalline water from the top down. Relative motion on the surface, producing bumping and grinding at the pan edges, gradually shapes the increasingly-congealed slush into *pancake* ice. In turn, pancakes may collect and freeze together to produce an ice *floe*. Near the Antarctic ice edge, for example, the high turbulence associated with wave action maintains a frazil matrix which subsequently freezes between the pancakes to create the floe. Fig. 4.8 illustrates the result. Floes may cover a large surface area, but their

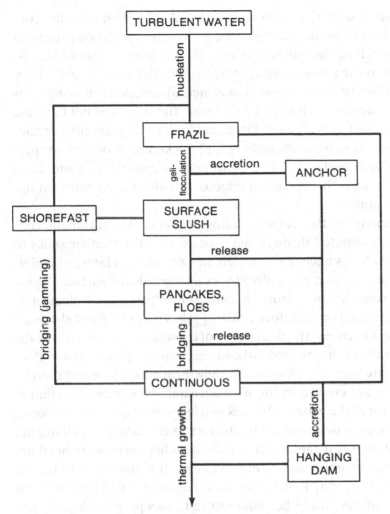

Fig. 4.10. Ice cover formation on turbulent water.

thickness tends to remain in the same order as that of the pancakes from which they were formed.

Whether the water is calm or turbulent, the first frozen layer constitutes *primary* ice which is seldom produced as a large continuous sheet, at least initially. Fig. 4.9 provides an outline of the processes as they occur on calm water. Crystal growth usually, but not always, generates shorefast ice first, with thin sheets of skim appearing later. As the latter multiply and grow they begin to interact with the shorefast ice and with each other. This process of growth followed by interaction may take place several times in a season, depending upon the prevailing winds and currents. Winds, for example, may disturb the surface enough to break the skim, or plate ice, into smaller pieces usually known as *flake* ice. As the wind dies, the continuing loss of heat from the water surface revives the nucleation process in the presence of which the flakes may consolidate into continuous sheets. River currents, on the other hand, tend to carry the skim or plate fragments downstream until an obstruction is met, whereupon following ice is stopped further upstream. Early in the season, the skim may not have the mechanical strength to withstand large end forces and may disintegrate. Eventually, however, the encroaching and thickening shorefast ice permits enough local bridging for a continuous ice sheet to form from obstructed plate ice. The upstream edge of this sheet then starts an upstream progression.

Fig. 4.10 illustrates the corresponding sequence for turbulent conditions. Frazil distributed through the supercooled water either sticks to submerged surfaces as anchor ice, which may be released later, or undergoes geli-flocculation as it gradually moves up towards the surface. Again the process of consolidation, from slush to pans to pancakes to floes, may be interrupted several times before bridging takes place between shorefast and floating ice. On rivers, the final stages of consolidation may include the spectacular results of thermohydraulic interaction. As pancakes and floes congregate upstream of an obstruction they form an ice jam and thereby impose a significant change in the hydrodynamic boundary condition at the upper surface of the water. This has two immediate effects: it implies a leading edge for each new piece of floating ice to encounter; and it implies an increased stream resistance with a corresponding increase in head upstream. Depending upon the stream velocity (or more precisely, the Froude number (Ashton, 1986)), pieces of floating ice colliding with the upstream edge of the jam either come to rest, thus propagating the jam upstream, or are underturned, thus being added to the under side of the jam. The forces acting on each piece of ice, new or old, are not well understood, and therefore it is difficult to describe in detail the processes by which new ice is added beneath, or by which ridges are formed (Flato, 1987). Uncertainty surrounding the cohesion of the loose aggregate and the boundary stresses to which it is subjected make stability analysis of the jam particularly difficult. Suffice it to say that should the stability limit be exceeded, as it frequently is, the jam breaks up sending a disruptive surge of water and ice downstream where it will either be dissipated on another jam or destroy it, and perhaps other jams further downstream, in a growing wave of destruction (Joliffe and Gerard, 1982).

The same processes of flocculation and consolidation may be seen in the creation of primary sea ice, although spatial scales are larger and velocity scales are usually smaller (Weeks and Ackley, 1982). Otherwise, the freezing of sea water is similar to the freezing of fresh water, as described in Figs. 4.9 and 4.10. The principal difference arises because of the expulsion of salt from solution at an equilibrium freezing temperature near −1.9 °C. Since this occurs above the density maximum temperature (Pounder, 1965) it reinforces the natural turnover caused by cooling of the surface layers from above. This mixing tends to maintain a deep isothermal layer close to the equilibrium freezing temperature, and on the top of which nucleation begins. In calm weather, the production of frazil crystals which are able to float and grow on the surface leads to the formation of a thin elastic sheet of ice known as *nilas*. This is the equivalent of freshwater skim. Initially, this sheet is very dark in colour, presumably because the *c*-axes are predominantly vertical, but gradually lightens as it thickens. If the calm is shortlived, the cover will be broken up into pieces of *rind* which are subsequently frozen together after some juxtaposition; this resembles the formation of flake and plate ice.

   More commonly, the appearance of frazil marks the beginning of geliflocculation and slush formation. Shorefast ice is usually the first to form and last to leave, if it leaves at all, while on open water the accumulation of a thick cover of slush eventually creates *grease* ice, with its characteristic damping effect and matte appearance. In this state, the ice behaves much like a large membrane free to stretch and flex with swell and wave, but as time passes, oscillatory tensile stresses (essentially in the horizontal plane) disrupt the cover, causing it to break into pans. Presumably, it is the same, or similar, oscillations which create the jostling action by which pancake ice is formed and compacted. Of the two motions which produce the roughly circular shapes of pancakes with their characteristically raised edges, bumping and sloshing appear to be more important on lakes and seas, while rotation and grinding assume greater importance where a shoreline influences a current and produces a sustained horizontal velocity gradient over the water surface. It is, of course, the consolidation of pancakes into floes which eventually leads, often after several false starts, to a continuous sheet of primary sea ice.

4.3 Growth of an ice cover

   Once a thin, but continuous, cover of primary ice is in place the stage is set for further growth (Michel and Ramseier, 1971). This takes either or both of two forms: *secondary* ice is produced on the bottom of the cover either by direct freezing of the water adjacent to the interface, or by the accretion of ice particles (usually frazil) followed by freezing of the interparticulate water; *superimposed* ice, as its name suggests, forms on the top of the cover when it is inundated with water as a result of direct flooding or indirect submersion. In the course of a season, either or both of these forms of growth may occur more than once, and may be attributable to either or both of their respective sources. A vertical slice taken through the cover reveals the history of growth (Michel and Ramseier, 1971). Examined in white light or polarized light, the slice generally provides useful insight into the details of crystal growth with its many and varied hydrometeorological causes (Palosuo, 1961; Ragle, 1963; Shumskii, 1964; Knight, 1962; Adams, 1981).

4.3.1   Additive growth: snow, frazil and pans

As noted in the previous section, the consolidation and jamming of primary river ice may cause sudden flow surges which can lead to inundation of an intact cover. Later in the season, the appearance of cracks, or the seepage of neighbouring groundwater, may lead to local flooding of the cover. This form of superimposed ice, described as a naled, is treated in detail in Section 6.2. The other major source of superimposed ice is snow.

In general, snow participates in the formation and growth of an ice cover in several ways. Falling on open water, the snow crystals provide numerous nucleation sites for the production of frazil, and should the snowfall be heavy enough their agglomeration on the surface can lead to a slush cover from which primary ice may be formed directly. Later in the season, snow falling on discrete pieces of surface ice – skim, plates, pancakes, floes – introduces additional mass. This porous addition may eventually undergo metamorphosis, as later described in Section 6.2, and add directly to the body of the ice cover. More frequently, the additional weight of the snow causes the ice to submerge thereby permitting water to rise up into the snow cover by capillary action. For shorefast ice, the added weight may be insufficient to permit this, but for floating ice a simple force balance reveals that a layer of snow of thickness Δ_s falling on ice of thickness Δ_i would depress the top of the ice down to the water level if

$$\frac{\Delta_s}{\Delta_i} = \frac{\rho_w - \rho_i}{\rho_s}$$

Water then added to the snow through capillarity, obviously increases the density of the superimposed ice by partially or completely filling the interstitial spaces, which otherwise contain air and water vapour. Subsequent freezing, which is usually from the top down, creates *snow* ice, which is

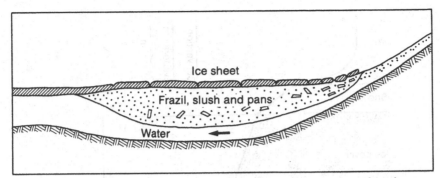

Fig. 4.11. Schematic of frazil, slush and pan deposition creating a hanging dam (following Michel (1971)).

often called *white* ice, partly because of the light scattering effect of trapped air pockets, and partly because of the random crystal orientation created by the initial snow fall (Adams, 1981). The entire sequence of events involving snow is indicated in Fig. 4.9. For the sake of clarity, these events are omitted from Fig. 4.10 which applies, for example, to the Antarctic coast where the high oceanic heat flux keeps the ice thin (60 cm) and the heavy snowfall is known to yield abundant snow ice (Wadhams, Lange and Ackley, 1987).

Bottom ice growth by addition, although essentially independent of conditions on the top of the cover, is strongly influenced by the temperature and velocity fields beneath. Early in the season, rapid growth often results from the addition of discrete pieces or particles of ice carried along on the current. Whether these are large pans or blocks swept under the leading edge of a jam or distributed crystals and floes of frazil, their natural buoyancy coupled with the bottom surface resistance attributable to irregular geometry frequently creates conditions in which they are likely to stick to the cover, subsequently being frozen in place. This is particularly true in supercooled water.

The behaviour of distributed frazil crystals is especially interesting and complex because the under surface of the ice cover is seldom smooth and the deposition process is a special case of sedimentation–saltation in which the particle density is close to that of the carrying fluid and the particles are inclined to stick to each other and to the surface. It has been observed that this deposition–accretion process in rivers is only effective for bulk water velocities below a critical value of 0.5–$0.8\,\mathrm{m\,s^{-1}}$ (Michel and Drouin, 1981). Subcritical water velocities permit the continual accretion of frazil over periods of time long enough to create substantial amounts of secondary ice. Nowhere is this demonstrated more dramatically than in a river in

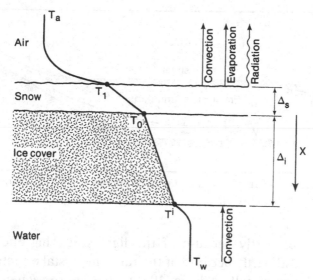

Fig. 4.12. Temperature distribution for simplified growth model.

which a continuous ice cover overlies a slow section located downstream of a more rapid section capable of generating large amounts of frazil slush and pans. The accreted frazil then forms a cohesive body of slush thereby building a hanging dam which is capable of reducing the flow cross section to a small fraction of its initial value. This process is noted in Fig. 4.10 and depicted schematically in Fig. 4.11.

4.3.2   Thermal growth: gelation

Once a continuous sheet of ice covers a large area of a sea, a lake or a river, the process of secondary growth may depend entirely on the removal of latent heat upwards through the cover. The problem then reverts to the Stefan form in which the progress of the interface may be described as thermal growth or gelation[6]. This is shown in Figs. 4.9 and 4.10 as the final process, which it frequently is, but thermal growth and additive growth often occur intermittently in alternations which are clearly revealed in photographs through vertical sections (Ashton, 1986).

In general, the cover loses heat from its upper surface and may or may not gain[7] heat at its lower surface. The net rate of heat loss represents a removal of both sensible and latent heat, the latter normally being the greater. Environmental conditions may vary from point to point over the surface of the cover, but such variations are usually gradual enough to permit the heat balance to be written in one-dimensional form. In essence, the problem is a variation of the basic Stefan problem with prescribed heat fluxes on the upper and lower surfaces.

The simplest formulation of this problem was given in Section 3.1 for conditions where the sensible heat may be ignored i.e., *Ste* $\ll 1$. Fig. 4.12 shows the temperature distribution in and near such an ice cover while it grows into water which is above the equilibrium freezing temperature T^i. Ignoring convection in the water for the moment, the effective resistance to conduction through the snow combined with convection in the atmosphere may be calculated quite simply. Letting R be the thermal resistance to heat flowing steadily between the upper ice temperature T_0 and the air temperature T_a, it may be shown that

$$T^i - T_0 = (T^i - T_a)/(1 + Rk_i/\Delta_i)$$

so that the interface equation may be written

$$d\Delta_i/dt = k_i(T^i - T_a)/\rho_i\lambda_{iw}(\Delta_i + Rk_i) \tag{4.5}$$

When there is no ice initially, this integrates to give

$$\Delta_i = \left[\frac{2k_i(T^i - T_a)t}{\rho_i\lambda_{iw}} + (Rk_i)^2\right]^{\frac{1}{2}} - Rk_i$$

For a snow resistance alone, $T_i = T_a$, and $R = \Delta_s/k_s$, so that

$$\Delta_i = \left[\frac{2k_i(T^i - T_a)t}{\rho_i\lambda_{iw}} + \left(\frac{\Delta_s k_i}{k_s}\right)^2\right]^{\frac{1}{2}} - \frac{\Delta_s k_i}{k_s}$$

whereas for a convective resistance alone, $T_i = T_0$, $R = 1/h_a$, and hence

$$\Delta_i = \left[\frac{2k_i(T^i - T_a)t}{\rho_i\lambda_{iw}} + \left(\frac{k_i}{h_a}\right)^2\right]^{\frac{1}{2}} - \frac{k_i}{h_a}$$

When both resistances are present

$$\Delta_i = \left\{\frac{2k_i(T^i - T_a)t}{\rho_i\lambda_{iw}} + \left[\frac{(1 + h_a\Delta_s/k_s)k_i}{h_a}\right]^2\right\}^{\frac{1}{2}}$$
$$- \frac{(1 + h_a\Delta_s/k_s)k_i}{h_a} \tag{4.6}$$

The effect of convection in the water may be estimated in much the same way. Equation (4.5) is expanded to the form

$$\frac{d\Delta_i}{dt} = \frac{\alpha}{\Delta_i + \gamma} - \beta \tag{4.7}$$

where $\quad \alpha = k_i(T^i - T_a)/\rho_i\lambda_{iw}$, $\quad \beta = h_w(T_w - T^i)/\rho_i\lambda_{iw} \quad$ and $\quad \gamma = (1 + h_a\Delta_s/k_s)k_i/h_a$. Putting $\Delta = \alpha - \beta(\Delta_i + \gamma)$, the interface equation transforms to

$$\frac{\mathrm{d}\varDelta}{\mathrm{d}t} = \frac{\beta^2 \varDelta}{\varDelta - \alpha}$$

which integrates to give

$$t = \frac{\varDelta}{\beta^2} - \frac{\alpha}{\beta^2} \ln \varDelta + \text{constant}$$

Again assuming that $\varDelta_i = 0$ when $t = 0$,

$$t = \frac{\rho_i \lambda_{iw}}{h_w(T_w - T^i)} \left(\frac{k_i(T^i - T_a)}{h_w(T_w - T^i)} \ln \left\{ 1 \middle/ \left[1 - \frac{\varDelta_i}{k_i(T^i - T_a)/h_w(T_w - T^i) - (1 + h_a \varDelta_s/k_s)k_i/h_a} \right] \right\} - \varDelta_i \right) \quad (4.8)$$

The final ice thickness may also be found by equating the growth rate to zero in equation (4.7) thus giving

$$\varDelta_i^f = \frac{\alpha}{\beta} - \gamma = \frac{k_i(T^i - T_a)}{h_w(T_w - T^i)} - \frac{(1 + h_a \varDelta_s/k_s)k_i}{h_a} \quad (4.9)$$

The advantage of the above approach is that it may be used to estimate and compare various effects when the ice latent heat dominates the sensible heat in both snow and ice: this is especially true of the thermal resistances above and below the ice cover. Equations (4.6), (4.8) and (4.9) are most accurate in temperate or subpolar regions, where we would expect that $c_{pi}(T^i - T_a) \ll \lambda_{iw}$. However, they assume that snow and ice properties are fixed, and that the upper boundary condition may be written in the form

$$\dot{q} = h_a(T_1 - T_a)$$

thus implying that the net effects of convection, evaporation and radiation shown in Fig. 4.12 may be lumped together in a fixed overall heat transfer coefficient h_a. If necessary, these boundary conditions may be treated separately (Lock, 1969a).

More generally, the surface heat and mass transfer processes must be treated in greater detail in order to accommodate a time-dependent surface heat flux with components from long and short wave radiation, convection and evaporation. The thermal growth of sea ice provides a good example (Maykut, 1986). The net rate of heat gain may be written

$$\dot{Q} = \dot{Q}_R + \dot{Q}_L + \dot{Q}_C \quad (4.10)$$

where $\dot{Q}_R$ is the net radiative gain, $\dot{Q}_L$ the net evaporative gain and $\dot{Q}_C$ the net convective gain. The evaporative and convective contributions[8] may be prescribed globally from climatological data, or may be specified more precisely in micrometeorological terms as functions of local terrain, windspeed, air temperature and humidity. The radiative heat flux may be specified in the form

$$Q_R = (G_A - E_A) + (1 - \rho)G_I - I_1$$

where E_A is the Stefan–Boltzmann (long wave) emission rate, G_A the incident long wave (atmospheric) radiation, G_I the incident short wave (solar) radiation, and ρ is the surface reflectance (albedo): I_1 is the short wave radiation transmitted into, and subsequently through, the snow as a distributed heat source. Given each of the terms in equation (4.10) (see Section 6.1 for a fuller discussion of surface heat balance) the heat flux on the upper surface may be specified as a time-dependent boundary condition.

When a snow cover is present, this upper boundary condition (at 1) may be used in solving the heat transfer problem within the snow; the other snow boundary condition merely states the continuity of heat flux between ice and snow i.e.,

$$k_s \left(\frac{\partial T_s}{\partial X} \right)_0 = k_i \left(\frac{\partial T_i}{\partial X} \right)_0$$

where X is measured downwards. Within the snow pack, heat transfer is by conduction[9] and is thus governed by the equation

$$\mu_s \frac{\partial}{\partial t} (c_{ps} T_s) - \frac{\partial}{\partial X} \left(k_s \frac{\partial T_s}{\partial X} \right) + \gamma_s I_1 \exp(-\gamma_s X) \qquad (4.11)$$

where the heat generation term arises from the application of Beer's law with an incident penetrating radiative flux of I (Maykut and Untersteiner, 1971). In solving this conduction problem the simplest approach is to assume that the thickness of the snow cover and its thermophysical properties remain constant, but it will become apparent later (Section 6.2) that deposition and metamorphosis of snow are dynamic processes (Langham, 1981; Crocker, 1984). As noted, the thermal conductivity is a function of the density, which is known to be dependent upon the formative wind and the subsequent ageing process; the latter, in turn, reflects the thermal history of the snowpack.

The heat conduction problem within the ice is no simpler. Once the cover has formed from primary ice it begins to thicken as described in the previous section. Constitutional supercooling tends to orientate the *c*-axes nearer and nearer to the horizontal plane in a wedging-out process which limits horizontal growth along the *a*-axes, but permits it vertically (Pounder, 1965; Weeks and Ackley, 1982). After a transition region, which is 5–10 cm thick, the downward dendritic growth develops into columnar, polycrystalline ice, but in so doing tends to trap brine pockets and air bubbles in the interdendritic gaps, thus altering the ice properties. Gas, liquid and solid inclusions readily lend themselves to a simple averaging analysis provided they are uniformly distributed and are roughly spherical in shape. However, the brine pockets in sea ice tend to be elongated parallel to the direction of growth. Furthermore, they not only introduce anisotropy but change their size, shape and location according to the temperature profile in the ice. Typically about 0.05 mm in width, they migrate up the temperature gradient, and have a tendency to enlarge and interconnect at temperatures above $-15\,°C$ (Pounder, 1965). The effect of temperature is felt in other ways, most notably through the release of latent heat over a temperature range. As noted in Section 3.3, the effective specific heat must be written as

$$c_p' = c_p + \lambda \; dm_w/dT$$

where m_w is the mass fraction of the unfrozen water. This requirement does not change the form of the transient term in the heat conduction equation but it may alter its relative importance.

In sea ice, the volumetric heat capacity $c = \rho c_p$ may be represented (Maykut and Untersteiner, 1971) by

$$c(S, \theta) = c_i + \alpha_c S(X)/\theta^2 \tag{4.12}$$

thus reflecting the effect of temperature, here measured in degrees Celsius, and the spatial variation of salinity $S(X)$. Thermal conductivity takes the form

$$k(S, \theta) = k_i + \alpha_k S(X)/\theta \tag{4.13}$$

Using equations (4.12) and (4.13), heat conduction in the ice may be described in the form of equation (4.11) which, if the properties are taken to be independent of time, reduces[10] to

$$c(S, \theta)\frac{\partial T_i}{\partial t} = k(S, \theta)\frac{\partial^2 T_i}{\partial X^2} + \gamma_i I_0 \exp{(\gamma_i X)} \tag{4.14}$$

for sea ice reached by a penetrating radiative flux I_0. A similar form would apply to lake ice or river ice. The interface equation may be written in the form

$$\rho_i \lambda_{iw} \frac{d\Delta_i}{dt} = k\left(\frac{\partial T_i}{\partial X}\right)_I - h_w(T_w - T^i) \tag{4.15}$$

where h_w is determined from the convective heat transfer system immediately beneath the interface.

In a nominally quiescent body of lake water, h_w represents a natural

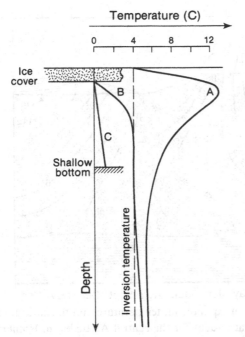

Fig. 4.13. Representative temperature profiles in a northern lake.

convection flow, although when $T_w < 4\,°C$, natural convection is replaced by a weak transient condition. Fig. 4.13 shows representative temperature profiles in a northern lake: curve B corresponds to rapid, early ice growth on deep water while curve C represents advanced growth on shallow water. The latter situation implies that heat is transferred from the bottom to the interface by pure conduction. Unless the local geothermal gradient is unusually high, or a substantial amount of heat is released from the bottom sediments, this conduction process produces water very close to the equilibrium freezing temperature (Ashton, 1986). Curve A (shortly before freezing) shows a strong adverse temperature gradient near the surface while curve B (after freezing) suggests a weaker adverse gradient at depth. Adverse gradients above $4\,°C$ ensure that vertical mixing is facilitated and thus lead to an almost uniform temperature field of $4\,°C$ shortly before freezing. After ice has formed on a lake, heat transfer near the interface is commonly limited to conduction alone, but the temperature distribution about the inversion point and the effect of rivers entering or exiting the lake may change the situation and create substantial variations in the local magnitude of h_w. In a river, the bulk flow usually dictates that fully mixed forced convection predominates. In sea water, the salinity and temperature profiles play an important role in stability and natural convec-

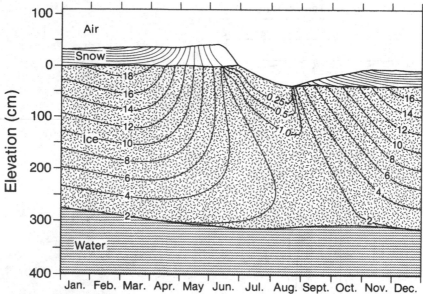

Fig. 4.14. Predicted values of equilibrium temperature and thickness for sea ice, based on Fletcher's heat budget for the central Arctic basin. Isotherms in the ice are labeled in negative degrees Celsius; isotherms in the snow (unlabelled) are drawn at 2 °C intervals. To distinguish between movements of the upper and lower boundaries, they are plotted without regard to hydrostatic adjustment. The vertical coordinate therefore corresponds to ice thickness only before the onset of ice ablation at the upper surface. (Following Maykut and Untersteiner (1971)).

tion, as do tides and currents in the corresponding forced convection system.

Bulk flow beneath an ice cover affects interface growth through advective transport processes. It may simply displace other water with a different bulk temperature or salinity, and thus change the overall potentials driving the convective system. In sea water, for example, the *c*-axis at the interface lies not only in the horizontal plane but tends to be aligned with the direction of the prevailing current (Weeks and Ackley, 1982; Wadhams, 1981; Langhorne, 1982). No such effect has been observed in lake ice (Adams, 1981), presumably because it would require the *a*-axis propagation of the interface associated with sustained supercooling. Section 4.5 offers a more detailed discussion. Currents may also carry large quantities of frazil which deposits itself by an adhesive accretion process. As noted earlier, this process is capable of producing a much thicker cover leading to the hanging dam depicted schematically in Fig. 4.11. A similar process has been noted with young sea ice (Wadhams, 1981).

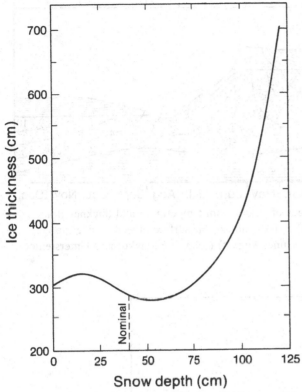

Fig. 4.15. Average equilibrium thickness of Arctic sea ice as a function of maximum annual snow depth (following Maykut and Untersteiner (1971)).

It is evident that thermal growth of an ice cover may be modelled in thermodynamic terms to include all of the major and most of the minor factors which affect such growth. It is also clear that within the limitations of one-dimensionality the growth will depend strongly upon local conditions. The sea ice model of Maykut and Untersteiner offers valuable insight into the relative importance of several major factors which will now be reviewed using their results. A representative solution based upon typical input data for the central Arctic basin is shown in Fig. 4.14 for both growth and decay during an annual cycle: a periodic state was reached 38 years after a prescribed initial state. The equilibrium thickness of about 3 m is consistent with field observations.

The effect of snow cover thickness is typically quite small, as indicated by Fig. 4.15 which shows the average equilibrium ice thickness as a function of maximum annual snow depth: the nominal snow thickness is 40 cm. Two important points emerge from this curve. Firstly it is clear that while snow depths of less than twice the nominal value produce little change in

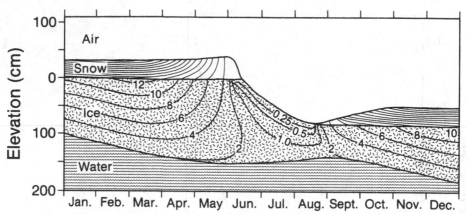

Fig. 4.16. Predicted values of equilibrium temperature and thickness for Arctic sea ice, given a 10% reduction in the surface albedo during the summer ablation season (June–August) (following Maykut and Untersteiner (1971)).

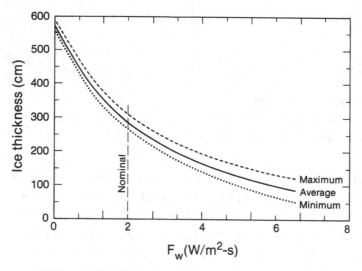

Fig. 4.17. Equilibrium thickness of Arctic sea ice as a function of the annual value of the oceanic heat flux (F_w) (following Maykut and Untersteiner (1971)).

ice thickness, the effect is not monotonic, as might be expected. The undulation reflects a changing balance between both freezing and top ablation which tend to offset each other. Secondly, it is apparent that for snow depths in excess of 80 cm, the decreased ablation of snow and surface ice eventually dominates, thus producing a substantial increase in average ice thickness.

Snow cover may also affect ice thickness through a change in albedo but

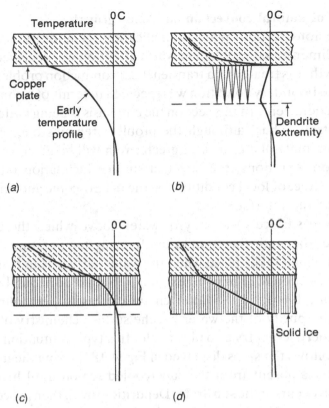

Fig. 4.18. Nucleation and initial growth of ice beneath a rigid, horizontal substrate: (*a*) initial supercooled condition; (*b*) formation of dendrites; (*c*) formation of polycrystalline ice; (*d*) final stable condition (following Gunderson (1966)).

such values usually remain high (0.8–0.9) when snow is present. Changing the nominal value of the ice albedo (0.64) by 10% during the melt season produced the results shown in Fig. 4.16 which provides a strong contrast with the representative results shown in Fig. 4.14. It is immediately apparent that a small change in the summer ice albedo has a large effect on the average ice thickness. Maykut and Untersteiner estimate that a 20% reduction would eliminate the ice cover completely within three or four years.

The influence of convection at the ice–water interface is shown in Fig. 4.17 in which the expected monotonicity is readily apparent. Field data on convection coefficients are not very well established for the wide range of conditions which actually occur, and therefore the consequences of this curve are hypothetical. None the less, it does make clear that the absence of the oceanic heat flux would double the ice thickness while a four-fold increase in the flux would virtually eliminate the ice cover.

4.4   The effect of natural convection on planar growth

In the previous section, planar solidification (gelation) was considered as a one-dimensional problem in which attention was focussed on an ice sheet. Growth was treated as a transient heat conduction problem in which the interface boundary condition was specified in terms of a convective heat transfer coefficient. In this section the emphasis is being switched to the interface itself. Thus, although the problem remains one of ice formation at a nominally plane surface, greater rein will be given to two very important considerations: the effect of surface inclination on the interface, and the effect of local conditions on the macroscopic and microscopic geometry of the interface.

Consider the free surface of a body of water above which the temperature is lowered beneath the equilibrium freezing temperature. The formative process which takes place when the water surface is in contact with cold air (see Section 4.2) may be extended to rigid horizontal substrates (Gunderson, 1966). In the latter circumstances, supercooling is again limited by the purity of the water but the surface chemistry of the substrate and its microgeometry also play a role. In a typical situation, the freezing process follows the steps displayed in Fig. 4.18. During the initial cooling period, ice is absent from the supercooled region until heterogeneous nucleation occurs at the substrate. Dendritic growth then proceeds fairly rapidly, the crystal growth rate gradually diminishing, and even reversing, as latent heat is released into progressively less supercooled water which is also being heated by the bulk below. A planar interface then forms and persists, although its subsequent shape may vary with local convective conditions. Should the substrate or the water contain a sufficient number of effective nucleators, the dendritic transition period is not observed. When it does occur, its duration increases monotonically with the substrate cooling rate.

In a more detailed study of dendritic growth, Gilpin (1976) demonstrated that natural convection and dendritic growth interact in a manner which depends upon surface orientation and the degree of supercooling. For a vertical surface in bulk supercooled water, the dendrites grow horizontally but their growth rate (and hence their length at any given time) is not uniform over the surface if the supercooling is less than about 2 °C. In this range, natural convection occurs between, and immediately beyond, the dendrites thus altering their local growth rate. The release of latent heat produces interdendritic warming of the water, making it denser. The denser water flows in a downward direction between the dendrites and, where possible, escapes downward and flows back into the supercooled bulk; continuity demands that this interdendritic water be replaced by supercooled water from further up the surface. The dendrites thus increase both in length and vertical width (Miksch, 1969) with increased elevation along the surface. Dendrites may exceed 30 cm in length.

In general, the thermal diffusion velocity and the dendrite tip velocity will be different so that at any time their respective penetration distances from the substrate will also be different. Treating a simple one-dimensional situation, the diffusional penetration distance is of the order of $(\varkappa_w t)^{\frac{1}{2}}$ while the (Stefanian) ice penetration distance is of the order of $(k_w \theta_w t / \rho_i \lambda_{iw})$, where θ_w is the amount of supercooling. Hence the diffusion–crystallization ratio of the two penetration zones is $(\rho_i / \rho_w Ste_w)^{\frac{1}{2}}$, where $Ste_w = c_{pw} \theta_w / \lambda_{iw}$ is the water Stefan number. In fact, the dendrite tip velocity, which is along the a-axes, would be much greater than the Stefanian velocity and the penetration ratio would be reduced accordingly. This ratio reveals the central importance of supercooling in determining the relative extent of the dendrites and the associated natural convection system. When the supercooling was small ($\theta_w \lesssim 2°C$), Gilpin found that the convective zone was more extensive than the dendrite zone; it is then appropriate to regard the natural convection as an external determinant of dendrite growth and shape. When $\theta_w \approx 2°C$, the convective and dendritic zones were approximately coextensive. For $\theta_w \gtrsim 2°C$ the dendrite velocities were evidently high enough to dominate the convective system, while for still greater supercooling the convection system ceased to have any effect on dendrite growth; bulk supercooling is then the sole determinant, under which conditions a uniform bulk temperature yields uniform growth over the substrate surface.

In the same work, Gilpin also studied dendritic growth over horizontal surfaces. With the surface facing down, interdendritic warming produced a Rayleigh instability resulting in an array of plumes falling through and ahead of the dendrites, and thereby enhancing heat transfer and growth in an uneven manner. Again the effect decreased when $\theta_w \gtrsim 2°C$ and dendrite growth became more uniform. It is interesting to note that downward growth in a supercooled under-ice melt pond is governed by similar behaviour, at least during its initial stages (Martin and Kauffman, 1974). With the surface facing up, the Rayleigh instability no longer occurs but growth is complicated by the buoyancy of the dendrites themselves: when $\theta_w \lesssim 1°C$, they tend to detach and float upwards, thus producing a form of frazil. For $\theta_w \gtrsim 1°C$, dendrite growth is essentially uniform across the surface. Using simple physics, Gilpin was able to show that the ratio of a representative advective velocity V_c to the dendrite tip velocity V_I is given by

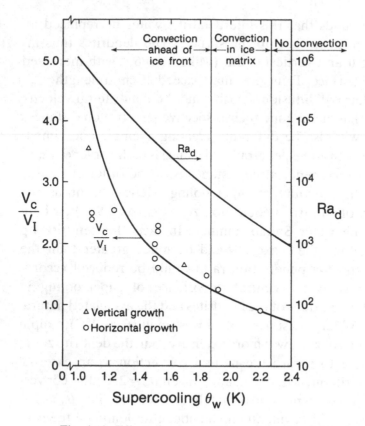

Fig. 4.19. Effect of supercooling on the ratio of the advective velocity to the dendrite tip velocity (following Gilpin (1976)).

$$\frac{V_c}{V_I} \approx \frac{1}{8} Ra_d^{\frac{1}{4}} \qquad (4.16)$$

where $Ra_d = \beta g \theta_w d^3 / \nu \varkappa$: d is the interdendritic gap, an inverse function of θ_w. This result is shown along with experimental data in Fig. 4.19 which reveals that for $\theta_w \gtrsim 2.3\,^\circ\text{C}$, $Ra_d \lesssim 10^3$, a value low enough to prevent instability on a downward facing surface and to produce only a very weak flow near a vertical surface.

Whenever the water is superheated rather than supercooled, the interface assumes a smooth form devoid of dendrites. Fig. 4.20(*a*) illustrates a situation in which a cold vertical substrate removes heat by conduction through polycrystalline ice: latent heat at the interface and sensible heat in the water are thus withdrawn. The temperature profile in the thin film of cooled water adjacent to the ice both causes fluid motion and is influenced by it. For pure water, the density maximum at $4\,^\circ\text{C}$ creates two overlapping flow regimes, as indicated in the illustration. When the bulk water

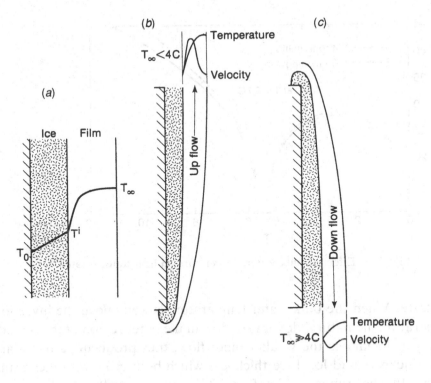

Fig. 4.20. Freezing on a plane vertical surface in the presence of fresh water.

temperature is equal to or less than 4°C, the flow is upwards at every point within the boundary layer (Fig. 4.20(b)). When the bulk temperature is substantially greater than 4°C, e.g., 10°C or more, the flow is downwards (Fig. 4.20(c)). However, for bulk temperatures in the vicinity of 4°C, i.e., 4–7°C, the flow is bidirectional, with the cooler water nearer the ice tending to move up while the warmer water further out tends to move downward.

Ice growth under these circumstances is governed by the interface equation (4.15) in which the heat transfer coefficient is determined by the natural convection system adjacent to the ice. Often, this may be treated as a boundary layer flow in which longitudinal (vertical) temperature gradients are negligible in comparison with the lateral gradients. When the water temperature is nowhere greater than the inversion temperature, the ascending boundary layer exhibits a heat transfer coefficient increasing with depth; if the substrate temperature is uniform, this generates an ice thickness which decreases with depth, as the figure indicates. Only near the bottom of the substrate will the parabolic form of the boundary layer equations be inaccurate and fail to generate the slight growth below the

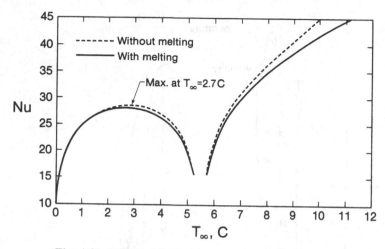

Fig. 4.21. Effect of bulk water temperature on heat transfer rate.

substrate. When the bulk water temperature is well above the inversion temperature the natural downward flow in the outer region of the boundary layer dominates the weaker inner flow, thus producing a boundary layer thickness and local ice thickness which both grow with increasing depth. In the range $4°C < T_\infty < 7°C$, and especially in the range $5°C < T_\infty < 6°C$, the opposition of the inner and outer regions of the boundary layer tends to produce a very weak flow with a consequent reduction in the average heat transfer coefficient and a corresponding increase in ice thickness. Fig. 4.21 shows how the Nusselt number varies with the bulk water temperature (Vanier and Tien, 1968). It is thus apparent that the mean ice thickness does not decrease monotonically with bulk water temperature, but exhibits a local maximum when T_∞ is between 5°C and 6°C.

Much of the above discussion also applies to ice growth from horizontal substrates immersed in superheated water. There are, however, two important qualifications. Firstly, boundary layer natural convection over nominally horizontal surfaces is much weaker, being driven not by the local temperature difference but by the local horizontal gradient of the hydrostatic pressure, as controlled by the temperature field. Near an isothermal surface, such a flow is very weak and would therefore generate very low heat transfer coefficients together with ice thicknesses greater than those found on the corresponding vertical surface. The second qualification arises from the possibility of a Rayleigh instability.

For pure water below a finite, growing ice sheet, the boundary layer would tend to flow outwards towards the edges of the sheet when

$T_\infty < 4\,°C$ and inwards when $T_\infty \gg 4\,°C$. The reverse would be true for water above an ice sheet. It is quite likely that when $4\,°C < T_\infty < 7\,°C$ the convective system would be bidirectional and thus create an exceptionally small heat transfer coefficient. The existence of a thermal instability would greatly modify the boundary layer flow and may in fact destroy it. Water above ice, but with a temperature below the inversion point, is stable provided that the Rayleigh number based on the boundary layer thickness does not exceed a critical value. A similar statement applies to pure water below ice when $T_\infty \gg 4\,°C$. For either of these situations, the surpassing of the appropriate critical Rayleigh number produces a dramatic alteration in the flow, varying from the appearance of longitudinal rolls superimposed on a boundary layer flow to Bénard cells in water which is otherwise quiescent.

In the absence of a leading edge, the effect of a horizontal, natural convection boundary layer is usually negligible if the neighbouring substrate is isothermal. The freezing problem then reverts to a quasi-one-dimensional form in which Rayleigh instabilities may induce cellular convection near the interface. For example, if pure water with a bulk temperature in excess of $4\,°C$ lies above an ice sheet, a potentially unstable region exists immediately adjacent to the interface. On the other hand, the same water lying beneath the ice sheet is stable, at least in the region extending down to the $4\,°C$ isotherm. It has been noted that this isotherm provides a natural dividing surface between the stable and unstable regions in fresh water (Forbes and Cooper, 1975).

The laboratory work of Yen and Galea (1969) has demonstrated the existence of Rayleigh instabilities in pure water and shown their effect on the nominally horizontal interface. In quiescent water, the appearance of Bénard cells changes the local heat transfer coefficient across the surface in a regular way, thereby producing a scalloped interface which is depressed near the points towards which relatively warm water penetrates. Critical Rayleigh numbers, based on the depth of water separating the ice from a warm solid substrate, are found to be a function of the substrate temperature, as might be anticipated. The Rayleigh numbers fall in the range 10^3–10^5.

In an aqueous solution, the above situation is modified by the rejection of solute at the interface. During the freezing of sea water from above, for example, the adverse temperature and concentration gradients create an instability (Farhadieh and Tankin, 1975) when the Rayleigh number exceeds 5×10^3. Bénard cells once again cause sculpting of the interface, and lead to the formation of salt plumes which originate at local 'peaks' in the interface and descend to significant depths in the solution. It is interesting to note that the cellular convection system thus produced has the initial effect of raising the interface temperature above the equilibrium freezing temperature.

4.5   The effect of forced convection on planar growth

In an engineering context, there are many examples of ice formation on a subcooled substrate which is planar (or nearly so) and near which the effect of natural convection is either absent or dominated by forced convection. In the production of 'Flakice', for example, a large, internally-refrigerated drum rotates slowly about a horizontal axis while partly submerged in chilled water. The thin layer of ice appearing on the drum surface as it passes through the water constitutes an essentially one-dimensional growth in the presence of weak forced convection. The emerging ice layer is peeled from the top of the drum to produce the 'Flakice'. Stronger forced flows are frequently encountered on a larger scale, as when a river flows past an engineering structure which removes large amounts of heat either naturally, by means of conduction into sub-zero air, or artificially, by means of refrigeration. Once again, ice growth may be considered planar so long as the ice thickness is small compared with the overall dimensions of the structure.

On a natural scale, quasi-one-dimensional ice growth is more often the rule than the exception, and there are many situations in which forced convection exerts a significant influence. The decelerating inflow to a lake, and the accelerating outflow, each give rise to alterations in the local ice thickness depending not only on the bulk water temperature but on the local details of the water flow. Much the same may be said of rivers generally, but the level of turbulence, reinforced by wind-induced wave action, has the initial effect of delaying the appearance of a solid cover while a layer of frazil slush is developed, as noted in Section 4.2. However, the persistence of slush in early winter invariably leads to the formation of a continuous solid ice cover and a substantial change in the hydrodynamic boundary condition and heat transfer coefficient at the upper water surface.

Should the water contain frazil particles which originated upstream, further effects of advection may be seen. For example, the under surface of river ice may grow by accretion, a feature which may not only lead to the formation of a hanging dam mentioned earlier (see Fig. 4.11) but may alter the shape of the leading edge of the ice cover, thereby creating a hydro-dynamic force which tends to submerge it. Similar phenomena may occur with sea ice when it is relieved by stretches of open water described as *polynyas* or leads. The latter tend to be long and narrow stretches opened by divergent or shear stresses while the former are randomly shaped and tend to occur in the lee of coastlines or fast ice edges. As discussed earlier, the combined effect of winter wind and wave action on this open water may generate frazil in large amounts. Wind-generated currents are then capable of building the ice cover along the windward side by frazil accretion at the edge and underneath. In passing, it is interesting to note that frazil formed in supercooled sea water by this and other means is considered to be an important factor in the collection and establishment of algal communities (Horner, 1985). While it is known that the communities thrive in the bottom ice created from the frazil, it is not yet certain whether the algae actually initiate the nucleation or are simply the beneficiaries of it by becoming attached to the rising crystals which are subsequently integrated into a habitat (Ackley, 1982).

It is evident from these few examples that the growth of an ice sheet or film in the presence of forced advective effects is a common problem the precise form of which varies widely. In this section the discussion will be limited to simple, particle-free, forced convection over a nominally plane ice–water interface. Although such a treatment automatically excludes many other important and more complex situations, it should serve to establish the basic characteristics of the problem as they relate to the growth rate and shape of the interface in water which is superheated or supercooled.

Consider first the growth of ice in flowing water which is supercooled. Unless the supercooling is very small e.g., $\theta \leqslant 0.3\,°C$, crystal growth will be dendritic and therefore a useful starting point is the study of dendrite growth under forced convection conditions. Given the very small cross section of dendrite crystals, it is most likely that local Reynolds numbers based upon their width would also be small, thus implying laminar flow. For water flowing normal to the c-axis, Miksch (1969) has demonstrated that dendritic branches grow faster on the upstream edge and slower on the downstream edge than when the flow is absent. Although this is not too surprising, it was also observed that the streamwise temperature variation caused the 'spine' of the dendrite to change direction. The angular alteration was found to decrease with an increase in supercooling: i.e., with a decrease in the relative importance of latent heat release. These findings are consistent with observations and speculations on sea ice growth in the presence of constitutional supercooling. Although very different situations, they both lead to the general conclusion that crystal orientation is strongly influenced by current-induced alterations to the local temperature (and concentration) field. Removal of latent heat by convection establishes a principal growth direction in which this removal is optimal. The preferred direction is always normal to the c-axis but the angle between this and the water flow direction varies with the degree of mutual interference between neighbouring dendrites. With the c-axis normal to a horizontal flow direction, horizontal a-axis growth directly into the flow optimizes the upstream (and minimizes the downstream) growth rate of a single crystal. The bottom (or top) tip, however, is partly in the wake of the upstream edge and therefore grows more slowly. A parallel stack of such crystals in a horizontal stream of supercooled water would grow even more slowly on their lower (or upper) surface because an undisturbed thermal boundary layer would tend to develop in the direction of flow. If the flow were parallel to the c-axis it would run across dendrite tips rather than between them and would thus induce higher heat transfer coefficients at the tips. This provides a plausible explanation of the alignment between c-axes and flow direction observed on the bottom of sea ice (Weeks and Ackley, 1982; Langhorne, 1982).

From these observations and conclusions it is possible to anticipate the growth behaviour of dendrites when nucleation takes place on a planar substrate. Growth is essentially outward but slower lateral growth eventually leads to the substrate being completely covered in ice, after which point the dendrite 'roots', then in solid ice, begin moving away from the substrate. In a small volume of water, dendrite growth may continue until the ice volume reaches the limit established by the original amount of supercooling i.e., by the Stefan number. In a semi-infinite water volume, growth could continue indefinitely, with the rate of growth of individual dendrites, upstream and cross-stream, varying in the direction of the water flow. Upstream release of latent heat tends to retard downstream growth and therefore the dendrites would be thickest and protrude furthest at a substrate leading edge. The abrupt change at the leading edge itself may have a considerable effect on the local flow field. Flow within and around such a dendrite matrix does not appear to have been the subject of much study and therefore the possibility of mainstream separation immediately behind the leading edge, or of local separation in a dendrite wake, has not been explored. Although Reynolds numbers may not be large, much would depend upon the level of free stream turbulence and the precise shape of the substrate.

When the water is superheated, the interface takes on a smooth glassy[11] appearance. Under these conditions interface growth depends upon the

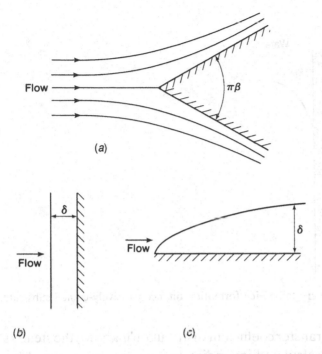

Fig. 4.22. Laminar boundary layer flow around a wedge: (a) wedge;
(b) stagnation ($\beta = 1$); (c) Blasius ($\beta = 0$).

inequality of the heat fluxes normal to it. If the substrate temperature is T_0, the gradient in the ice may be approximated by

$$\nabla T \approx \frac{T^i - T_0}{\Delta_i(X)} \tag{4.17}$$

when $Ste \ll 1$ and $\Delta_i \ll L$, the extent of the substrate in the longitudinal X-direction. In the water, the gradient may be determined from the heat flux represented by

$$\dot{q} = h_w(X)(T_\infty - T^i) \tag{4.18}$$

where the heat transfer coefficient derives from the local flow. Under steady conditions it follows that

$$k_i \frac{(T^i - T_0)}{\Delta_i} = h_w(T_\infty - T^i)$$

and hence

$$\epsilon \, Bi = 1 \tag{4.19}$$

where $\epsilon = (T_\infty - T^i)/(T^i - T_0)$ is the superheat ratio, and $Bi = h_w \Delta_i / k_i$ is the Biot number of the ice layer. It therefore also follows that $\Delta_i(X) \propto 1/h_w(X)$ once growth has ceased. This last result dictates that

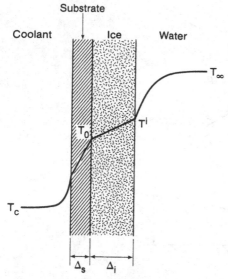

Fig. 4.23. Polycrystalline ice formation on a convectively-cooled substrate.

whenever the heat transfer coefficient distribution is known the steady state ice profile may be calculated immediately.

Heat transfer over a plane isothermal surface situated in a mainstream has been widely studied and may be broadly represented in two dimensions by the wedge flow depicted in Fig. 4.22. For any wedge angle $\pi\beta$, it is known (Schlichting, 1968) that the laminar boundary layer thickness δ is given by

$$\delta \propto X^{(1-m)/2}$$

where $m = \beta/(2-\beta)$, and X is the distance from the leading edge (or forward stagnation point). The heat transfer coefficient is inversely proportional to the boundary layer thickness and hence

$$h_w(X) = A X^{(m-1)/2}$$

for laminar conditions. Of special interest are the limiting situations which occur when the surface is either parallel to the main flow or normal to it. For the former, often called the Blasius problem, $\beta = 0$, $m = 0$ and hence $h_w \sim X^{-\frac{1}{2}}$; for the latter, a stagnation flow, $\beta = 1$, $m = 1$ and hence h_w is independent of X.

Given the understanding that variations in heat transfer coefficient strictly imply a non-planar interface, but that this effect is felt only in a small region near the leading edge, the growth of ice on a plane substrate may be calculated in the quasi-one-dimensional form depicted in Fig. 4.23. The interface equation

$$\rho_i \lambda_{iw} \frac{d\Delta_i}{dt} = \frac{k_i (T^i - T_0)}{\Delta_i} - h_w (T_\infty - T^i)$$

takes on precisely the same form as used in Section 4.3 if the substrate wall is substituted for the snow and the coolant replaces the atmosphere. The ice thickness is therefore given by the implicit form

$$t = \frac{\rho_i \lambda_{iw}}{h_w (T_\infty - T^i)} \left(\frac{k_i (T^i - T_c)}{h_w (T_\infty - T^i)} \ln \left\{ 1 / \left[1 - \frac{\Delta_i}{k_i (T^i - T_c)/h_w (T_\infty - T^i) - (1 + h_c \Delta_s/k_s) k_i/h_c} \right] \right\} - \Delta_i \right) \quad (4.20)$$

in which the subscripts c and s refer to the coolant and substrate, respectively. The final ice thickness is given by

$$\Delta_i^f = \frac{k_i (T^i - T_c)}{h_w (T_\infty - T^i)} - \frac{(1 + h_c \Delta_s/k_s) k_i}{h_c} \quad (4.21)$$

thus extending equation (4.19) to include the effects of coolant and substrate.

The function $h_w(X)$ determined for an ice-free substrate may thus be substituted directly into the equations above so long as curvature of the interface is unimportant. As the effect of curvature near the leading edge becomes more important the form of $h_w(X)$ must be modified through a process relating it to ice shape. In these circumstances, the problem becomes fully two-dimensional and has been studied by Hirata, Gilpin and Cheng, (1979a) under steady flow conditions, beginning with equation (4.19) in the form

$$\frac{X}{\Delta_i} = \frac{k_w}{k_i} \epsilon \, Nu_X$$

where Nu_X is the local Nusselt number. For a flat plate lying parallel to a laminar free stream,

$$Nu_X = 0.332 \, Pr^{\frac{1}{3}} \, Re^{\frac{1}{2}}$$

in which Pr is the Prandtl number. Thus

$$\frac{X}{\Delta_i} = 0.332 \, Pr^{\frac{1}{3}} \frac{k_w}{k_i} \epsilon \, Re_X^{\frac{1}{2}} \quad (4.22)$$

which suggested to the authors two important possibilities:

(1) since the material properties are essentially fixed, $\epsilon Re_X^{\frac{1}{2}}$ may be considered as the single correlating parameter, and

(2) since the equation implies that $\Delta_i \propto X^{\frac{1}{2}}$, the shape of the leading edge of the ice may be approximated by a parabola.

The second possibility led the authors to re-phrase the problem by

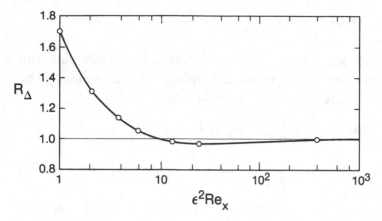

Fig. 4.24. Comparison of ice thickness predicted by the two-dimensional and one-dimensional theories (following Hirata *et al.* (1979a)).

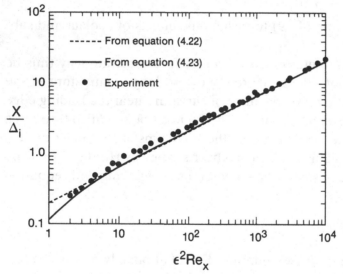

Fig. 4.25. Comparison of predicted and measured ice layer profiles on a constant temperature plate (following Hirata *et al.* (1979a)).

transforming it from a rectangular Cartesian coordinate system to a parabolic system. In the transformed coordinate system the convection problem becomes a stagnation flow for which the potential velocity field is known (Milne-Thomson, 1968). From this field may be found the variation of the free stream velocity in the physical convection problem which may then be solved to determine the variation of heat transfer coefficient over the parabolic ice–water interface. Under the steady conditions assumed, the heat flux to the water must equal that from the ice at each point on the interface. Equating the two interfacial heat fluxes yields

$$\frac{X'}{\Delta_i} = 0.730 \, \frac{k_w}{k_i} \, \epsilon \, Re^{\frac{1}{2}} \, (1 + f) \qquad (4.23)$$

where $X' = X_o + X$, X_o is the ice extension upstream of the plate leading edge, and f is a correction depending solely on X'/Δ_i. Comparing equations (4.22) and (4.23), it is evident that they are essentially the same[12] when $f = 0$ i.e., far from the forward stagnation point.

The ratio R_Δ of two-dimensional and one-dimensional estimates of ice thickness determined from equations (4.23) and (4.22) is shown plotted in Fig. 4.24 as a function of $\epsilon^2 Re_X$. The one-dimensional approximation is evidently satisfactory provided $\epsilon^2 Re_X$ is more than about 10. In water at the equilibrium freezing temperature, $\epsilon = 0$ and the effect of leading edge curvature may not then be neglected. However, since $Re_X > 10^3$ for boundary layer flow it is clear that the effect of leading edge curvature may safely be neglected when $\epsilon \gtrsim O(1)$; indeed over the entire laminar boundary layer regime ($10^3 < Re_X < 5 \times 10^5$) the leading edge will usually play only a minor role. The plot of X/Δ_i versus $\epsilon^2 Re_X$ shown in Fig. 4.25 offers an alternative[13] assessment of the limits of one-dimensional theory. The theoretical inadequacy of equation (4.22) occurs at values of $\epsilon^2 Re_X < 10$, but this figure suggests that, in practice, the equation is still operative for $\epsilon^2 Re_X \approx 2$. As the authors point out, however, good agreement over the range $2 < \epsilon^2 Re_X < 50$ is fortuitous and reflects the cancellation of two opposing effects: flow acceleration and streamwise heat conduction.

The utility of equation (4.22) under steady conditions suggests that a quasi-one-dimensional model may also be useful under transient conditions if $Ste \ll 1$. In general, therefore, equations (4.20) and (4.21) constitute a valuable starting point for laminar flows provided $\epsilon^2 Re_X$ is not too small. If it is too large, however, other factors come into play and the boundary layer may become unstable. In the absence of ice, boundary layer instability on a horizontal substrate is normally attributable to the appearance of Tollmein–Schlichting[14] waves but the existence of a vertical temperature gradient may alter the situation. For example, an upward facing surface at a temperature of 0°C in fresh water with a bulk temperature of 4°C or less may generate Taylor–Goertler waves (Gilpin, Hirata and Cheng, 1978a; Gilpin, Imura and Cheng, 1978b) which are capable of producing longitudinal grooves in an upward facing ice interface. For a downward facing interface, the same temperature distribution would tend to stabilize the boundary layer and thus delay transition to turbulent flow. Under a smooth sea ice cover, the adverse vertical salinity gradient would also tend to destabilize a boundary layer flow.

In general, laminar–turbulent transition over an ice sheet with a leading

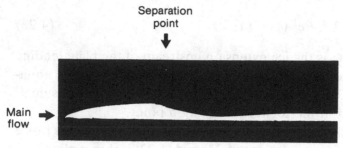

Fig. 4.26. Photograph of abrupt transition with $U_\infty = 78\,\text{cm s}^{-1}$ and $T_\infty = 1\,°\text{C}$.

edge is not a simple process (Hirata, Gilpin and Cheng, 1979b). As equations (4.19) and (4.21) indicate, the steady state ice thickness is inversely proportional to the water heat transfer coefficient, and since turbulent heat transfer is much more efficient than laminar heat transfer it follows that ice thickness should decrease beyond the transitional regime. However, the transition may create either of two distinct variations in the shape of the interface: gradual and abrupt, both of which have a tendency to migrate upstream. In the first of these, thinner and flatter ice becomes unstable and produces a rippled surface (Gilpin, Hirata and Cheng, 1980) on which separation, re-attachment and re-laminarization are repeated regularly between the ripple peaks. An abrupt change in surface shape, on the other hand, usually occurs on thicker and more curved ice at a point where separation is not followed by re-laminarization, at least not immediately. Erosion in the separated zone causes the 'step' thus created to move upstream until the heat transfer coefficient immediately downstream of the 'step' is no longer much greater than the original local value. A silhouette of this ice shape is shown in Fig. 4.26. Re-writing equation (4.22) in the form

$$\Delta_i = \alpha X^{\frac{1}{2}}$$

where

$$\alpha = \left(\frac{\nu}{U_\infty}\right)^{\frac{1}{2}} k_i / (0.332\ Pr^{\frac{1}{3}} k_w\ \epsilon)$$

and U_∞ is the free stream velocity, Hirata *et al.* 1979b noted that the range of the two types of interfacial effect during transition could be delineated by the quantity α which strongly influences both the flatness and the thickness of the ice. Abrupt transitions occurred for $\alpha > 6 \times 10^{-2}\,\text{m}^{\frac{1}{2}}$ while gradual transitions occurred for $\alpha < 3 \times 10^{-2}\,\text{m}^{\frac{1}{2}}$. Between these two limits both types were observed.

For a gradual transition, the critical Reynolds number is found to be less than if the plate were ice-free (Hirata *et al.* 1979b). This is attributed to

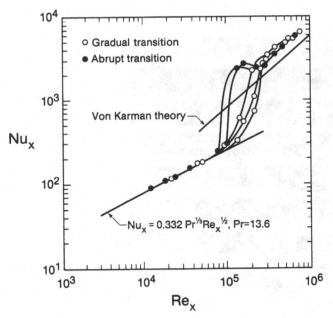

Fig. 4.27. Heat transfer data for forced convection over an ice sheet (following Hirata *et al.* (1979b)).

the fact that downstream reductions in local ice thickness introduce a mild adverse pressure gradient which moves the transitional point (and the ripple) upstream. For an abrupt transition, this same effect is even more dramatic. From the initial transition point the 'step' is observed to migrate upstream to a point where the local Reynolds number may be an order of magnitude less than the transitional value for an ice-free substrate. Well downstream of the transitional zone the flow becomes fully turbulent. As the heat transfer data in Fig. 4.27 reveal, laminar conditions are well represented by well-established flat plate theory, but the transitional and turbulent data are not. The two separate routes through the transitional regime reflect a hysteresis phenomenon suggesting bistable behaviour when $3 \times 10^{-2} \, m^{\frac{1}{2}} \lesssim a \lesssim 6 \times 10^{-2} \, m^{\frac{1}{2}}$. In the turbulent region, the data are about 20% above the prediction for a flat plate, and this may be attributed to large scale turbulence generated by alterations in ice topography.

As a general rule, the growth of perturbations on an initially plane ice–water interface is believed to be stimulated by convective heat transfer on the water side and damped by conductive heat transfer on the ice side. Using linear stability analysis under turbulent conditions, Gilpin *et al.* (1980) demonstrated that stability during freezing required $\epsilon Bi < 0.4$, yet

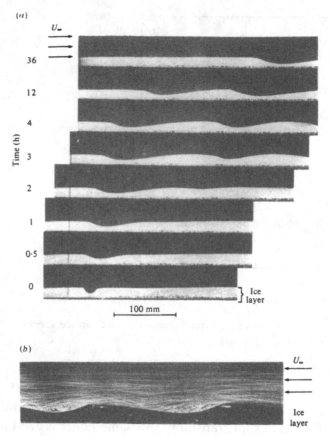

Fig. 4.28. (*a*) Photographs of developing ice surface wave. (*b*) Flow visualization over rippled ice surface (following Gilpin *et al.* (1980)).

for the steady state they concluded that $\epsilon > 1/12$ (or $Bi < 12$) was necessary to guarantee stability. Any small surface irregularities, natural or artificial, grow and migrate according to the particular flow conditions existing at the time. With respect to turbulent behaviour, Fig. 4.28(*a*) illustrates the evolution of an initial recess, showing both formative growth and subsequent movement *downstream*. In general, good agreement has been observed (Gilpin *et al.*, 1980) between the wave numbers predicted and those measured. Early growth develops a smooth wavy pattern which decays, stabilizes or grows, depending upon the values of ϵ and Bi. Unstable growth leads to waves of sufficient amplitude to generate separation in the wake of the 'crest', thus producing a rippled surface. A typical rippled surface with separated flow is shown in Fig. 4.28(*b*).

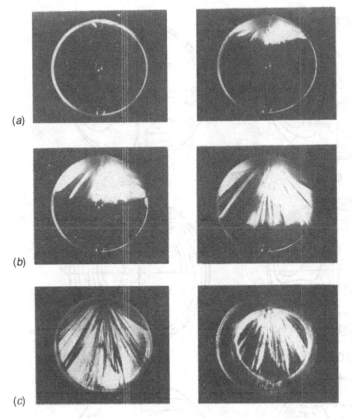

Fig. 4.29. Appearance of ice in a water pipe: (*a*) just before and just after nucleation; (*b*) during dendritic growth; (*c*) during annular growth (following Gilpin (1976)).

4.6   Freezing inside cavities and conduits

In the previous two sections consideration was given to freezing on plane substrates under a variety of conditions. This basic geometry enabled the growing ice to be modelled as a one-dimensional or quasi-one-dimensional layer whose precise local shape and thickness were determined from local conditions, more or less; the multi-dimensionality implicit in such non-planar ice interfaces was often found to have a limited role. Once the substrate geometry is two- or three-dimensional such simplification must be abandoned, unless the ice remains very thin. There are, of course, a great many practical substrate shapes, but the geometries most commonly encountered may be divided into two categories: those for which the ice growth is wholly or partly confined by the substrate; and those for which growth is entirely external to the substrate. The former group will be the subject of this section, and the latter will be discussed in the following, and last, section.

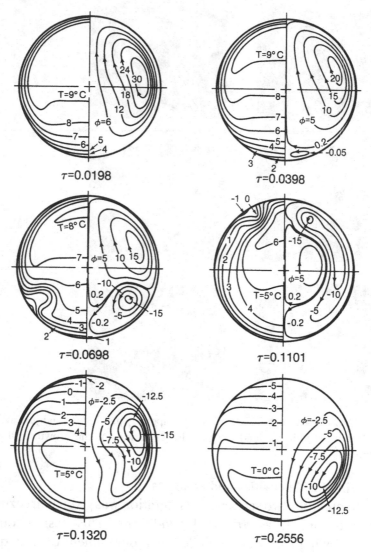

Fig. 4.30. Transient streamline patterns and isotherms in cooling water: non-dimensional time $\tau = kt/R^2$ (following Cheng et al. (1978)).

The circular cylinder is one of the simplest and commonest of cavity geometries, especially when a large length–diameter ratio converts it into a pipe. Freezing inside horizontal water pipes has been the subject of much study, especially by Gilpin and his coworkers. Fig. 4.29 shows the various stages of freezing in a pipe if there is no main flow (Gilpin, 1976, 1977, 1978a,b). There are, in general, four stages: cooling and supercooling[15] of the water; nucleation and dendritic growth, possibly filling the pipe;

annular growth gradually replacing a modified dendritic growth; and sub-cooling of the complete ice plug. The second and third of these will now be discussed in detail.

As the water cools through 0°C it does so with its coldest region closest to the pipe wall, thus establishing a natural convection system in which water temperatures range, in general, above and below the equilibrium freezing temperature. This circulation is nominally an annular flow but may be complicated by the spanning of the inversion temperature. Cheng, Takeuchi and Gilpin (1978) give the results of a numerical study of natural convection in a convectively-cooled horizontal pipe with the initial water temperature set at 10°C and the environmental temperature equal to −10°C. Fig. 4.30 shows the evolution of temperature and flow, revealing three stages. Initially, a predominantly positive thermal expansion coefficient ensures that the central core flow is upwards, whereas much later in the process, and especially when the water is substantially supercooled, the expansion coefficient is negative and the flow direction reversed. Between these extremes is a transitional stage in which an incipient flow reversal appears at the bottom of the pipe (Fig. 4.30 top right) and gradually spreads upwards. Such alterations in the flow pattern correspond to alterations in the temperature field. In the early stages, the coldest water remains in the bottom half of the pipe until the equilibrium freezing temperature is reached at the pipe wall. This 0°C isotherm moves gradually up around the wall while a small supercooling appears in the upper half of the pipe (middle right). With the flow now reversing, a substantial degree of supercooling begins to develop in the upper half of the pipe. It is thus evident that dendritic growth will be delayed until the third, reversed-flow stage and will begin by spreading throughout the upper half of the pipe, as demonstrated in Fig. 4.29.

In general, the temperature field in the water will be dependent on the thermal boundary conditions, and is therefore influenced by the precise details of the external convection system, thermal insulation (if any), and the thermal resistance of the pipe wall, both radially and circumferentially. The greater the total radial thermal resistance (pipe wall, insulation and external convection) the lower will be the cooling rate, and the deeper the supercooling will penetrate down into the lower half of the pipe. Lowering the circumferential resistance tends to even out the temperature distribution in the water and creates a similar effect. It is thus evident that dendritic growth throughout the entire 'quiescent' water volume is most likely to occur in small, insulated metal pipes. Larger, uninsulated plastic pipes in the same thermal environment may produce just enough dendritic ice to freeze the remaining water close to the wall, and thereby create conditions for annular growth.

Annular growth may in fact be the natural growth form if the water is initially superheated enough and the cooling rate is high enough. Any supercooled region close to the pipe wall then resembles a plane substrate wrapped into a cylindrical shape, with initial dendrite growth gradually transforming into polycrystalline ice, as described in Fig. 4.18. In the absence of a main flow, it is quite likely that dendritic growth will not only be extensive but will continue after annular growth has begun, as Fig. 4.29 indicates. The rather flimsy dendritic structures are often numerous enough to demand a significant pressure gradient before a main flow can be re-established; start up will then produce an ice slush formed from the fragmented dendrites. Should the dendrites be allowed to remain in place, they gradually become thicker and grow (within the limit imposed by the initial Stefan number) into webs or sheets of ice anchored in the annular ice. The start up pressure gradient is thereby increased substantially.

Annular growth will always occur when the water is not supercooled. If superheat is also absent, the system follows the dictates of the Clausius–Clapeyron equation relating the equilibrium freezing pressure to the equilibrium freezing temperature. In a pipe which is shut off at both ends, the density difference between ice and water is capable of inducing very high freezing pressures, depending on the rigidity of the pipe wall (Sugawara, Seki and Kimoto, 1983). As freezing takes place, the equilibrium freezing temperature falls, and it is then possible to achieve stable equilibrium states in which the pipe and its contents are uniformly at a temperature well below the initial freezing temperature but yet all of the water is not frozen. A sudden release of pressure would then create supercooled water into which dendrites would presumably grow from the polycrystalline ice already present.

The problem of freezing in superheated water is different again in type and significance. Metastable behaviour and its effect on nucleation is no longer important and interest is focussed instead on the formation of polycrystalline ice under a variety of convective conditions both inside and outside the pipe. Under steady conditions, the system differs in only one important respect from ice-free pipe flows, which have been studied extensively. This difference is, of course, the presence of an ice–water interface with a shape which is unknown *a priori*; interface shape influences convective behaviour which, in turn, determines the interface shape. Interface–convection interaction is seen in almost any polycrystalline ice

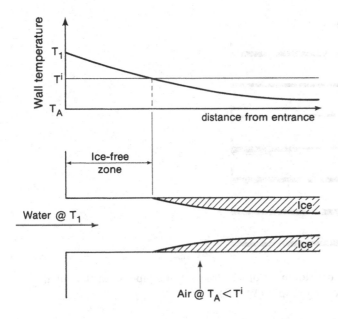

Fig. 4.31. The ice-free zone in a convectively cooled pipe.

growth problem, as the previous section has made clear, but it is a major characteristic of internal flows.

Convection in the water closely resembles a classical thermal entrance flow and thus poses a modified Graetz[16] problem (Zerkle and Sunderland, 1968). When the pipe wall temperature is uniform, and less than the equilibrium freezing temperature, the prescription of the water inlet velocity and temperature is sufficient for the development of the solution to the variable-section Graetz problem which may then be matched to the ice conduction problem at the interface. However, for the more practical outer boundary condition of uniform convective cooling depicted in Fig. 4.31, the situation becomes slightly more complex for two reasons: firstly, the pipe wall temperature does not fall below the freezing point until the water has travelled through an ice-free zone, the length of which must first be determined from the solution of the modified Graetz problem posed therein; secondly, the ice-free zone solution provides the upstream boundary condition at entry to the subsequent variable-section Graetz problem (Lock, Freeborn and Nyren, 1970; Hwang and Yih, 1973).

Several experimental and analytic investigations have been devoted to the study of polycrystalline ice growth in a conduit under laminar or turbulent flow conditions (Mulligan and Jones, 1976; Cheng and Wong, 1977; Shibani and Ozisik, 1977; Thomason, Mulligan and Everhart, 1978).

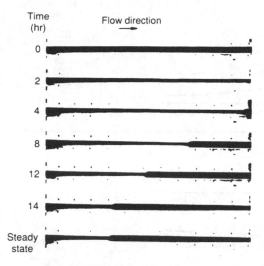

Fig. 4.32. Transient development of ice structure in a pipe when the initial flow is laminar (following Gilpin (1981a)).

By and large, these studies suggest that a thermal entrance model, suitably modified by the presence of the ice, adequately describes heat and fluid flow. Satisfactory agreement has been reported between theory and experiment over a practical, if restricted, range of conditions. In particular, the influence of ice growth on pressure drop appears to have been adequately modelled in both the laminar (Mulligan and Jones, 1976) and turbulent (Thomason *et al.*, 1978) regimes. Despite these findings, however, a full range of variables has yet to be explored and the models themselves then subjected to refinement or replacement. It is only recently, for example, that the fundamentals of ice–water stability – hydrodynamic, thermal and interfacial – have been studied in any depth.

The interfacial stability of annular pipe ice affords an excellent example of unusual, and perhaps unexpected, behaviour being explained in fundamental terms. This behaviour was first discovered by Gilpin (1979b) for flow conditions near the point of transition[17] from laminar to turbulent flow. In the absence of ice, this transition had been revealed almost a century earlier (1883) by Osborne Reynolds. The presence of ice might not be expected to alter the transitional phenomenon greatly but Gilpin discovered an entirely different type of behaviour which, in extending over a Reynolds number range of 374–3025, evidently precludes the existence of the usual type of transition, at least when the cooling rate is sufficiently low.

As mentioned above, several previous studies of pipe icing suggested that annular ice merely introduces a geometrical complication: reported

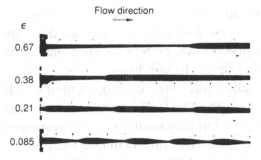

Fig. 4.33. Steady-state ice profiles for successively lower superheat ratios (following Gilpin (1981a)).

pressure drop measurements, visual and profilometric studies, all combine to support such a model. However, most of these results reflect a fairly high (low value of ϵBi) cooling and freezing rate, for which the time scales may not have been long enough to detect the type of interfacial instability discussed in the previous section, and which, although much slower, is clearly persistent. Fig. 4.32 shows a series of pipe ice profiles photographed over a 16 hour period during which the steady conditions imposed on the controlling variables (i.e., water flow rate, wall temperature and water inlet temperature) produced transient interfacial behaviour: the initial flow was laminar. This can only be explained in terms of an interfacial instability which Gilpin correctly attributed to the local heat flux imbalance resulting from changes in the flow cross section.

It is well known that a converging channel accelerates a fluid flow, and it is therefore to be expected that ice growing inside a uniformly cooled pipe would produce the same effect because the progressive loss of sensible heat from the water as it moves downstream implies a progressively lower water heat flux and therefore an ice annulus which thickens with distance along the pipe. At exit from the pipe, however, the presence of a relatively abrupt enlargement may create a decelerating zone in which the flow is likely to separate and thereby exhibit a vigorous convection: i.e., the water heat fluxes immediately upstream and downstream of such a well-defined local maximum in the ice thickness will be quite different. As noted earlier, the higher downstream flux leads to local ablation of the interface, and the abrupt change in shape slowly propagates upstream. This gradual backward progression, which may not occur in rapid cooling experiments, continues until a point is reached where the radial heat flux in the ice matches the enlargement-increased water heat flux. Downstream of this stationary point the water heat flux may be great enough to cause the ice to melt away completely, as demonstrated in Fig. 4.32. Ice then remains only on the upstream side, unless and until nucleation re-occurs well down-stream. With new nucleation, polycrystalline annular ice again develops downstream, following which the process may be repeated in a succession of upstream migrations, depending upon the conditions.

This succession of events gradually leads to the development of steady state profiles such as those shown in Fig. 4.33. The results shown were obtained with an initially laminar flow but very similar results occur when the flow is turbulent. It has been suggested that turbulent flow will destabi-lize an initially planar ice interface when $\epsilon Bi \gg O(1)$. A similar situation may occur inside a pipe. Waves begin to appear, grow in amplitude, and finally transform themselves into the banded structure evident in Fig. 4.33. The shape of the bands determines the local water heat flux distribution and vice versa. The final steady state profiles imply that $\epsilon Bi \simeq 1$ at each axial location, keeping in mind that ϵ is a decaying function of distance downstream.

It is clear from Figs 4.32 and 4.33 that interfacial instabilities generate an ice profile markedly different from that supposed in the thermal entrance model. The ramifications of these periodic growths, or Gilpin bands, are several. Not least, it is to be expected that flow resistance will change with the geometrical alterations. Without the bands, the monotonically in-creasing ice thickness ensures an accelerating flow which tends to maintain laminar conditions or suppress turbulence; even so, the water may still be fully turbulent. Once the bands appear, the situation changes substan-tially. In the accelerating portion, the narrowing of the water flow path is fairly rapid, and the acceleration parameter K may be high enough[18] to re-laminarize the flow after its separation in the neck of the band. When this occurs, separation and re-laminarization not only control heat transfer and drag over and between the bands but imply that flow conditions at the inlet to the pipe will then have little influence downstream.

Assuming complete re-laminarization, and the presence of a boundary layer over the interface, Gilpin (1979b) used a simple interfacial heat balance to show that

$$L/D\epsilon^2 = 0.0725 \, Re_D \ln^2(D/d) \tag{4.24}$$

where L is the band length (or distance to the first neck), D is the pipe diameter and d is the neck diameter: the coefficient, 0.0725, was obtained empirically. It is obvious that the neck Reynolds number Re_d may greatly exceed the pipe Reynolds number which, in the work discussed, extended over the range $3.7 \times 10^{-2} < Re_D < 1.4 \times 10^4$. Since the banded struc-ture was always observed, the usual mechanism for transition to

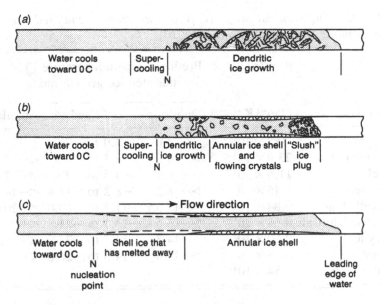

Fig. 4.34. Ice growth mechanisms during pipe filling: (*a*) dendritic mode; (*b*) mixed mode; (*c*) annular mode (following Gilpin (1981b)).

turbulence does not generally apply. A recent review of Gilpin bands is given by Hirata (1987).

Thus far, the discussion has focussed on freezing in cavities and conduits already filled completely with water. The section ends with a brief description of freezing phenomena which arise during the filling process itself. Subcooling of an empty pipe and any surrounding material such as frozen soil creates a sensible heat deficit which, if great enough, may overcome the superheat of the incoming water and thereby give rise to the possibility of ice formation (Gilpin, 1981b). Fig. 4.34 suggests three possible freezing regimes, each dependent on initial temperatures and, to a lesser extent, on flow rate. The illustration is based on observations made when the inlet water temperature was above the equilibrium freezing temperature by less than 1 °C.

For pipe wall temperatures only slightly below 0 °C, ice does not form or, if it does, it is quickly melted. The sensible heat deficit in the pipe wall is not sufficient to produce more than a thin film of ice. When the wall temperature is lower the effects of supercooling increase and dendritic growth begins to occur. This is illustrated in Fig. 4.34(*a*) and is characterized by dendrites large enough and numerous enough to create a significant increase in resistance to bulk flow. Should the upstream pressure not increase to compensate for this additional resistance, the pipe will freeze

Table 4.1. *The material property, β, and its effect on the predicted ice growth mode (following Gilpin (1981b))*

Material	$\beta \times (\varrho Ck)^{\frac{1}{2}}$ ($\mathrm{J\,m^{-2}\,K^{-1}\,s^{-\frac{1}{2}}}$)	Predicted pipe temperature (°C) for indicated ice growth mode		
		Dendritic	Mixed	Annular
Copper	185×10^3	$>-\ 4.0$	-4.0 to -7.0	$<-\ 7.0$
Aluminium	23.9×10^3	$>-\ 4.3$	-4.3 to -7.5	$<-\ 7.5$
Steel	11.6×10^3	$>-\ 4.5$	-4.5 to -7.9	$<-\ 7.9$
Concrete	1.46×10^3	$>-\ 8.2$	-8.2 to -14.4	<-14.4
Polyethylene	0.84×10^3	>-11.3	-11.3 to -20.0	<-20.0
Polyvinylchloride	0.64×10^3	>-13.6	-13.6 to -23.8	<-23.8
Acrylics	0.60×10^3	>-14.3	-14.3 to -25.0	<-25.0
Wood	0.58×10^3	>-14.6	-14.6 to -26.6	<-26.6
Water	1.54×10^3			

shut, either suddenly or after a period of intermittent stick-slip displacements dislodging and re-locating the mass of slush ice behind the leading edge of the water.

When the pipe wall temperature is much lower e.g., less than $-20°C$, dendritic growth is not noticeable but the pipe wall is then coated with polycrystalline ice. This annular growth is represented in Fig. 4.34(c) and is characterized by an increase in flow resistance which is negligible, at least initially. In fact, the sensible heat in the water which follows later tends to melt back the upstream section of the annulus formed earlier. Between these two modes is a third, a mixed mode, which combines the features of the other two. Fig. 4.34(b) indicates that the mixed mode includes the relatively benign occurrence of polycrystalline ice but the effect of supercooling is also present: upstream, in the forming and fracturing of small dendrites; and downstream, in the geli-flocculation of dendrites within the water front slush.

Freezing could be avoided during filling, at least in principle, by simply superheating the water sufficiently or by pumping fast enough to introduce substantial viscous and inertial effects. It is also possible that nucleating agents, including hoarfrost on the empty pipe wall, for example, could limit the ice to annular form. However, in the absence of such measures the actual behaviour of the ice and water are controlled essentially by the temperatures and materials present. Using such a basic prescription, Gilpin estimated the range of the three modes for various pipe materials. The results, which are consistent with field experience and his experimental findings, are given in Table 4.1. Perhaps surprisingly, it is found that a pipe cooled only slightly below the equilibrium freezing temperature is more likely to become blocked than one cooled to much lower temperatures. This stems from the fact that a limited degree of supercooling allows dendritic growth to spread, and should the growth become extensive it presents a much greater hydraulic resistance than an annular coating of comparable mass. Even physical breakage of the dendrites may not reduce the resistance more than temporarily: a supercooled slush, in which dendrite fragments multiply, grow and sinter, is not likely to be less 'viscous' than pure water, and will have a tendency to catch on small protuberances near which it may re-freeze. The annular growth mode, although not without its disadvantages, is clearly the lesser of two evils. The comparatively small, initial increase in hydraulic resistance may be reduced by subsequent melt-back, and the time scale during filling is not likely to be long enough to permit interfacial instabilities to develop a roughened surface. Growth and decay after the filling period are governed by the considerations discussed earlier.

4.7   Freezing on submerged bodies

As a starting point, ice formation on submerged bodies may be treated in much the same way as suggested earlier for plane surfaces. Given the velocity and temperature field in water flowing around the ice-free substrate, the appearance and growth of ice may be considered as a perturbation whose characteristics are essentially determined from the convection system it disturbs. For example, moderate cooling of a substrate immersed in unsuperheated water may generate dendritic ice crystals whose shape, orientation and rate of growth depend upon the local flow field. On the other hand, superheated water flowing over a colder substrate is more likely to generate a continuous coating of polycrystalline ice which will respond to local convection conditions in a different way.

In such an approach, and especially for a continuous coating, the ice is treated one-dimensionally or quasi-one-dimensionally. This has obvious analytic and computational advantages, and will be adequate for many practical circumstances in which the ice coating introduces no significant change in the velocity field: any change is either small enough to be ignored or may be incorporated in a simple iterative scheme. Under such conditions, known solutions for heat and fluid flow about immersed bodies provide approximations to the local water heat flux over the interface and

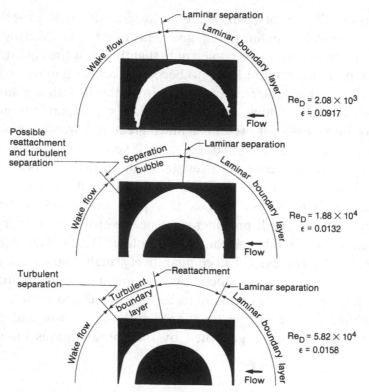

Fig. 4.35. Photographs of an ice layer (light region) around a cylindrical substrate at various Reynolds numbers. Flow regions indicated were obtained by an interpretation of the local heat transfer coefficient (following Cheng *et al.* (1981)).

thus determine the local ice growth rate. This is especially convenient for bodies of revolution. For example, the wedge flow mentioned in Section 4.5 may be replaced by flow around a cone, thus making possible the study of ice growth on surfaces ranging from circular discs to sharply-pointed needles. However, this approach may not be used when the external flow field is radically altered by the presence of the ice. External ice, like internal ice, does not generally respond passively to the flow field: laminar flow may become turbulent, turbulent may become laminar; unseparated flow may separate, detached flow may re-attach; unrestricted flow may be blocked.

Cross flow over a circular cylinder has received much attention in the literature, which has established a variety of flow regimes delineated by ranges of the Reynolds number. Broadly speaking, these regimes are also observed on iced cylinders but the details may be noticeably different (Cheng, Inaba and Gilpin, 1981). Variations in the heat transfer coefficient

around an initially circular substrate create variations in the radial growth
rate of the ice which thus becomes non-circular. This departure from
circularity influences the local heat transfer coefficients, and therefore the
interface enters first into dynamic equilibrium, and ultimately into static
equilibrium, with the surrounding flow. Fig. 4.35 shows how the flow par-
ticulars vary over a range of Reynolds numbers. Although the thermal
conditions differ slightly throughout the figure it is clear that the transition
to a turbulent boundary layer flow, conventionally taken to be around
$Re_D = 4 \times 10^5$, occurs at a much lower Reynolds number. The transition
is evidently marked by the appearance of a separation zone, much like
transition on the nominally plane ice surface discussed in Section 4.5, and
again illustrates how a sharp change in flow and heat transfer coefficient
may generate an interfacial instability leading to a change in interface
shape.

The heat transfer correlation for an iced cylinder may be expected to be
a function of the Prandtl number, Reynolds number (based on cylinder
diameter), and superheat ratio. Cheng *et al.* (1981) have suggested that

$$Nu_D = 1.46 \, Re_D^{0.457}/\epsilon^{0.311}$$

when $0.05 < \epsilon < 0.5$ and $Re_D < 5 \times 10^3$,

$$Nu_D = 0.289 \, Re_D^{0.637}/\epsilon^{0.151} \tag{4.25}$$

when $0.013 < \epsilon < 0.16$ and $5 \times 10^3 < Re_D < 5 \times 10^4$, and

$$Nu_D = 0.0756 \, Re_D^{0.792}/\epsilon^{0.112}$$

when $0.015 < \epsilon < 0.025$ and $Re_D > 5 \times 10^4$. They also give the latent
heat storage Q' of the ice as

$$Q' = 11.2/(\epsilon^2 \, Re_D)^{0.612}$$

when $3 < \epsilon^2 Re_D < 50$. Sensible heat is comparatively small: $Ste \leqslant 1$.

This last result is important in the design of water chillers where banks of
tubes build a body of ice later allowed to thaw in order to keep the surrounding water temperature near the freezing point. It is also a useful
starting point in estimating the amount of energy required to build an ice
veil or 'dam' across a river (Lock and Kaiser, 1985). However, the application of single-tube data to rows or bundles of tubes is rather limited. The
effects of mutual proximity are capable not only of modifying the overall
heat transfer rate and thereby the ice growth rate, but may lead to a
fundamental alteration in the nature of the flow (Lock and O'Callaghan,
1989).

The flow and heat transfer characteristics of a row of tubes in a cross flow
differ only slightly from single-tube behaviour when the gap–diameter

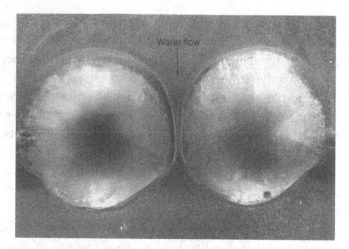

Fig. 4.36. Occluding ice shapes: end view (following Lock and Kaiser (1985)).

ratio is not too small, e.g., greater than 0.5:1. For any given Reynolds number, the flow behaviour changes as the gap–diameter ratio is decreased. In particular, the transitional Reynolds number is greatly reduced. As the gap shrinks to zero, which it must if complete occlusion of the water stream is to be achieved, the flow at exit from the gap, and in the wake, behaves in a complex manner, becoming unstable or bistable with respect to directionality. Water may thus issue either to the right or the left of the gap rather than in the expected downstream direction. In keeping with earlier observations on interfacial stability, such alterations in the flow must be reconciled with consequent alterations in the growth rate and shape of the ice coating. Mutual proximity of neighbouring interfaces creates a local increase in velocity which tends to flatten the ice profile, as illustrated in Fig. 4.36. Flow and heat transfer in the nearly-parallel gap thus produced may once more be modelled as a Graetz problem in which the shrinking gap, combined with the contraction region immediately upstream, tends to maintain or re-establish laminar conditions. Immediately downstream of the gap, in a separation zone, the interface is noticeably scalloped.

As mentioned earlier, the ice–water interface couples the temperature field and the velocity field under forced convection conditions. This coupling is even stronger when the external flow field is driven by thermal buoyancy, and may be further complicated by the phenomenon of inversion. The complex flow behaviour of fresh water near the inversion temperature has been mentioned several times, notably in Sections 4.4 and 4.6. On the outside of a horizontal cylinder, the flow field again exhibits

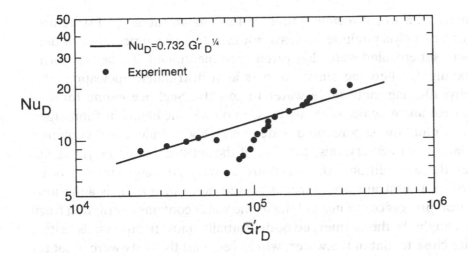

Fig. 4.37. Heat transfer relation for a horizontal cylinder causing freezing in cold water (following Cheng *et al.* (1987)).

three regimes (Cheng, Inaba and Gilpin, 1987) which parallel those described in Fig. 4.20 for a vertical surface. When the bulk water temperature is less than the inversion temperature, the water flow near the interface is essentially upwards: this generates an ice profile thickening around the cylinder from bottom to top. The reverse is true when the bulk water temperature is much greater than the inversion temperature. Between these regimes is a third in which $4°C \lesssim T_\infty \lesssim 6°C$, and the ice thickness tends to be greater and more uniform.

The characteristics of the local heat flux are reflected in the average Nusselt number which, when plotted against bulk water temperature, exhibits the familiar minimum in the vicinity of $T_\infty = 5°C$. In turn, this minimum is reflected in the heat transfer correlation, written in terms of a modified Grashof number defined by

$$G'r_D = \frac{ag\,D^3}{v^2}\,\phi\,(T_\infty - T^{i})^2\,[1 + b\phi(T_\infty - T^{i})]$$

where a and b are empirical coefficients determined from the density–temperature curve, and $\phi = (T_{in} - T_\infty)/(T^{i} - T_{in})$, in which T_{in} is the inversion temperature. Experimental data are shown in Fig. 4.37 along with an empirical curve designed to fit data outside of the inversion region. The minimum (average) Nusselt number, corresponding to the thickest ice, is generated near $G'r_D \simeq 7 \times 10^4$. The effect of the superheat ratio is largely contained in $G'r_D$.

Finally, it is worth recalling that ice forming on submerged structures may not be polycrystalline, at least not initially. The appearance of dendrites in supercooled water has often been mentioned. These may form, for example, when the substrate acts as a distributed nucleation site, thereby allowing metastable water to solidify. Such ice is one form of anchor ice and may occasionally be seen on aquatic plants or submerged rocks. Should the supercooled water be highly turbulent and contain a population of frazil crystals, the role of the substrate changes yet again. Under these conditions, the frazil are actively growing, and will often adhere to a substrate whether it is coated in ice or not. This adhesion–accretion process continues as long as the water contains distributed frazil to be 'caught' by the submerged body. Initially, most frazil crystals, with a density close to that of the water, would veer past the body were it not for the capricious behaviour of turbulent eddies, but the cumulative deposition of frazil gradually becomes an efficient collector. Such deposits constitute the major form of anchor ice.

5

Ice and air

Ever since the pronouncements of Aristotle, it has been surmised that the occurrence of ice in air is the direct result of temperatures low enough to cause solidification of water vapour. Equally, it has long been known that the distribution and concentration of this water vapour is strongly influenced by the aerodynamic and thermal history of the air containing it. Whether the temperatures leading to precipitation originate within the air itself or on a solid surface, it would seem natural to view the resultant phenomena, collectively, as the central core of *aeroglaciology*.

Broadly speaking, the formation of ice from water vapour follows two distinct paths. In the first, the vapour condenses directly into the solid phase, whereas in the second there are two successive stages: condensation into liquid water followed by freezing. In general, therefore, the study of aeroglaciology divides naturally into two parts, and the organization of this chapter reflects that fact. The first and second sections treat the distinct physical processes of deposition and accretion, more or less independently, while subsequent sections use this basic knowledge in a particular context. Section 5.3 deals with the various natural ice forms, and ice forming processes, which are of special interest to the meteorologist. The remaining sections largely concentrate on the accumulation of ice on man-made structures: in the air, over water and on land.

5.1   Deposition of water vapour

Deposition of water vapour takes place only at temperatures and pressures below the triple point, and under conditions of stable equilibrium the state must lie on the deposition–sublimation ruled surface discussed in Chapter 2. In practice, the vapour may be supercooled ($T < T^i$) or supersaturated[1] ($P > P^i$), either event, or both, implying a metastable condition. The form of the deposit depends principally on the temperature of the substrate, as Table 5.1 indicates, but the rate of cooling, the vapour

Table 5.1. *Summary of some of the experimental results on the structure of solids formed from the deposition of water vapour (following Hobbs (1974)).*

Experimental method	Substrate temperature (C)						
	-180	-160	-140	-120	-100	-80	-60
X-ray diffraction	vitreous or amorphous				semi-crystalline		hexagonal
Calorimetric	amorphous				crystalline		
X-ray diffraction	small crystals		intermediate range not investigated				hexagonal
Electron diffraction	small crystals			cubic			hexagonal
Calorimetric	amorphous				crystalline		
Calorimetric	amorphous				crystalline		
Electron diffraction	crystal growth poor		cubic				hexagonal
X-ray diffraction	Mixture of cubic and hexagonal						hexagonal
Electron diffraction	amorphous or small crystals		cubic	hexagonal and cubic		hexagonal	
X-ray diffraction	vitreous	vitreous and cubic			hexagonal		
Electron microscope and electron diffraction				cubic		hexagonal	
X-ray diffraction	vitreous				cubic		hexagonal
Calorimetric	vitreous (glass transition at -134 C)				cubic		hexagonal
Electron microscope and electron diffraction	amorphous			cubic		hexagonal	
X-ray diffraction	(diffuse rings)				cubic		hexagonal
Electron microscope and electron diffraction	minute cubic crystals		cubic	hexagonal and cubic		hexagonal	

pressure and the nature of the substrate also influence the outcome.

It is evident from Table 5.1 that deposition divides into three overlapping regions. For $-100\,°C \le T < -60\,°C$ (and above), nucleation leads to the growth of the hexagonal crystals of ice Ih, while for $T \le -138\,°C$ the deposit is vitreous or amorphous. Between these regions is a third in which the deposit is again crystalline but most often has the cubic lattice of ice Ic.

In the absence of a substrate, a natural ice crystal is most likely to take on its most preferred shape under stable equilibrium conditions (Hobbs, 1974): hexagonal prisms, hexagonal bipyramids or hexagonal prisms with pyramidal ends, each of which is illustrated in Fig. 5.1. The three distinct faces of these forms have all been observed, but the pyramidal face, which tends to grow fastest and disappear, is by far the rarest. The hexagonal basal face, normal to the *c*-axis, and the rectangular prism face, normal to

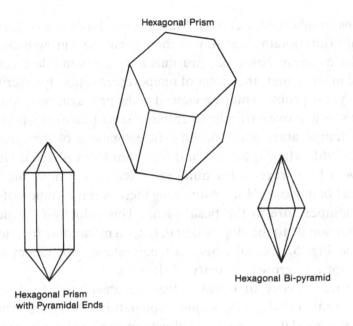

Fig. 5.1. Forms of natural ice crystals growing from water vapour.

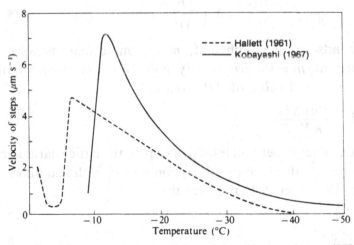

Fig. 5.2. Temperature variation of the velocity of steps (250 Å high) on the basal plane of ice crystals growing epitaxially on covellite. Excess vapour densities 2.5×10^{-7} mg m^{-3} (following Hobbs (1974)).

the *a*-axes, are usually the dominant faces and it is therefore their relative growth rates which determine the primary geometry, or habit, of the crystal.

Growth at an ice crystal surface may be viewed in two ways. Macroscopically, the problem may be treated as one of uniform deposition over a simplified shape simulating the crystal (Mason, 1953; Fukuta, 1969). However, the results thus obtained, although they reveal mass growth rates of the same order as those observed, are unable to accommodate certain fundamentals; in particular, the effect of temperature is poorly described. Microscopically, the problem may be viewed as the propagation of surface steps which arise and move naturally over the crystal surface in response to the prevailing temperature and humidity. Investigations of step propagation (Hallett, 1961; Mason, Bryant and Van Den Heuval, 1963; Hobbs and Scott, 1965) have revealed a number of features compatible with observed crystal behaviour. Not least among these is the variation of step velocity with temperature in the basal plane. This velocity U, which is inversely proportional [2] to the step height h, is not a monotonic function of temperature, as Fig. 5.2 reveals. Such a temperature dependency is reflected in the normal[3] growth velocity of the crystal.

The propagation theory of spiral defects, as developed by Burton, Cabrera and Frank (1951), offers one explanation of crystal growth. According to this model, the normal velocity of the crystal face may be expressed as

$$V_I = \frac{\alpha}{\rho_i} \left(\frac{m}{2\pi kT} \right)^{\frac{1}{2}} \frac{(\Delta P)^2}{\Delta P_1} \tanh \left(\frac{\Delta P_1}{\Delta P} \right) \tag{5.1}$$

where α is the adsorption coefficient, m the molecular mass, k is Boltzmann's constant, ρ_i is the ice density, and ΔP is the excess vapour pressure: ΔP_1 is a critical value of ΔP given by

$$\Delta P_1 = 2\sqrt{2} \, \frac{\pi \eta a_0^2 P_0}{k X_S T} \tag{5.2}$$

where η is the ledge energy per unit length of step, a_0 the lattice parameter, X_S the mean migration distance of adsorption and P_0 is the equilibrium vapour pressure. Equation (5.1) requires that

$$V_I \propto (\Delta P)^2 \tag{5.3}$$

when $\Delta P_1 \gg \Delta P$ and

$$V_I \propto \Delta P \tag{5.4}$$

when $\Delta P_1 \ll \Delta P$. Such behaviour has been noted by Lamb (1970). Moreover, spiral growth theory (Section 2.9) implies that the step velocity is proportional to the normal face velocity (assuming the distance separating the steps is greater than the collection distance, otherwise the steps interfere with each other), and thus requires the temperature dependence of the former (Fig. 5.2) to be reflected in the latter. Alternatively, growth may stem from two-dimensional nucleation, as noted in Section 2.9.

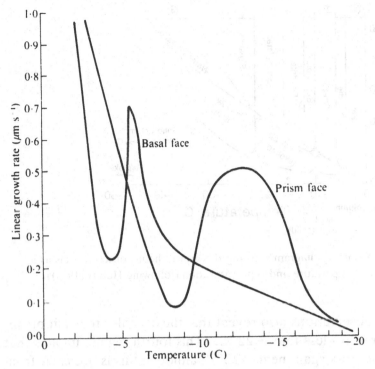

Fig. 5.3. Best fit curves to experimental measurements of the linear (normal) growth rates on the basal and prism faces, at a constant excess vapour pressure of 1.3×10^{-5} bar, as a function of temperature. (From Lamb and Hobbs (1971) with changes.)

Measurements of normal, or linear, face velocities are shown in Fig. 5.3 (Lamb and Hobbs, 1971), from which it is immediately evident that the dependency of the basal face normal velocity reflects the basal face step velocity data of Hallett in Fig. 5.2. It is also clear that there is a qualitative correspondence between the basal face velocity and the prism face velocity, so far as the effect of temperature is concerned. The relative lateral displacement of the curves provides a means of predicting crystal habit (Hobbs, 1974). Since the two curves in Fig. 5.3 intersect at $T \approx -5.3\,°C$ and $-9.5\,°C$, the temperature dependence of crystal shape should divide into three regions. Between these temperatures, the basal face velocity exceeds the prism face velocity and therefore hexagonal prisms would tend to be longer than they are wide. On either side of these temperature limits, the reverse holds true and the prisms would tend to grow as hexagonal plates. Laboratory observations on crystal habits (Kobayashi, 1957, 1961; Hallett and Mason, 1958) confirm that they tend to be plate-like from $0\,°C$ to $-4\,°C$, prism-like from $-4\,°C$ to $-10\,°C$, and plate-like from $-10\,°C$

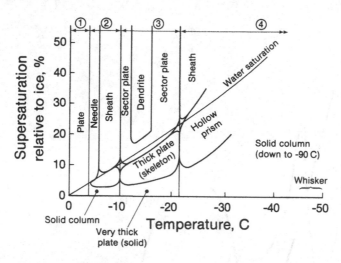

Fig. 5.4. Summary diagram showing the growth habits of ice crystals as a function of temperature and supersaturation (following Hobbs (1974)).

to $-22\,°C$; the observations also reveal that the crystals are again prism-like for temperatures less than $-22\,°C$. (This fourth regime implies that the curves must cross again near $-22\,°C$, although this is not clear from Fig. 5.3.)

There is thus both theoretical and experimental evidence on the role of temperature in determining the primary geometry of ice crystals grown from the vapour phase. But temperature is only one of two principal variables which affect growth, the other being the ambient vapour concentration or density. The role of the vapour density may be presented in a number of ways, the most common being through the vapour pressure in relation to the saturation pressure. Fig. 5.4 is a map of crystal habits in relation to both of the principal variables. The broad effect of temperature is evident as the four regions mentioned above but superimposed upon them are a variety of secondary effects attributable to the degree of super-saturation defined by

$$\phi = \frac{P - P_{sat}(T)}{P_{sat}(T)} \tag{5.5}$$

where P_{sat} is the saturation equilibrium phase change pressure, which may be chosen with respect to either water or ice[4]. The latter is used in Fig. 5.4 so that the water saturation curve may be shown separately, thus enabling behaviour relative to water saturation to be seen at a glance. Below the water saturation curve, water evaporates while water vapour deposits on ice; above the curve, condensation and deposition are both possible, but

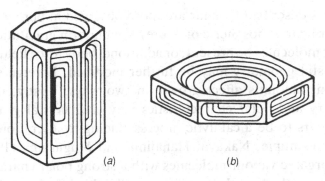

Fig. 5.5. The hopper structure of ice crystals (*a*) hollow prismatic columns and (*b*) a dish-shaped hexagonal plate (following Mason *et al.* (1963)).

wherever ice and water actually coexist the former will grow at the expense of the latter.

The secondary growth features of ice crystals may be quite substantial, as Fig. 5.4 indicates. When the supersaturation is lower than about 3% the primary geometry is displayed almost intact, and this continues without much change until the supersaturation has reached levels corresponding to the water saturation curve. When the supersaturation is above these levels, the secondary effects become increasingly pronounced. For example, in region 3, between −10°C and −22°C, the primary solid plate-like structure is transformed into a branched dendritic structure; and if the temperature is within a few degrees of these limiting values the structure frequently develops sectored plates around its tips, as seen in Fig. 1.12. In region 2, the range −4°C to −10°C, the primary solid prism-like structure grows instead as a needle (near −4°C) or a hollow sheath (near −10°C); the same tendency towards hollowness near the regional boundary is also observed for temperatures immediately below −22°C.

As mentioned earlier, the primary structure of an ice crystal is determined by normal facial velocities, assuming these to be generated more or less uniformly across the entire face. In contrast, secondary features are associated with the facial edges and corners of the crystal where concentration and temperature gradients are highest. Steps appearing along these edges run towards the centre of the face, thus forming a family of steps down towards the face centre. This behaviour is most clearly seen at high values of supersaturation, when the so-called hopper development is observed. The resulting shapes are illustrated in Fig. 5.5 which indicates that the steps remain parallel to the edge from which they were initiated except where the intersection of neighbouring edges introduces a modifying curvature.

The crystal habits described thus far are those observed in pure water vapour (in air at or near atmospheric pressure) in a uniform gravitational field. Should other molecules be present, or additional body force fields be operative, the crystal geometry may be further modified. Foreign substances in sufficient amounts affect growth in two ways. Firstly, their molecular geometry and polar characteristics may alter crystal growth through what appears to be a catalytic process leading to alterations in crystal shape. For example, Nakaya, Hanajima and Muguruma (1958) found that many organic vapour molecules with a strong polar character tend to accelerate growth along the *c*-axis. Secondly, the diffusion of heat and water vapour through foreign materials may alter both the rate of primary growth and the secondary growth features (Hallett and Mason, 1958; Gonda and Komabayashi, 1971).

The effect of an electric field has also been noted (Bartlett, Van Den Heuval and Mason, 1963). Some general influence on the polar water molecule is perhaps not surprising but the effects appear to be restricted to crystals with a needle or dendritic habit. Growth is accelerated (10–100 times) in the direction of the electric field thus producing 'electric' needles at the tips of which the local field is evidently intense enough to enhance surface diffusion properties.

Another important aspect of ice growth from vapour phase is the role of the cooled substrate. When the substrate is particulate, e.g., silver iodide aerosols, natural mineral dusts (e.g., sandstone), or industrial pollutants, it is their effect on nucleation which is usually of greatest interest. More will be said on nucleating particles in Section 5.3, and at this stage it is perhaps sufficient to note some of the principal characteristics which determine their effectiveness. Generally, they are insoluble in water, otherwise they would have difficulty in maintaining structural order, and they must be at least as large as the critical embryo dimension r^*. Typically, they exhibit surface hydrogen bonding similar to that of the H_2O molecule, and possess rotational symmetry. Most obvious is the need for a crystallographic structure similar to ice itself. Thermodynamic conditions (the state of the vapour, as defined by pressure and temperature) in general, and supersaturation in particular, may be unaffected by the particles, but the latent heat liberated by growing ice must still be dissipated in the environment. Since overall thermodynamic conditions determine the primary and secondary features of crystal growth, earlier comments apply to growth on particulates unless they act as cloud condensation nuclei with high number densities.

Growth on a continuous substrate having an extent much larger than the

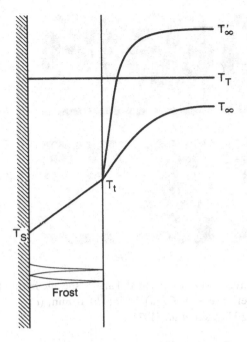

Fig. 5.6. Temperature field for a frost layer.

size of a typical ice crystal is very common on both natural and man-made structures. In certain natural circumstances, e.g., hoarfrost on a tree branch, the thermodynamic conditions are not unlike those discussed above, but in other circumstances, especially in engineering structures typified by an air-conditioning system, the conditions may be significantly different. For example, a cooling coil placed in moist air continually removes heat from the air. Should frost form on the coil, the latent heat thereby released may be removed entirely by the coolant flowing through the coil, and the surrounding air need not then be colder than the ice; in fact the ice is usually cooler than the air. This difference in thermodynamic conditions is important, because the crystal is no longer, of necessity, limited to growth in a supersaturated, metastable vapour; near the coil, the vapour must, of course, have passed through the saturation point but the bulk of the vapour need not be supersaturated. The nucleating surface may be supercooled before nucleation takes place, in which case the first stages of growth will be dendritic, but once ice has appeared on the substrate its subsequent growth is likely to take place under the condition of stable equilibrium. However, unlike stable growth from the liquid phase, the deposition of molecules at the ice–vapour interface is not characterized by the invariable occurrence of homogeneous, polycrystalline ice.

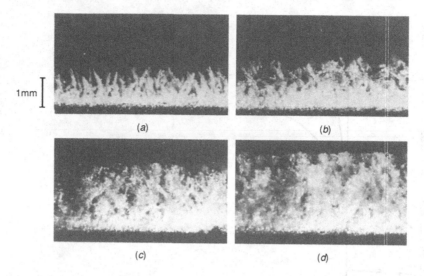

1mm

(a)   (b)

(c)   (d)

Fig. 5.7. Early growth of a frost layer: wind speed $4\,\mathrm{m\,s^{-1}}$, specific humidity $0.01\,(\mathrm{kg/kg})$, substrate temperature $-18\,^{\circ}\mathrm{C}$: (a) 3 min; (b) 10 min; (c) 15 min; (d) 20 min (following Hayashi *et al.* (1977)).

Microscopic studies of ice crystals growing on metal substrates in super-saturated vapour (Isono and Iwai, 1969; Lamb, 1970) suggest that their primary structure is isometric (hexagonal prisms, faceted spheres) and their secondary structure is weakly defined, if present at all. However, observations of hoarfrost, and of the interior of the domestic refrigerator, reveal that growing crystals commonly exhibit branching and mutual inter-ference which eventually lead to an intricate, porous matrix. The charac-teristics of this frost layer generally depend upon the prevailing conditions: the substrate temperature, the air (vapour) temperature and the humidity are obvious factors, but the details of convection immediately adjacent to the frost surface must also be considered.

Fig. 5.6 is a representation of the temperature field in and near a frost layer formed on a substrate at temperature T_s (a similar field exists for vapour concentration). At the outer surface of the layer, which is not generally well defined, the temperature is represented by T_t. When the ambient temperature T_∞ is less than the triple point temperature T_T, the initial effect of lowering the substrate temperature is to cool, and possibly supersaturate, the adjacent vapour which then deposits on the substrate. In the absence of supercooling, the vapour should, from a macroscopic standpoint, deposit uniformly to produce a solid, homogeneous layer of ice. The presence of slight supercooling, however, may help generate growth which, macroscopically at least, does not appear to exhibit primary or secondary features which change much with air temperature or humidity. Fig. 5.7 shows frost in the early stages of growth (Hayashi, Aoki and Yuhara, 1977) during which it is evident that individual needles are twiglike and have a tendency to grow and densify into a bushy matrix.

Conditions within the matrix are not usually known precisely. Fig. 5.6 shows a continuous temperature profile (shown linear as a convenience appropriate to a thin layer) between the roots at T_S and the tips at T_t, but this temperature is the local mean value, which is assumed to incorporate the ice temperature (and its cross sectional variations) along with the interstitial vapour temperature (and variations). It appears that deposition is usually slow enough to create quasi-steady conditions for which it may be assumed that the ice–vapour interface within the matrix is at the stable equilibrium deposition temperature T^i which varies between T_S and T_t. The local temperature in the ice interior is thus less than the local value of T^i (during deposition, latent heat is removed mainly through the ice by conduction).

So long as the frost remains porous, it will permit vapour to diffuse through its interior, adding mass by deposition. The local matrix density thus increases with time. Whether or not this corresponds to an increased density for the entire frost layer will depend upon the frontal area and the rate of growth of new ice depositing at the tips. Given local, stable equilibrium within the matrix it must be assumed that the features of crystalline growth are provided by Fig. 5.4, specifically along the horizontal axis. It therefore appears that the dominance of either basal or prism face growth will vary with the local temperature in the frost matrix; or, if the frost is very thin, will be a function of substrate temperature. Within the matrix, stable equilibrium precludes interfacial instability and implies slow growth which is not controlled by diffusion. At the tips, however, the vapour concentration gradient is much steeper[5] in the direction of outward growth, thus implying higher growth rates. These theoretical features are supported by macroscopic observations and enable the frost to be modelled as a continuous, variable property layer in which the distributed release of latent heat may be neglected in comparison with that liberated at its outer surface. This basic model will be employed in Section 5.6.

Fig. 5.6 also shows the air temperature profile when the ambient value T'_∞ is greater than the triple point temperature. The description of this situation is essentially the same as that given above with one important exception: the frost surface temperature may increase enough to reach T_T. For example, when the temperature and rate of heat withdrawal at the substrate are held constant, T_t must increase as the frost thickness

increases. Should T_t exceed T_T, deposition will be replaced by condensation, the resultant drops being drawn into the frost interstices by capillary
forces modified by the local effect of gravity. This interstitial water then
freezes (often unevenly) and thereby reduces the overall thermal
resistance of the frost layer. Given a constant rate of heat withdrawal, this
will result in the lowering of T_t below T_T, the consequent arrest of condensation, and a return to simple deposition, now on the frozen crust. This
behaviour may lead to oscillations in the frost surface temperature (White
and Cremers, 1981). It will also introduce a marked change in the mean
density and permeability of the layer, essentially converting the problem
into a Stefan problem with the ice–vapour interface fixed at the triple point
temperature.

5.2   Accretion

When vapour pressures are above the triple point value, the
lowering of the temperature usually produces the nucleation of water first,
and only if the temperature drops below the equilibrium freezing temperature will ice appear. In the absence of a substrate, condensation
produces droplets of water which, in a very large space such as that occupied by a cloud, may continue to exist in stable or metastable equilibrium
with their immediate environment. Should the temperature of this environment be lowered only slightly beneath the equilibrium freezing temperature the droplets will not likely freeze, unless they happen to contain
or contact active nucleants. In general, the smaller the droplet the greater
will be the degree of supercooling it can thus tolerate before spontaneous
(homogeneous) nucleation occurs. When droplets in this metastable state
collide with an equally cold object, such as an aircraft wing, they freeze.
The process whereby a substrate acquires ice from the freezing of supercooled droplets is commonly known as *accretion*.

Consider a single droplet emerging from the embryonic stage in an
infinite environment of water vapour existing in a metastable (supersaturated) state. The droplet assumes and retains the form of a sphere, more or
less, and grows at a rate determined by the prevailing thermodynamic
conditions. Under quasi-steady conditions, the droplet temperature
remains fixed above the surrounding vapour temperature, the rate of
sensible heat loss then being just sufficient to balance the latent heat gain
resulting from condensation on the droplet surface.

The rate of heat loss to the surroundings depends partly upon the precise shape of the droplet and partly on the amount of local ventilation,
but these tend to be minor contributions when the droplet is small. The

principal contribution is then steady, radial heat conduction, during which the temperature is inversely proportional to the radial distance R from a point in the centre of the droplet. It thus follows that the conductive heat flux $\dot{Q}$ from a droplet of radius r may be represented by

$$\lambda_{wv} dM/dt = \dot{Q} = -4\pi k r^2 (dT/dR)_r = 4\pi k r (T_r - T_\infty) \quad (5.6)$$

where M is the mass of the droplet and T_∞ is the ambient temperature. Similarly, the mass flux may be written as

$$dM/dt = \dot{M} = 4\pi D r^2 (d\rho/dR)_r = 4\pi D r (\rho_\infty - \rho_r) \quad (5.7)$$

where ρ_∞ is the ambient vapour density. Assuming the vapour behaves like an ideal gas, and the saturation pressure–temperature relation is governed by the Clausius–Clapeyron equation, equation (5.6) may be used (Mason, 1971) in showing that

$$\frac{\rho_s(r) - \rho_s(\infty)}{\rho_s(\infty)} = \frac{\lambda_{wv}}{4\pi k r T_\infty}\left(\frac{\lambda_{wv}}{RT_\infty} - 1\right)\frac{dM}{dt} \quad (5.8)$$

where ρ_s is the saturation value. Taking $\rho_r = \rho_s(r)$ and noting that

$$\frac{\rho_r - \rho_s(\infty)}{\rho_s(\infty)} = \frac{\rho_r - \rho_\infty}{\rho_s(\infty)} + (S - 1),$$

where $S = \rho_\infty/\rho_s(\infty)$ is the saturation ratio of the surroundings, equations (5.7) and (5.8) may be combined to give

$$\frac{dM}{dt} = \frac{4\pi r (S - 1)}{\dfrac{\lambda_{wv}}{kT_\infty}\left(\dfrac{\lambda_{wv}}{RT_\infty} - 1\right) + \dfrac{RT_\infty}{DP_s(\infty)}}$$

or

$$r\frac{dr}{dt} = \frac{S - 1}{\dfrac{\rho_w \lambda_{wv}}{kT_\infty}\left(\dfrac{\lambda_{wv}}{RT_\infty} - 1\right) + \dfrac{R\rho_w T_\infty}{DP_s(\infty)}} \quad (5.9)$$

This relation may have to be modified according to the prevailing conditions and the size of the droplet. For example, when the droplet is growing on a nucleus (of mass m_n and molecular weight M_n), and its radius is small enough to influence the surface energy, $\rho_r \neq \rho_s(r)$, in which case (Mason, 1971)

$$\rho_r = \rho_s(r)\left(1 + \frac{2\sigma}{\rho_w RT_r} - \frac{3i m_n M}{4\pi r^3 \rho_w M_n}\right)$$

where σ is the surface tension and i is Van't Hoff's factor. This modifies the right hand side of equation (5.9) accordingly. On the other hand, when the droplet radius becomes large enough to alter its shape or falling speed, and

Table 5.2. *Some cloud droplet characterstics (following Mason (1971))*

Cloud type	Droplet Radii (μm)			Droplet concentration (cm^{-3})	Water content ($g\,m^{-3}$)
	min	mean	max		
Altostratus (Germany)	1	5	13	450	—
Cumulonimbus (USA)	2	20	100	72	2.5
Maritime cumulus (Hawaii)	2.5	15	20	75	0.50
Continental cumulus (USA)	3	9	33	300	1.0

hence influence the amount of ventilation, the simple diffusive solutions implied in equations (5.6) and (5.7) must be modified. The surface heat and mass flux densities may be written more generally as

$$-k\,dT/dR = h_Q\,(T_r - T_\infty)$$

and

$$D\,d\rho/dR = h_M\,(\rho_\infty - \rho_r)$$

where h_Q, h_M are the heat and mass transfer coefficients, respectively. The extreme right hand sides of equations (5.6) and (5.7) must therefore be multiplied by the Nusselt ($Nu_r = h_Q r/k$) and Sherwood numbers ($Sh = h_M r/D$) respectively. For air in forced laminar flow, the heat and mass transfer processes are often taken to be analogous (i.e., the Prandtl number $Pr = \nu/\varkappa$ and the Schmidt number $Sc = \nu/D$ are both close to unity), in which case the Nusselt and Sherwood numbers may both be written in the same form. For example,

$$Nu_r = 1 + aRe_r^{\frac{1}{2}}$$

In the absence of a mean droplet velocity relative to the air, condensation is augmented solely by natural convection, for which

$$Nu_r = 1 + bRa_r^{\frac{1}{4}}$$

In these expressions, $Re_r = U_\infty r/\nu$ is the familiar Reynolds number for a relative (fall) velocity U_∞, and $Ra_r = \beta g(T_r - T_\infty)r^3/\nu\varkappa$ is the Rayleigh number. The coefficients a and b may be determined theoretically or

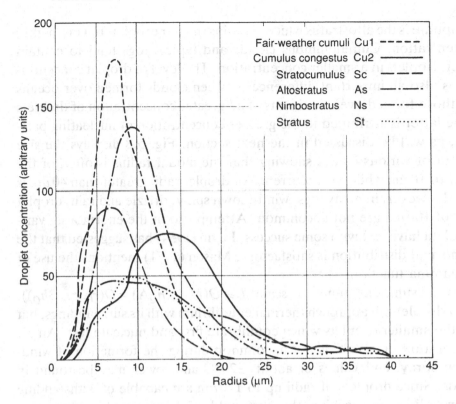

Fig. 5.8. The mean droplet-size distributions of various cloud types (following Diem (1948)).

empirically. Similar arguments and expressions apply to the growth of 'spherical' ice crystals when deposition is augmented by ventilation.

The droplets formed in cloud or fog begin their growth more or less as individuals but their mutual isolation may not last for long. The mass of the droplets increases with time, but the vapour from which they grow, however extensive it may be, is obviously finite; therefore droplets eventually find themselves in competition for the same vapour, and experiencing reduced growth rates. In addition, larger drops tend to fall faster than smaller drops, which are then gathered up by collision and coalescence, a process which may be slightly augmented by any small scale turbulence. It is thus evident that while bulk atmospheric motions will dictate the general environmental history of a droplet cloud, its temperature and humidity in particular, the various internal processes of collective interaction also exert a considerable influence on the droplet size spectrum.

Size spectra of cloud and fog droplets vary widely, as might be expected. Table 5.2 lists some typical characteristics. Among those clouds unlikely to precipitate is the altostratus which contains smaller droplets but in a higher concentration. Cumulonimbus clouds and typical fogs tend to contain larger droplets in a smaller concentration. The level of droplet concentration is also an important difference between clouds formed over oceans and those formed over continents. The greater concentration of droplets in the latter is attributed to the greater concentration of nucleating particles, as will be discussed in the next section. Fig. 5.8 displays the size spectra of various clouds showing that the modal radius is often of the order of $10\,\mu$m while the occurrence of droplet radii greater than $40\,\mu$m is rare. However, in many fogs, wind-blown spray, drizzle and rain, droplet radii of $100\,\mu$m are not uncommon. Attempts to fit the enormously variable data have met with some success. Levin (1954) has suggested that the log-normal distribution is satisfactory. Mason (1971) mentions the use of the gamma function.

Given sufficient time (a scale $t_c \geqslant O(Mc_{pw}/h_Q A) = O(r\rho_w c_{pw}/3h_Q))$), every droplet will approach thermal equilibrium with its surroundings, but it is the smaller droplets which equilibrate first and nucleate last. An air temperature of $-10\,°$C is not uncommon during the formation of wind-blown spray, and values of about $-20\,°$C and lower may be found in clouds. Since droplets of radii up to $120\,\mu$m are capable of withstanding supercooling approaching the threshold of homogeneous nucleation (Hobbs, 1974), it is clear that there are many natural situations in which much of the water content of cloud, fog or spray remains in a supercooled state. Under certain circumstances, spontaneous freezing does occur, producing a monocrystalline prism on the basal faces of which dendritic growth may occur (Weickmann, Katz and Steele, 1970; Pitter and Pruppacher, 1973). However, there are many other circumstances when freezing follows impact on a moving surface.

Droplets blown towards an object seldom continue in a straight path. As they travel with and through the air they are subjected to viscous and gravitational forces which tend to alter their trajectories. Thus for a droplet of mass M,

$$M\,d\mathbf{V}_d/dt = M\mathbf{G} - \tfrac{1}{2}\rho_a C_D A_d (\mathbf{V}_d - \mathbf{V}_a)\,|\,\mathbf{V}_d - \mathbf{V}_a\,|$$

is the equation of motion[6] in a body force field $\mathbf{G}$ where the droplet velocity is $\mathbf{V}_d$ and the local air velocity $\mathbf{V}_a$: C_D is the droplet drag coefficient and A_d its forward projected area. When the body force field may be neglected this reduces to

$$\frac{2}{9}\frac{\rho_w r^2}{\mu_a}\frac{d\mathbf{V}_d}{dt} = \frac{-C_D Re_r}{12}(\mathbf{V}_d - \mathbf{V}_a)$$

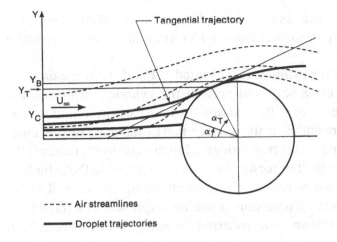

Fig. 5.9. Streamlines and trajectories.

where $Re_r = |\mathbf{V}_d - \mathbf{V}_a| r / \nu_a$ is the droplet Reynolds number and r is the droplet radius. In terms of normalized variables $\mathbf{v}_d = \mathbf{V}_d / U_\infty$ and $\tau = t U_\infty / R$, where R is the equivalent radius of the accreting body and U_∞ its speed relative to the surrounding air, this becomes

$$K \frac{d\mathbf{v}_d}{d\tau} = - \frac{C_D Re_r}{12} (\mathbf{v}_d - \mathbf{v}_a) \tag{5.10}$$

where the inertial parameter

$$K = \frac{2}{9} \left(\frac{r}{R} \right)^2 \frac{\rho_w}{\rho_a} Re_R,$$

and $Re_R = U_\infty R / \nu_a$ is the body Reynolds number.

In order to solve the Lagrangian form (5.10) it is necessary to know the precise velocity field of the air, but even in the absence of this field the equation still provides some insight. For example, when $K \ll 1$ the left hand side is essentially suppressed[7] implying that $\mathbf{v}_d \approx \mathbf{v}_a$: i.e., the droplets follow the air streamlines and do not impact on the body. Furthermore, if $Re_r < O(1)$, Stoke's law

$$C_D = 12/Re_r$$

will apply and equation (5.10) then reduces to

$$\frac{2}{9} \frac{\rho_w}{\rho_a} Re_R \frac{d\mathbf{v}_d}{d\tau} = - \left(\frac{R}{r} \right)^2 (\mathbf{v}_d - \mathbf{v}_a) \tag{5.11}$$

As the air changes direction while flowing over the body, the inertia of the droplet causes it to lag behind. Equation (5.11) reveals that droplet responsiveness to streamline curvature varies inversely with the droplet radius squared. Therefore, as a cloud of droplets approaches a body, the larger droplets will depart further from the air streamlines than the smaller droplets.

Consider flow over the cylinder represented in Fig. 5.9. Of the droplets passing through the cross sectional area of the cylinder, projected upstream, only the inner core will be unable to veer past the cylinder. The droplet trajectory tangential to the surface marks the edge of this inner core. The precise shape of this trajectory is a function of droplet radius; the smaller the radius, the further forward will be the tangent to the cylinder[8]. The droplets contained between the tangential trajectories will suffer collision, and the upstream area thus bounded, expressed as a fraction of the projected area, is known as the *collision efficiency* E_c. From Fig. 5.9 it is evident that

$$E_c = Y_c / Y_B$$

where Y_B marks the half width (radius) of the body and Y_c is the lateral distance to the edge of the impact core. If the local collision efficiency β at any point on the surface is defined by

$$\beta = \frac{\text{surface water flux density}}{\text{free stream water flux density}}$$

then the total water flux on the body when the water concentration is w_v, is given by

$$U_\infty w_v Y_c = \int_0^{\alpha_T} \beta U_\infty w_v Y_B \, d\alpha$$

Hence

$$Y_c = Y_B \int_0^{\alpha_T} \beta \, d\alpha$$

where α_T (or Y_T) defines the tangential point. Thus

$$E_c = \int_0^{\alpha_T} \beta \, d\alpha \tag{5.12}$$

in which it is important to note that β is generally a function of both droplet radius and position on the surface. Only if the air exerts no influence on the droplet trajectories will β be a function of position alone; at such a theoretical limit, $\beta = \cos \alpha$.

Collision with the surface is generally followed by a complex process

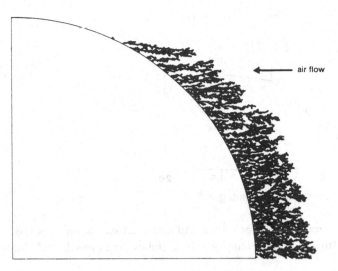

air flow

Fig. 5.10. Simulated atmospheric rime (following Gates *et al.* (1988)).

dependent upon the prevailing thermal and fluid flow conditions. Super-cooling of an amount $\theta = (T^i - T)$ implies a sensible heat deficit $c_{pw}\theta$ which may be made up wholly or partly by the latent heat released upon freezing. The ratio of these thermal energies, the Stefan number $Ste_w = c_{pw}\theta/\lambda_{iw}$, is thus an important parameter. Should $Ste > O(1)$, the droplet would have the ability to freeze completely, but this would require an ambient temperature of less than -80°C which is very rare under natural circumstances (recall that -40°C is the temperature associated with homogeneous nucleation in water). For temperatures well below the equilibrium freezing temperature, but above the homogeneous nucleation threshold, the droplet is still able to freeze quickly, and almost completely, because of additional heat loss after impact. Thus when $Ste_w \lesssim O(1)$, the accretion will likely be a porous structure in the form of stacked frozen spheres: this is often called *dry* accretion or rime. However, when $Ste_w \ll 1$, only part of the droplet may freeze, the remaining water being held in the ice matrix or spread across the freezing surface under the shearing action of the air stream. The resulting *wet* accretion tends to be a form of *glaze* ice. It is evident that aerodynamic and hydrodynamic conditions are important in the accretion process because they determine not only the general flow field around the body, and hence the collision and spectral distribution of droplets across its surface, but also control the various splashing, spreading, rippling and shedding phenomena associated with droplet impact and the subsequent development of a water film.

Dry accretion has been studied by many workers, and in recent years

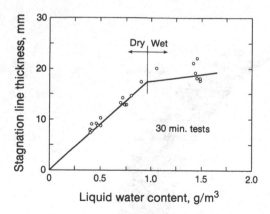

Fig. 5.11. Comparison of predicted and measured stagnation thickness: wind speed = $10\,\mathrm{m\,s^{-1}}$, average temp. = $-10°C$ (following Lozowski and Gates (1984)).

several attempts have been made to model it. The behaviour of a low concentration cloud of very small droplets represents one extreme. For comparatively large Stefan numbers, these droplets produce rime feathers, sometimes described as hoar[9] frost, in circumstances where the impact density may not be great enough to justify the use of an averaged flux density. Accordingly, Gates, Liu and Lozowski (1988) have developed a stochastic prediction in which a droplet, randomly chosen upstream, is assumed to follow straight or curved trajectories and to freeze completely upon impact. Fig. 5.10 shows the results of this model applied to a circular cylinder, clearly revealing the porous structure and distinct feathering, particularly in regions distant from the stagnation point.

When the droplet radius and concentration are greater, and the Stefan number is lower, the accretion process increases in complexity. The lower Stefan number implies that less of the droplet will freeze spontaneously and this, together with a larger radius, leads to the expectation that the droplet will not likely remain spherical upon impact. A greater droplet concentration reinforces this tendency to spread or splash because it reduces the time interval between successive impacts at any given point on the accreting surface. To a certain extent the radius and concentration effects may be embodied in the water content defined by

$$w_v = \frac{4}{3}\,n_v \pi r_m^3\,\rho_w$$

where n_v is the concentration, or number density, and r_m is the mean volumetric droplet radius. Increasing the water content from a very low value causes a corresponding increase in ice growth rate, as Fig. 5.11

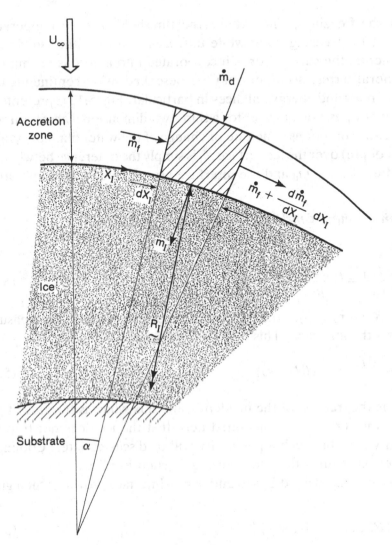

Fig. 5.12. Water balance in the accretion zone, where the inner boundary coincides with the outer boundary of previously formed ice.

indicates, but eventually the water is supplied at a rate greater than that at which it can freeze, thus producing a different (wet) regime (Lozowski and Gates, 1984). For a given water content, the same effect may be produced by varying the relative air speed over a sufficiently large range (Lozowski, Stallabrass and Hearty, 1983). There must always be a threshold water flux density above which some of the water will not freeze. Below this threshold, often called the Schumann–Ludlam Limit (Schumann, 1938; Ludlam, 1958), the rate at which heat is lost to the surrounding air is greater than the rate of latent heat release and the ice temperature therefore falls beneath the freezing point. Above the threshold, the ice temperature remains at the freezing point while unfrozen water is either shed, after flowing across the surface, or is incorporated into a porous ice matrix.

In general terms, accretion may be described using continuous concepts, the mass and energy balances in particular. Fig. 5.12 represents the source and disposition of water for a thin, two-dimensional accretion zone under steady conditions. The rate of change of the water film flow rate $\dot{m}_f$ (per unit depth) over the ice surface X_I is simply the difference between the droplet flux density $\dot{m}_d$ and the interface growth rate $\dot{m}_I$ (per unit area). Hence

$$\dot{m}_d - \dot{m}_I = \mathrm{d}\dot{m}_f/\mathrm{d}X_I$$

or

$$\frac{\mathrm{d}R_{nI}}{\mathrm{d}t} = \frac{\beta U_\infty w_v}{\rho_i} - \frac{1}{\rho_i}\frac{\mathrm{d}\dot{m}_f}{\mathrm{d}X_I} \tag{5.13}$$

where $\beta(X_I)$ (or $\beta(\alpha)$) is the local collision efficiency and R_{nI} is measured normal to the interface. This may also be written in the form

$$\rho_i \frac{\mathrm{d}R_{nI}}{\mathrm{d}t} = n(\beta U_\infty w_v) \tag{5.14}$$

where n is the fraction of the incident water flux which is freezing at that point on the surface. It is assumed here that the ice does not trap any unfrozen water, but such a possibility will be discussed later. Cooling of the accretion ensures that the freezing fraction $n \geqslant Ste_w$.

For dry growth, $\mathrm{d}\dot{m}_f/\mathrm{d}X_I = 0$ and $n = 1$. Interface growth is then given by

$$\mathrm{d}R_{nI}/\mathrm{d}t = \beta U_\infty w_v/\rho_i \tag{5.15}$$

where β is now the *collection* efficiency (defined as the product of the collision efficiency and the adhesion efficiency), the *adhesion* efficiency being 1.0. The dry growth rate at a stagnation point is thus proportional to both the air velocity and the water content as Fig. 5.11 suggests. For the entire body, equation (5.15) may be integrated to give the dry mass accumulation rate. Thus

$$\dot{M}_i = 2\rho_i \int_0^{X_T} \frac{\mathrm{d}R_{nI}}{\mathrm{d}t}\,\mathrm{d}X = 2U_\infty w_v \int_0^{X_T} \beta\,\mathrm{d}X = 2U_\infty w_v Y_B E_c$$

where X_T is half the extent of the icing surface. Under wet conditions, $n < 1$ and the rate of ice accumulation is given by

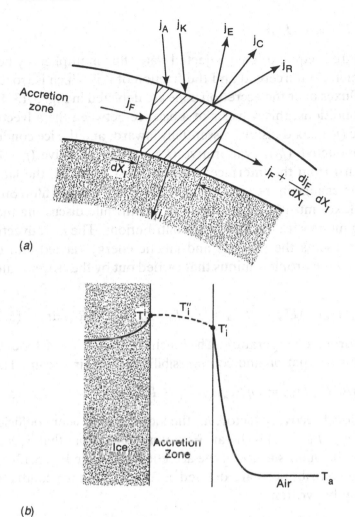

Fig. 5.13. Thermal conditions of the accretion zone: (*a*) energy flux densities; (*b*) temperature profile.

$$\dot{M}_{\mathrm{i}} = 2 U_\infty w_{\mathrm{v}} \int_0^{x_{\mathrm{T}}} n\beta \, \mathrm{d}X$$

In general, the freezing fraction n depends upon the energy balance in the accretion zone but its precise form is unknown *a priori* . If it is taken to have the uniform, and overall, value n_0,

$$\int_0^{x_{\mathrm{T}}} n\beta \, \mathrm{d}X = n_0 \int_0^{x_{\mathrm{T}}} \beta \, \mathrm{d}X = Y_{\mathrm{B}} E_{\mathrm{c}} n_0$$

and hence

$$\dot{M}_i = 2 U_\infty w_v Y_B E_c n_0$$

i.e., the icing rate is equal to the projected water flux multiplied by both the fraction which is intercepted and the fraction of this which is frozen.

The energy fluxes over the accretion zone are depicted in Fig. 5.13. For quasi-steady conditions, this consists of a balance between the advective (j_A) and kinetic (j_K) flux densities, essentially inward, and the ice conductive (j_i), air conductive (j_C), evaporative (j_E) and net radiative (j_R) flux densities outward from the interface: to these must be added the latent heat flux and the change in the sensible heat flux j_F of the water film. Since each of these fluxes must be prescribed it is worthwhile discussing them separately, beginning with the incoming contributions. The net advective flux density represents the sensible and kinetic energy carried into the accretion zone by the droplets minus that carried out by the frozen water. Hence

$$j_A = \tfrac{1}{2}(\beta U_\infty w_v) U_\infty^2 + \beta U_\infty w_v c_{pw} T_d - \rho_i c_{pw} T^i dR_{nl}/dt \quad (5.16)$$

where T_d is the droplet temperature. The kinetic[10] flux density includes the effects of viscous dissipation and compressibility of the air stream. Thus

$$j_K = \tfrac{1}{2}\eta h_Q U_\infty^2/c_{pa} = \tfrac{1}{2}\eta Ec_a h_Q (T_i' - T_a) \quad (5.17)$$

where η is the local recovery factor, h_Q the local heat transfer coefficient and $Ec_a = U_\infty^2/c_{pa}(T_i' - T_a)$ is the air Eckert number. Note that T_i' is the temperature on the outer surface of the accretion zone (see Fig. 5.13(b)).

The outgoing contributions are defined as follows. The ice conductive flux density may be written

$$j_i = (T^i - T_s)/R_i \quad (5.18)$$

where $(T^i - T_s)$ is the temperature difference between the inner surface of the accretion zone and the substrate and R_i is the thermal resistance of the ice (per unit interfacial area). The air conductive (or convective) flux density, the first of three losses to the environment, is given simply by

$$j_C = h_Q(T_i' - T_a) \quad (5.19)$$

whereas the evaporative[11] flux density is usually written as

$$j_E = \lambda_{wv} h_M \Delta C \quad (5.20)$$

where h_M is the local mass transfer coefficient and ΔC is the vapour concentration difference between the accretion zone surface and the air stream. The net outward radiative heat flux density is given by

$$j_R = \epsilon\sigma(T_i')^4 - \epsilon g \tag{5.21}$$

where ϵ is the effective emittance of the surface, σ is the Stefan–Boltzmann constant and g is the irradiation. Specularity and transmission through the ice have been neglected.

A quasi-steady state energy balance on the element dX_I of the accretion zone is thus written:

$$j_A + j_K - j_E - j_C - j_R - j_i = \frac{dj_F}{dX_I} - \lambda_{iw}\rho_i\frac{dR_{nI}}{dt}$$

In briefer form this may be written

$$\lambda_{iw}\rho_i\frac{dR_{nI}}{dt} = -\Delta j + \frac{dj_F}{dX_I} \tag{5.22}$$

where $\Delta_j = j_A + j_K - j_E - j_C - j_R - j_i$.

It is apparent from Fig. 5.13 that freezing of some of the droplets and runback water carried into the accretion zone will correspond to a distributed heat source. The details of this source are complex but the overall effect is clear: the average temperature of the freezing mixture T_i'' will be raised above T_a to a value approaching the equilibrium freezing temperature T^i, which defines the inside surface of the accretion zone. Heat loss ensures that the outer surface temperature T_i' lies between T_a and T^i. It is also apparent that liquid water is not the only fluid that enters into the accretion process. Water vapour, for example, continually leaves so long as there is an evaporative heat loss, although the magnitude is comparatively small. The role of the air is more complex. Much of the air separates from the droplet stream on impact but some of it may become trapped in pockets within the accretion. Strictly, therefore, j_F should be regarded as a mixture flux even though the water component is by far the greatest. For many practical purposes,

$$j_F = \dot{m}_f c_{pw} T_i''$$

and hence

$$\frac{dj_F}{dX_I} = \dot{m}_f c_{pw}\frac{dT_i''}{dX_I} + c_{pw}T_i''\frac{d\dot{m}_f}{dX_I}$$

For the special case of accretion forming on an adiabatic substrate under the sole cooling effect of the air conductive flux, equation (5.22) reduces to

$$\rho_i\lambda_{iw}\frac{dR_{nI}}{dt} = h_Q(T_i' - T_a) + c_{pw}\dot{m}_f\frac{dT_i''}{dX_I}$$

$$+ c_{pw}(T_i'' - T_d)\frac{d\dot{m}_f}{dX_I} + \rho_i c_{pw}(T^i - T_d)\frac{dR_{nI}}{dt}$$

if the advective flux contains only sensible heat and the water balance (equation (5.13)) is used. Hence, if the accretion zone is taken to be isothermal, with $T'_i = T''_i = T^i$, independent of X_I,

$$\rho_i \lambda_{iw} \frac{dR_{nI}}{dt} = h_Q(T^i - T_a) + \beta U_\infty w_v c_{pw}(T^i - T_d),$$

which has the simple interpretation that the latent heat not used in heating the cold droplet flux from T_d to T^i will be lost to the surrounding air. It is interesting to note that wherever runback is absent this expression may be written in the simpler form

$$\frac{dR_{nI}}{dt} = \frac{h_Q(T^i - T_a)}{\rho_i \lambda_{iw}(1 - Ste_d)} \tag{5.23}$$

where $Ste_d = c_{pw}(T^i - T_d)/\lambda_{iw}$. Since the droplet temperature is often close to the air temperature, this equation illustrates how the ice accumulation rate increases as the air temperature decreases: firstly, because the heat loss rate is increased and, secondly, because the Stefan number is increased.

In the sections which now follow, accretion will appear in a variety of contexts: hail, shipboard icing, aircraft icing, etc. Each of these will be given a more specialized treatment which relies upon the above discussion, the energy and mass balances in particular.

5.3   Atmospheric ice

It has been noted that ice forming from water vapour must follow one of two paths, the choice depending largely upon local conditions in relation to the triple point. Naturally occurring ice in the atmosphere may thus be subdivided into these same two categories in both of which the process usually begins with heterogeneous nucleation initiated by airborne particles, which are sometimes referred to as ice nuclei; less ambiguously they might be called ice *nucleants* (thus distinguishing between ice and other materials as potential nucleation sites).

The fact that clouds often exhibit temperatures in the neighbourhood of $-15\,°C$ confirms that supercooling and supersaturation are common, but this usually reflects nucleation inefficiency rather than the scarcity of nucleants. For example, it is estimated (Hobbs, 1974) that the number of nucleants n_i per cubic metre which are active at a temperature T is given by

$$n_i = 10^3 \exp[\alpha(T_c - T)] \tag{5.24}$$

where $T_c \approx -20\,°C$ is the temperature when n_i is one per litre, and $0.3 < \alpha < 0.8$ is an empirical constant. Comparing this number with a typical concentration of particles, which is of the order of 10^{11} per cubic metre, it is clear that very few particles are active (1 in 10^8). It is also clear that the concentration of active nucleants is very sensitive to temperature, increasing by an order of magnitude for a 4 °C drop.

Ice nucleants originate from many sources, the majority being associated with continentality. Natural sources include soils and vegetation. The former constitute the largest single source of nucleants, mainly from clay and mica (Mason, 1960). With sizes ranging up to a few microns, these dust particles may be found at altitudes up to 5 km, and may range over great distances, probably as a result of dust storms, and occasionally through volcanism. Decomposing leaves, poplar mulch in particular, are also an effective natural source (Schnell and Vali, 1972). Sea-salt nuclei, which originate when air bubbles burst on the sea surface producing minute droplets, which evaporate to leave the salt nucleus, have been suggested as a major source in a maritime climate (Blanchard and Woodcock, 1957; Mason, 1971), but this seems unlikely in a high humidity environment where they would soon be dissolved in water droplets. Industrial air pollution is another obvious source but, despite its concentration in certain areas, it only amounts to a few per cent of the total global source of nucleants.

Nucleants other than ice itself have certain surface characteristics, as noted in Section 5.1. As atmospheric particles, they act in essentially three ways. Firstly, they simply provide surfaces on which ice is deposited directly from the vapour phase. Secondly, they act within a supercooled droplet to initiate freezing. Thirdly, they exist outside of a supercooled droplet and cause nucleation on contact. The first mechanism operates below the extension of the water-vapour saturation curve beneath the triple point; it may also be operative above the extension. The second and third mechanisms are clearly restricted to the region above the triple point temperature, and simply distinguish between internal and external sources. The latter usually permits greater supersaturation or supercooling (higher Stefan numbers).

Self-generating ice nuclei may be produced in both the freezing and contact processes (Dye and Hobbs, 1968; Hobbs and Alkezweeny, 1968). When a droplet is frozen rapidly and uniformly, a thin shell of ice forms on the outside, thus constraining expansion of the water which continues to freeze on the inside. Unless sufficient time elapses to allow deformation of the shell, it will fracture and thereby produce ice fragments which become ice nuclei. The contact between an ice crystal and a supercooled droplet produces a very different effect but with a similar result. Such an event causes a sudden release of latent heat at the point of contact, thus subjecting the ice crystal to a thermal shock which is evidently capable of fracturing it, again producing fragments (Dye and Hobbs, 1968). However, both of these fragmentation processes are too inefficient to explain the high multiplication rates observed both in the laboratory and the field. These rates have been attributed instead to a splintering mechanism which occurs in clouds with temperatures between $-3\,°C$ and $-8\,°C$ when droplets larger than about $24\,\mu m$ collide with riming ice particles at a relative velocity near $2.5\,\mathrm{m\,s^{-1}}$ (Hallett and Mossop, 1974; Mossop and Hallett, 1974; Mossop, 1976). From Fig. 5.4 it is evident that riming near $-5\,°C$ is characterized by needle-shaped growths which might be expected to fracture easily upon collision with larger droplets. Appropriately, this form of multiplication has been called the Hallett–Mossop process.

Turning now to the development of atmospheric ice particles, consider first depositional growth from the vapour phase. As noted earlier, this growth proceeds only in the presence of supersaturation, which is often in the range 10–20% (with respect to ice; water saturation is a common condition). The analysis of droplet growth, as described in equations (5.6)–(5.9), applies equally to the diffusive growth of 'spherical' ice particles, and therefore, for given ambient conditions, it follows from equation (5.9) that

$$M \propto t^{3/2}$$

which has qualitative experimental support, at least for small particles (Fukuta, 1969; Mason, 1953). This suggests radial growth rates which are extremely high initially but which decay rapidly as the particle grows. On the other hand, it is evident from Sections 2.9 and 5.1 that linear growth rates, i.e., normal to the basal or prism faces, are independent of crystal size. For non-diffusive deposition, therefore, the mass would be proportional to t^3, the rate of change being proportional to the crystal surface area. Such growth must reflect the temperature dependency of the basal and prism face velocities, as displayed in Fig. 5.3, and this implies maxima in the growth rates at $-5.5\,°C$ and $-12\,°C$ for a simple, hexagonal crystal (Lamb and Hobbs, 1971). However, these growth rates, like diffusive growth rates, are too low to establish the depositional process as the cause of the larger ice particles often found in clouds. It must therefore be assumed that depositional growth is sometimes succeeded by the processes of aggregation and accretion.

Snowflakes provide a good example of additive growth processes. Broadly speaking, aggregation of snow crystals is similar to the aggregation of water droplets: heavier ones tend to fall faster than lighter ones thus bringing about the encounters which are the source of their growth.

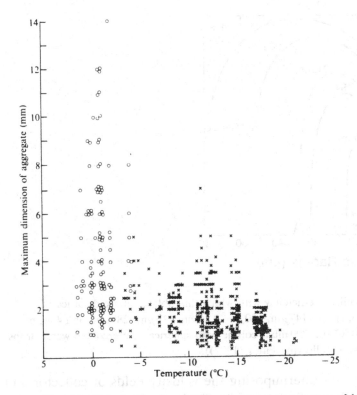

Fig. 5.14. Maximum dimensions of natural aggregates of ice crystals as a function of the temperature of the air where they were collected. Open circles: crystals collected from an aircraft; Crosses: crystals collected on the ground (following Hobbs (1973)).

Smaller ice crystals moving relative to larger crystals are subject to the viscous drag forces discussed in the previous section where it was noted that collision efficiency was dependent upon the relative size of the particles and their relative velocity. Taking both collision and adhesion efficiency to be fixed, Magono (1953) concluded that the volume V_s of a 'spherical' snowflake grows from the accretion of ice crystals (of volume V_i) at the rate

$$\mathrm{d}V_s/\mathrm{d}t = (9/16\pi)^{\frac{1}{3}} n_i V_i E V_s^{\frac{2}{3}} \Delta U_\infty \qquad (5.25)$$

where ΔU_∞ is the difference in their terminal velocities and n_i is the ice crystal concentration. Hence

$$V_s(t) = [V_s^{\frac{1}{3}}(0) + (9/16\pi)^{\frac{1}{3}} n_i V_i E \Delta U_\infty t]^3 \qquad (5.26)$$

which predicts snowflake dimensions compatible with the field observations plotted in Fig. 5.14 (Hobbs, 1973). More recently, Pitter and Pruppacher (1974) and Pitter (1977) have examined the effect of particle size on

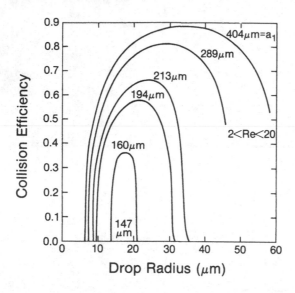

Fig. 5.15. Collision efficiency with which thin oblate spheroids of ice of semi-major axis $a_1 = 147\,\mu m$, $160\,\mu m$, $194\,\mu m$, $213\,\mu m$, $289\,\mu m$, and $404\,\mu m$, in air of $-10\,^\circ C$ and 700 mbar collide with spherical, supercooled water drops of various radii (following Pitter (1977)).

collision efficiency by superimposing the velocity fields of collector and collected while both are falling freely. Their results, applied to droplet accretion by ice particles, reveal important features of the riming process; these are illustrated in Fig. 5.15. Firstly, they suggest that ice particles with a radius below about 150 μm diameter are unable to grow by accretion and must therefore grow by deposition. Secondly, they suggest that droplets with a radius smaller than about 10 μm will not be collected. Thirdly, they suggest that although the collision efficiency begins to increase with droplet size, it exhibits a maximum before falling again to zero. All of these indications are in good agreement with the experimental observations of Ono (1969) and Hariyama (1975).

The collection efficiency is also influenced by temperature which alters the adhesion efficiency. If adhesion is taken to be a sintering[12] mechanism its effectiveness would tend to decrease with decreasing temperature. Observations of natural clouds indicate that the mechanism is not very effective below $-20\,^\circ C$, as Fig. 5.14 shows. The figure also reflects the temperature dependence of crystal habit which creates two local maxima, as noted earlier. It must be understood that the collision-based collection of ice crystals and water droplets may be augmented by other mechanisms which include wake capture, atmospheric turbulence, electrostatic force and the aerodynamic rotation of the aggregates themselves.

Table 5.3. *Empirical relationships between the terminal velocities V_0 (ms^{-1}) of solid precipitation particles and their maximum dimension D_m (mm) (following Locatelli and Hobbs, (1974)).*

	type of particle[a]	V_0–D_m relationship
Graupel and graupel-like snow	Cone-shaped graupel	$V_0 = 1.2D_m^{0.65}$
	Hexagonal graupel	$V_0 = 1.1D_m^{0.57}$
	Graupel-like snow of lump type	$V_0 = 1.1D_m^{0.28}$
	Graupel-like snow of hexagonal type	$V_0 = 0.86D_m^{0.25}$
Unrimed aggregates	Combination of sideplanes, plates, bullets and columns	$V_0 = 0.69D_m^{0.41}$
	Sideplanes	$V_0 = 0.82D_m^{0.12}$
	Radiating assemblages of dendrites or dendrites	$V_0 = 0.8D_m^{0.16}$
Densely-rimed aggregates	Radiating assemblages of dendrites or dendrites	$V_0 = 0.79D_m^{0.27}$
Densely-rimed columns		$V_0 = 1.1D_m^{0.56}$

[a] Based on Magono and Lee's (1966) classification.

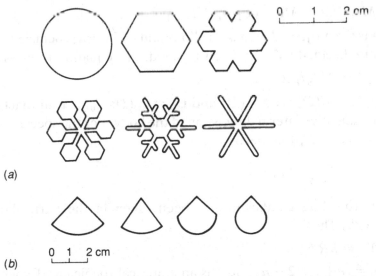

(a)

(b)

Fig. 5.16. Snow crystal and hailstone models: (*a*) Snow crystal models: top row, left to right: disc, hexagonal plate and broad-branched crystal: bottom row, left to right: stellar crystal with plates, dendrite and stellar crystal. (*b*) Conical modes: left to right: 90° cone-spherical sector, 70° cone-spherical sector, 90° cone-hemisphere and 90° teardrop (following List and Schemenauer (1971)).

Continued aggregation obviously leads to much larger ice particles which, broadly speaking, fall into two categories. When the crystals retain their basic habits, and conglomerate into feathery masses, the resulting structure is generally described as *snow*, even though it may have grown in part by droplet accretion. Alternatively, when environmental conditions produce more or less isometric crystals and/or allow extended accretion of supercooled droplets, the resulting structure will tend to be denser. Loosely speaking, the ice particles thus produced may be described as *hail*, but such a broad category needs several subdivisions (International Association of Hydrology classification (1951)). *Graupel* are ice crystals which are heavily rimed: they include soft hail, small hail and snow pellets. *Ice pellets* are small transparent spheroids of ice which are sometimes called sleet. *Hailstones* are larger (usually greater than 5 mm diameter), laminated lumps of ice having a smooth glassy surface, usually caused by melting near the ground. Each of these types deserves a separate discussion of its origin and growth, but first it is worthwhile giving some general consideration to the falling motions of the variously-shaped bodies representative of these groups.

In its simplest form (a more general form was discussed in Section 5.2), the equation describing the translational, falling motion of a body of mass M, is given by

$$M \, \mathrm{d}\mathcal{V}/\mathrm{d}t = Mg - \tfrac{1}{2} C_\mathrm{D} \rho_\mathrm{a} \mathcal{V}^2 A \qquad (5.27)$$

where A is the projected area of the body and C_D its drag coefficient. When the terminal velocity $\mathcal{V}_0$ has been reached, this equation reduces to

$$\mathcal{V}_0^2 = 2Mg/\rho_\mathrm{a} C_\mathrm{D} A \qquad (5.28)$$

in which $M = M(D)$, $A = A(D)$ and $C_\mathrm{D} = C_\mathrm{D}(D, \mathcal{V}_0)$ are all functions of D, a representative lateral dimension of the body. For the special case of a sphere of radius R, this becomes

$$\mathcal{V}_0^2 = \frac{8\rho_\mathrm{i} g R}{3a \rho_\mathrm{a}} Re_\mathrm{R}^n$$

in which the drag coefficient has been taken in the particular form $C_\mathrm{D} = a/Re_\mathrm{R}^n$. Thus

$$\mathcal{V}_0 = bR^m \qquad (5.29)$$

where $m = (n + 1)/(2 - n)$ and b is an empirical coefficient. Experiments on the effect of body dimensions on falling velocity generally conform to equation (5.29). Some empirical findings are given in Table 5.3.

The model experiments of List and Schemenauer (1971) (see Fig. 5.16) generally reveal the expected behaviour of a drag coefficient decreasing

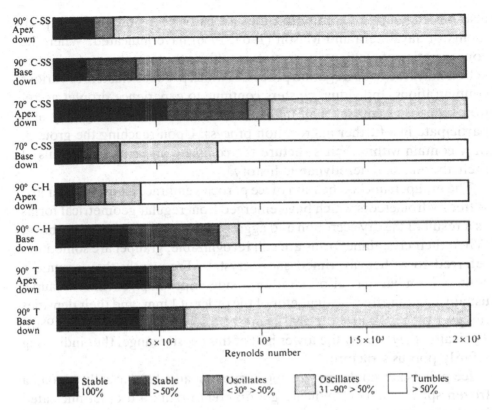

Fig. 5.17. Degree of oscillation of conical models as a function of Reynolds number: C–SS cone-spherical sector; C–H cone-hemisphere; T teardrop (following List and Schemenauer (1971)).

with increasing Reynolds number, although the solid hail-like particles do begin to exhibit the opposite tendency when the Reynolds number approaches and exceeds 10^3. This anomalous behaviour reveals the limitations of equations (5.27) and (5.28) when the mass of the body may no longer be treated at a point, and the motion is not steady. Minima in the $C_D - Re$ curve mark a transition to unsteady behaviour characterized by two basic types of superimposed oscillatory motion: lateral translation of the centre of mass, and tilting of the vertical axis. Fig. 5.17 provides details of the transitions for specific types, and reveals that oscillations may lead to a tumbling motion.

Stable, oscillatory, spiralling and tumbling motions clearly influence the accretion characteristics of ice particles, and emphasize the essential aerothermodynamic interaction: shape determines motion which controls growth; growth creates shape change which alters motion. Motion may also be altered by proximity to other ice particles, as the model experiments of Jayaweera and Mason (1965, 1966) have indicated. When the concentration of elementary crystals is sufficiently great this mutual interaction may lead to aggregation processes which produce clusters of various configurations. Individual clusters continue to experience droplet accretion (which, of course, may influence their motion) but they may also participate in a further aggregation process. Upon reaching the ground, they contain within their structure the complete and complex details of their thermal and aerodynamic history.

The graupel subclassification of ice particles embraces those smaller and softer hydrometeors which have emerged from regular geometrical forms as a result of the dry accretion and aggregation processes discussed above. When their crystalline forms are still recognizable, graupel are sometimes referred to as heavily rimed snow crystals. Frequently they assume a conical form but may also be rather nondescript lumps of rime. Graupel usually have an effective diameter of the order of 1 mm, and their densities range from $50 \, \mathrm{kg \, m^{-3}}$ to $890 \, \mathrm{kg \, m^{-3}}$ (Pruppacher and Klett, 1980). Typically, they occupy the lower half of this density range, thus indicating a fairly porous structure.

Ice pellets are similar in size and shape to graupel but usually contain a frozen spheroidal core indicating a different origin: a droplet nucleated internally may initiate the accretion process, thus accounting for most ice pellets; a droplet slowly frozen from the outside may account for those pellets with an unfrozen core. The core, solid or liquid, tends to create higher densities, typically of the order of $900 \, \mathrm{kg \, m^{-3}}$ (List, 1965).

Hailstones originate in a number of ways but it is their sheer size, from millimetres to centimetres, which characterizes them. They may, for example, be formed from graupel where an increasing fall velocity gradually converts a dry accretion process into one of wet accretion; this generates excess water which, through capillary action, enters the porous graupel structure where it freezes. The frozen water adds to the particle density and weight, thereby accelerating the conversion process. Accretion on the lower face often causes an asperity which may render it aerodynamically unstable, as Fig. 5.17 would suggest for Reynolds numbers in excess of 10^3. In the event of an instability, the hailstone is free to rotate, thus exposing a different face to the 'wind'. The rotational process facilitates the development of a fairly uniform structure and a roughly spherical shape. However, instability is by no means an invariable occurrence, as the oblate shape of many large hailstones testifies.

The observations of Macklin and Ludlam (1961) indicate that the

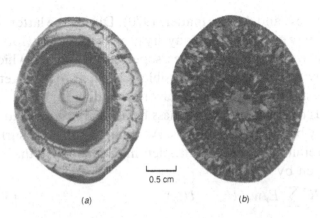

Fig. 5.18. Thin section of a natural hailstone: (*a*) normal transmitted light; (*b*) polarized transmitted light (following Knight and Knight (1968)).

Reynolds number of a hailstone is typically in the range 3×10^4–2×10^5. This corresponds to subcritical behaviour for a smooth rigid sphere with a fairly constant drag coefficient of 0.4–0.5 (Schlichting, 1968). Such a C_D range is comparable with the observed values of 0.5–0.7 for hailstones, bearing in mind that the surface is usually both rough and wet, and it is seldom perfectly spherical.

A typical hailstone is shown in cross section in Fig. 5.18. In normal light, the hailstone reveals a laminated structure with surface asperities which, on occasion, may become lobes as a result of preferential growth (because of a higher local collision efficiency) in relation to surrounding areas. It has been noted (Knight, 1968; Knight and Knight, 1968) that the details of lobe growth and shape depend upon whether the accretion is dry or wet, and it has been suggested that these differences may be used to describe the observed variations in sponginess. The layers mark the various conditions of growth, and thus reflect the types of accretion. Opacity is attributable to the presence of air bubbles seen in both dry and wet regimes: transparency, on the other hand, indicates the absence of air bubbles, as would occur under just wet conditions.

In polarized light, the layering is seen to consist of changes in the size, shape and orientation of crystallites. Opaque layers, with high air bubble concentrations tend to contain numerous small, and randomly orientated, crystallites; transparent layers tend to have larger crystallites with their c-axes parallel to the direction of growth (List, MacNeil and McTaggart-Cowan, 1970). Wind tunnel studies suggest that the transparent layers with radial c-axes are formed under slow, dry growth conditions, while faster, wetter growth tends to produce radial a-axes (Levi, Achaval and

Aufdermauer, 1970; Levi and Aufdermauer, 1970). During the latter conditions excess water may penetrate into any dry growth previously formed and, in any event, the water may constitute a supercooled film into which a dendritic matrix may grow. It is thus possible for a hailstone to retain unfrozen water interstitially, and thereby avoid shedding it.

Using the methods of Section 5.2, the mass balance appropriate to dry hailstone growth may be written as follows. Assuming a spherical surface which captures and retains all the water droplets incident upon it, the mass rate of growth is given by

$$\mathrm{d}M/\mathrm{d}t = \pi R^2 \sum_r E_r w_r (U_\infty - U_{\infty r}) \tag{5.30}$$

where R is the hailstone radius and E_r represents the overall collision efficiency for droplets of radius r moving with a terminal velocity $U_{\infty r}$. (Typically $r \ll R$, $U_{\infty r} \ll U_\infty$ and $\sum_r E_r w_r = Ew$, where E and w are the overall efficiency and water content, respectively.) This is an equivalent form of equation (5.15) for dry growth, under which conditions the energy balance plays an insignificant role, at least from the standpoint of mass addition. The effective hailstone radius would then increase linearly with time if the collision efficiency and fall speeds were fixed. If, however, E were to increase with the collector size, as is suggested in Fig. 5.15 for ice particles less than 0.4 mm in radius, the rate of growth would change dramatically. For example, if the droplet terminal velocity is neglected and the hailstone terminal velocity is assumed to be fixed, a linear relation between E and R requires that $\mathrm{d}R/\mathrm{d}t \propto R$ and the growth would then be exponential. This behaviour may be compared with the earlier result for diffusive growth (following from equation (5.9)) for which $\mathrm{d}R/\mathrm{d}t \propto 1/R$.

Under wet conditions, a complete description of the growth process must include the energy balance described by equation (5.22). When cooling is mainly attributable to convection (including both sensible and evaporative contributions), the overall balance is given by

$$\rho_i \lambda_{iw} \frac{\mathrm{d}R}{\mathrm{d}t} = h_Q \theta_a + \lambda_{wv} h_M \Delta C + \rho_i c_{pw} T^i \frac{\mathrm{d}R}{\mathrm{d}t}$$
$$+ (1 - n_0) \frac{E}{4} U_\infty w_v c_{pw} T'_a - \frac{E}{4} U_\infty w_v c_{pw} T_a$$

if the accretion zone is isothermal[13] at T^i, $\theta_a = T^i - T_a$, T_a is the air (and droplet) temperature, T'_a is the temperature of shed water, n_0 is the overall freezing fraction, and

$$E = \frac{1}{\pi R^2} \int^A \beta \, \mathrm{d}A$$

is the overall collection coefficient for a sphere of surface area A. The corresponding water balance is

$$\rho_i \, dR/dt = E U_\infty w_v n_0/4$$

which enables the energy balance to be re-written

$$\rho_i \lambda_{iw} \frac{dR}{dt} (1 - Ste_d) = h_Q \theta_a + \lambda_{wv} h_M \Delta C$$

$$+ (1 - n_0) \frac{E}{4} U_\infty w_v c_{pw} (T'_a - T_a) \quad (5.31)$$

where $Ste_d = c_{pw}(T^i - T_a)/\lambda_{iw}$. Thus the total convective heat loss rate balances the rate of generation of latent heat (less the sensible heat required to raise the droplets to the freezing temperature), except for the fact that the shed water carries away a sensible heat gain. This last effect is zero when all or none of the droplet flux freezes, i.e., either $n_0 = 1$ or $dR/dt = 0$, the latter occurring when $T^i = T'_a = T_a$, and each term in equation (5.31) vanishes identically under steady conditions. Between these extremes, the contribution of the shed water is often small, in which case the energy balance may be approximated by

$$\rho_i \lambda_{iw} \frac{dR}{dt} (1 - Ste_d) = h_Q \theta_a + \lambda_{wv} h_M \Delta C \quad (5.32)$$

The first of the heat losses on the right hand side of equation (5.32) may be written in terms of the heat transfer correlation for a sphere. Thus

$$\frac{2 h_Q R}{k_a} = Nu_D = 2 + a Re_D^b \quad (5.33)$$

where a and b are empirical constants. For a sphere in air[14], when $10^3 < Re_D < 10^5$, $a = 0.31$ and $b = 0.57$ (Whitaker, 1972). By analogy, the evaporative loss may be based on the same form

$$\frac{2 h_M R}{D_a} = Sh_D = 2 + a Re_D^b \quad (5.34)$$

The concentration difference $\Delta C = C_I - C_a$ is expressible in a number of ways. Water vapour diffusing through air may be treated as a dilute mixture for which $C = \rho$, the vapour density.

Combining equations (5.32), (5.33) and (5.34), yields

$$\rho_i \lambda_{iw} \frac{dR}{dt} (1 - Ste_d) = \left(\frac{2 + a Re_D^b}{2R} \right) [k_a \theta_a + \lambda_{wv} D_a (\rho_I - \rho_a)]$$

from which the growth of the hailstone may be found through integration. An approximate solution is readily available if it is assumed that changes in the stone (relative) velocity U_∞ are small enough to ignore, in which case Re_D varies with R only and the equation is conveniently re-written in the form

$$\frac{dR}{dt} = \frac{K(2 + aRe_D^b)}{R}$$

where

$$K = \frac{k_a \theta_a + \lambda_{wv} D_a (\rho_I - \rho_a)}{2\rho_i \lambda_{iw}(1 - Ste_d)}$$

may be treated as a constant in a given environment. It is now evident that when ventilation is substantial, i.e., $2 \ll aRe_D^b$, $dR/dt \propto 1/R^{0.43}$ which may be compared with growth rates under other conditions. In general, $dR/dt \propto R^p$. For diffusive deposition, it was shown earlier that $p = -1$ while for non-diffusive deposition $p = 0$. It also appears that p may be greater than zero for dry accretion on some small hail and graupel.

Finally, a few words on spongy hail, in which ice and water are found in stable equilibrium at the equilibrium freezing temperature T^i. It has been noted that the water may have entered a previously formed, porous, dry growth, but such growth would occur at a temperature T_0 below T^i and therefore not all of the water entering the pores may remain unfrozen. In fact, if the pores are very small the surface of the hailstone may be sealed by the first coating of water, thus preventing further penetration into the core unless and until melting occurs. However, if sealing does not occur, the original sensible heat deficit of the stone, $c_{pi}(T^i - T_0)$ per unit mass of ice, is eventually balanced by latent heat released as water penetrates throughout the pores. (It is assumed that the pores are initially full of air and finally full of water: $c_{pa} < c_{pi}$.) Thus, if the porosity of a dry growth of volume V is $\mathcal{N}_p$,

$$\rho_i(1 - \mathcal{N}_p)Vc_{pi}(T^i - T_0) = \rho_i \lambda_{iw} V(\mathcal{N}_p - \mathcal{N}_w)$$

where $\mathcal{N}_w$ is the water fraction remaining in the stone unfrozen. Hence

$$\mathcal{N}_w = \mathcal{N}_p - Ste(1 - \mathcal{N}_p)$$

and the resulting ice fraction is given by

$$f_i = 1 - \mathcal{N}_w = (1 - \mathcal{N}_p)(1 + Ste)$$

Alternatively, or additionally, unfrozen water may be retained during the accretion process itself, as might be expected when a supercooled water film permits the formation of an ice matrix (Lock and Foster, 1989). The water balance must then be re-written to accommodate the fact that

$$1 = n_i + n_w + n_s$$

where the subscripts refer to the fraction of the incident water flux which becomes ice, retained water and shed water, respectively. Hence

$$\frac{E}{4} U_\infty w_v (1 - n_s) = \rho_i \frac{dR}{dt}$$

where the growth rate now applies to the spongy composite within which the ice fraction $f_i = n_i/(n_i + n_w)$, and at the surface of which the latent heat flux density is $\rho_i \lambda_{iw} f_i dR/dt$. With these changes, the energy balance, equation (5.31), becomes

$$\rho_i \lambda_{iw} \frac{dR}{dt}(f_i - Ste) = h_Q \theta_a + \lambda_{wv} h_M \Delta C$$

$$+ n_s \frac{E}{4} U_\infty w_v c_{pw} (T_a' - T_a) \quad (5.35)$$

with the usual interpretation: namely, that the latent heat liberated by the ice fraction f_i, less the sensible heat required to raise the spongy material temperature from T_a to T^i, must be carried away convectively along with the sensible heat gain of the shed water. It is evident that sponginess will increase the rate of growth of the accretion, especially when $f_i \to Ste$.

5.4 Icing of airborne structures

A brief discussion of cloud droplet features was given in Section 5.2 where it was noted that they depend upon the movement and thermal history of the cloud air mass. The principal characteristics – temperature, water content and droplet size distribution – are known to vary considerably from place to place and from time to time, and therefore it is difficult to specify precisely the potential icing hazard faced by aircraft. None the less, modelling the droplet size distribution is essential if the details of subsequent accretion, and its severity, are to be understood.

Comparing three common distributions, Takeuchi, Johnson, Callander and Humbert (1983) suggested that the gamma function offers the best prospects for the number density n_D. Written in the form

$$n_D = n_0 \left(\frac{D}{D_+} \right)^\alpha \exp(-\lambda D), \quad (5.36)$$

where D is the droplet diameter and n_0, α, λ and D_+ are appropriate parameters, this equation has the advantage of goodness of fit to both natural and wind tunnel droplet data. It also benefits from simplicity, especially with respect to the forms of commonly required quantities. For example, the total number density is given by

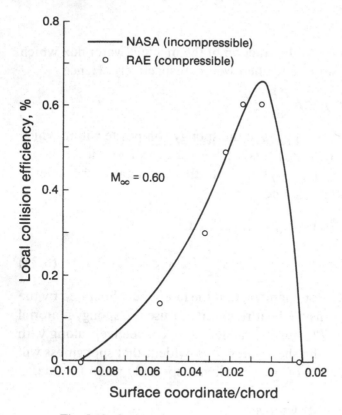

Fig. 5.19. Comparison of National Aeroneutics and Space Administration/ Royal Aeronautical Establishment (NASA/RAE) water droplet trajectory analyses. NACA 0012 airfoil chord = 21 in; $\alpha = 4°$; drop diameter = 12 μm (following Shaw (1984)).

$$n_T = \frac{n_0}{D_+^\alpha} \frac{\Gamma(\alpha + 1)}{\lambda^{\alpha+1}}$$

while the fractional water volume is defined by

$$v_w = \frac{\pi n_0}{6D_+^\alpha} \frac{\Gamma(\alpha + 4)}{\lambda^{\alpha+4}}$$

thus yielding the liquid water content $w = \rho_w v_w$.

Icing on aerofoil sections has been studied for decades, but it is only in recent years that the increased capabilities of digital computers and experimental facilities have made it possible to model the accretion process with sufficient accuracy to permit a wide range of comparisons between prediction and observation; hitherto, only the simplest situations could be treated with confidence. These advances have been felt in a number of ways (Shaw, 1984): experimentally, in the simulation of full scale effects; and theoretically, in the calculation of the velocity field, the corresponding

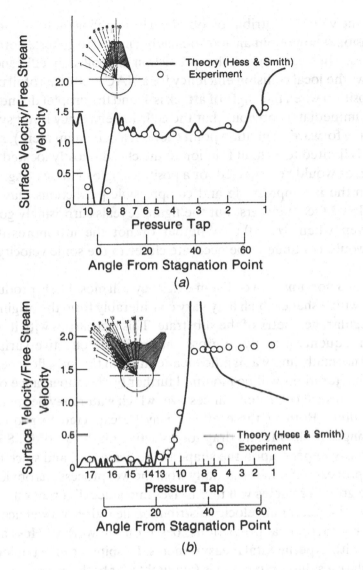

Fig. 5.20. Potential surface velocity as a fraction of free stream velocity for dry and wet accretion: (*a*) rime accretion, (*b*) glaze accretion.

prediction of droplet trajectories, and the consequent predictions of heat transfer and drag.

Typically, the Reynolds number of an aerofoil is large enough to ensure the existence of a boundary layer on the outside of which the velocity may be calculated from a potential flow field. Several schemes are available for the calculation of potential flow fields, among which the methods advocated by Hess and Smith (1967) and Gent and Cansdale (1985) are noteworthy. The velocity distribution obtained by applying these methods to two-dimensional flow about an aerofoil has been used to generate droplet trajectories, thus leading to the prediction of collision efficiency. Fig. 5.19 shows the local collision efficiency for a NACA 0012 aerofoil as a function of position when the angle of attack is 4° and the droplet diameter is 12 μm. It is immediately obvious that the collision efficiency is not symmetric about the forward stagnation point (nor is it a maximum there), and that collision is limited to a small fraction of the chord, mostly located on the underside, as would be expected for a positive angle of attack. Agreement between the incompressible and compressible predictions may be expected for low Mach numbers, but the figure reveals surprisingly good agreement even when $M_\infty = 0.6$. Whether or not the incompressible formulation would continue to be accurate closer to the sonic velocity is unknown.

It is clear that a non-uniform collision efficiency will most likely produce an ice surface with a shape which may vary considerably from the original, and usually regular, geometry of the substrate. This fact carries with it two important consequences: firstly, a gross change in the effective surface geometry will normally imply a significant alteration in the local flow field; secondly, such alterations will be modified further by the appearance of a multitude of local surface protuberances over which water may be splashing and spreading. Both of these effects may be expected to produce significant changes in the lift and drag forces acting on the aerofoil. Such changes obviously depend upon the shape of the accretion, and since the dry accretion process differs substantially from the wet process it is obvious that these two growth regimes will themselves introduce different effects. Fig. 5.20 shows the potential velocity distribution near dry and wet accretions, comparing theoretical predictions, based on the work of Hess and Smith (1967), with experimental measurements. Despite the effect of local surface irregularities the comparison is favourable for both dry (rime) and wet (glaze) growth, although separation behind the lobes in the latter destroys the validity of the potential flow method there. Evidently, the flow field may be modelled fairly accurately, at least for two dimensions.

It was noted in Section 5.2 that the calculation of ice growth rates combines information on droplet trajectories and local collection efficiency (recall that 'collection' includes both collision and adhesion) with heat and mass transfer balances in the region of impingement. The balances must, of course, be applied locally, and this creates considerable difficulty with a body shape which is neither constant nor geometrically simple. For a typical leading edge, represented by a cylinder in cross flow for example,

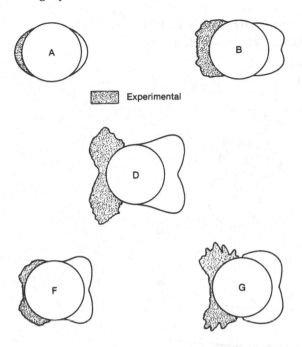

Experimental

Fig. 5.21. A comparison between some representative experimental accretion profiles and model profiles: (A) $-15°C$, $30.5\,m\,s^{-1}$, $0.40\,g\,m^{-3}$, $5.0\,min$; (B) $-15°C$, $122\,m\,s^{-1}$, $0.44\,g\,m^{-3}$, $2.5\,min$; (D) $-8°C$, $110\,m\,s^{-1}$, $0.65\,g\,m^{-3}$, $5.0\,min$, (F) $-5°C$, $91.5\,m\,s^{-1}$, $0.38\,g\,m^{-3}$, $4.0\,min$, b; (G) $-5°C$, $91.5\,m\,s^{-1}$, $1.17\,g\,m^{-3}$, $4.0\,min$ (following Lozowski *et al.* (1983)).

the problem is straightforward provided the ice thickness is small compared with the cylinder radius, but even then the effect of surface roughening is great enough to modify the heat transfer rate (Achenbach, 1977; Makkonen, 1984b). For a cylinder Achenbach gives the empirical form (valid up to $Re_D = 2.6 \times 10^5$ with a roughness parameter value of 9×10^{-3})

$$Nu(\alpha) = Re_D^{\frac{1}{2}} \{2.4 + 1.2 \sin [3.6(\alpha - 0.44)]\} \qquad (5.37)$$

where α is the angle measured in relation to the forward stagnation point. Lozowski *et al.* (1983) have used this, along with the analogous evaporative form and Seban's (1960) expression

$$\eta = 0.75 + 0.25 \cos 2\alpha \qquad (5.38)$$

for the local recovery factor, to compute ice growth around a cylinder. Some of their predictions are shown in Fig. 5.21 together with their measurements for a few minutes of exposure: air temperature and speed, water content and test duration are given for each comparison. It is evident that predictions for dry conditions (A) are excellent, as might be expected

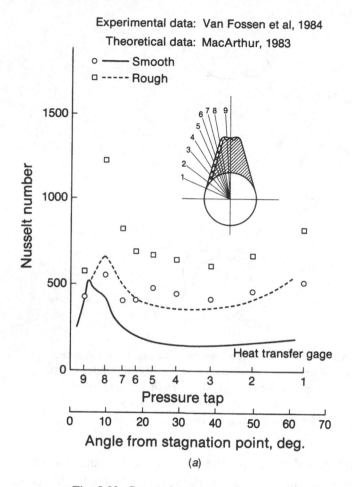

Fig. 5.22. Comparisons of heat transfer coefficient predictions with experimental measurements for smooth and rough ice shapes: (*a*) rime accretion;

when the effect of the water film is absent. When the growth is wet, however, and the prescription of local conditions becomes increasingly inaccurate as the shape departs from circularity, close agreement could not be expected. None the less, the features of stagnation point depression and lobe growth are predicted qualitatively, and the total mass of ice grown is still estimated with fair accuracy.

The effects of accretion on lift and drag, and on the local values of collection efficiency and heat transfer coefficient, are not yet well established, although some progress has been reported (MacArthur, 1983; Van Fossen, Simoneau, Olsen and Shaw, 1984). Fig. 5.22 provides a comparison of heat transfer results for the rime and glaze shapes of Fig. 5.20.

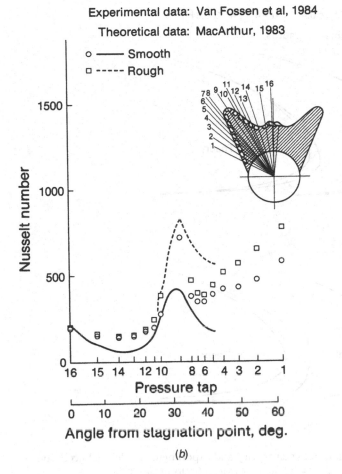

Experimental data: Van Fossen et al, 1984

Theoretical data: MacArthur, 1983

Fig. 5.22.(*b*)

(*b*) glaze ice accretion (following Shaw (1984)).

As expected, the results differ substantially from those for a circular cylinder, thus emphasizing the importance of shape. Perhaps surprisingly, the rime data are no better reconciled than the glaze data. The effect of roughness is also seen to be important except over the frontal face.

Despite the fact that aerofoil icing is understood and predictable qualitatively, it is evident that quantitative predictions are still wanting, except for simple situations. It is also clear that our limited ability, demonstrated in the handling of two-dimensional fixed aerofoil sections, will become more acute when applied to rotating sections such as propellers, turbine rotors and helicopter rotors. In general, these elements must be treated three-dimensionally in a speed range where the effects of compressibility

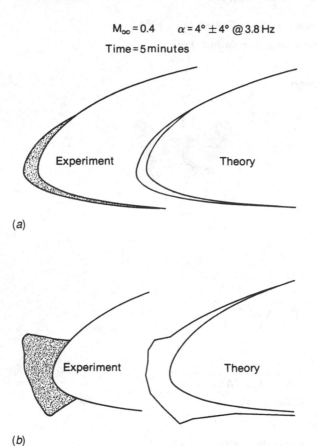

$M_\infty = 0.4$ $\alpha = 4° \pm 4°$ @ 3.8 Hz

Time = 5 minutes

(a)

(b)

Fig. 5.23. Comparison between theory and experiments for 'cyclic' accretion model; (a) rime: $T_\infty = -18\,°C$, $w_v = 0.16\,\mathrm{g\,m^{-3}}$; (b) glaze: $T_\infty = -12\,°C$, $w_v = 0.54\,\mathrm{g\,m^{-3}}$ (following Gent and Cansdale (1985)).

may be important. Along the helicopter rotor, for example, not only may the linear tip speed approach the sonic velocity, but centrifugal effects and spanwise variations of rotor speed create spanwise variations in accretion conditions and therefore in the shape and growth rate of the accretion itself. Several preliminary studies of helicopter rotor accretion have been reported (Stallabrass and Lozowski, 1978; Gent and Cansdale, 1985; Shaw and Richter, 1985) and have drawn attention to the difficulty of the problem. None the less, it appears that modelling experience with two-dimensional aerofoils provides a suitable point of departure for the rotating aerofoil, and may even be useful in simulating the cyclic pitch and velocity variations which occur in forward flight. Fig. 5.23 compares theoretical and experimental data for accretion under cyclic pitch conditions over a period of 5 min (Gent and Cansdale, 1985). The model, like that mentioned earlier, is evidently capable of quantitative prediction under dry conditions but only qualitative prediction when the ice is wet.

The icing behaviour of axial compressor blades is very similar to that of the aerofoils discussed above, the two main differences being the effect of the shorter span and mutual interaction between the blades themselves and between blades and walls. Centrifugal rotors obviously constitute a separate and more complex study, as do the multitude of intakes and nacelles, although the latter may often be simplified through the use of axi-symmetry. To a first approximation, a nacelle entry may be treated as a thick aerofoil wrapped spanwise into the form of a ring, but the velocity field within the ring will generally differ from that on the outside, thus influencing droplet trajectories and collection efficiency (Schmidt, 1965).

In concluding this section it is perhaps worth mentioning briefly an active thermal method whereby icing of airborne structures may be reduced, and perhaps avoided completely. Icing may always be reduced through reduced exposure but, given a period of unavoidable exposure, the tolerance of icing conditions may be increased further by heating the substrate from beneath. In general terms, this has the effect of reducing the conductive heat loss through the ice and thus generates a reduced accretion rate. In particular, should it raise the substrate temperature close to 0°C, the ice may not be able to form at all or, if it does, the adhesive bond with the substrate will be weakened (Jellinek, 1959) and may not be strong enough to withstand the shearing action of the surrounding air stream. The melting or removal of ice, once formed, will be discussed in more detail in Chapter 8.

5.5 Icing of waterborne and offshore structures

The discussion of icing in the previous section was based on the underlying assumption of a steady or quasi-steady state; for most aircraft flying through, or hovering in, extensive cloud formations this assumption is a reasonable, first approximation. For most waterborne and offshore structures, the steady state condition is at best a crude approximation to the truth, although a slowly shifting fog bank may lend itself to such an approach. Gusting wind, breaking waves and vessel motion all ensure that, in general, the impinging water spray has time-dependent characteristics; and matters are complicated further by spray-free periods during which the accretion may experience the effects of a chilling wind or a deluge of water.

Marine icing originates from a number of sources: vessel-generated

Table 5.4. *Characteristics of icing sources in the atmospheric surface layer (following Makkonen (1984a)).*

| Source | Droplet diameter (μm) | | Liquid water content (g m^{-3}) |
	Range	Mean	
Sea spray on a moving ship	1000–3500	2400	0–219
Sea spray in first 10 cm	10–1000	200	
Sea spray on a stationary ship			
$V > 15$ m s^{-1}, $h = 2$ m	3–2000	5–30	0.03
$V > 15$ m s^{-1}, $h = 7$ m	3–90	5–30	0.00
Marine advection fog		8–16	0.03–0.17
Coastal fog	4–20		0.01–0.16
Evaporation fog	6–120	13–38[a]	0.01–0.30
Evaporation fog			0.04–0.14
Evaporation fog		8–10	0.20
Arctic fog	7–130	16	0.00–0.15
Arctic fog	2–75	18	0.02
Arctic fog	6–60		0.04–0.17
Continental winter fog		10	0.00–0.45
Mountain fog		7–23[a]	0.05–0.30
Low stratus	2–43	5	0.05–0.25

[a] median volume diameter

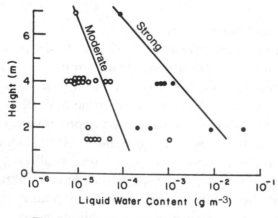

Fig. 5.24. Vertical distribution of spray water content over the sea surface at moderate (6–10 m s^{-1}) and strong (15–25 m s^{-1}) winds (following Preobrazhenskii (1973)).

spray, freezing rain, fog, wind-generated spray (spindrift), snow and hoarfrost. The discussion in Section 5.1 suggests that only the lightest accretion would result from hoarfrost; the same expectation would apply to dry or wet snow, which does not generally affect vertical surfaces and tends to occur at temperatures where its removal usually creates little difficulty. For wind speeds greater than about $5\,\mathrm{m\,s^{-1}}$ wind-generated spray occurs on the open sea with droplet diameters in the order of $10-1000\,\mu\mathrm{m}$. The corresponding water content is shown in Fig. 5.24 (Preobrazhenskii, 1973) from which it is evident that not only does the water content decrease rapidly over the first few metres, but it is also characterized by magnitudes very much lower than those typical of other sources. This is clear from Table 5.4. In general, therefore, wind-generated spray is capable of producing only light accretion.

Table 5.4 also reveals that fogs usually have higher water contents, typically in the range $0.01-0.20\,\mathrm{g\,m^{-3}}$; their droplet spectra, however, are not substantially different from those of wind-generated spray, particularly at elevations above $2\,\mathrm{m}$. Because of their life span, fogs are likely to be found in metastable equilibrium with the surrounding air and therefore exhibit a degree of supercooling indicated by the air temperature. Potentially, fog is a major source of marine icing, as is freezing rain, for which droplets are typically of the order of 1 mm and water contents may be high. Much the same comment applies to vessel-generated spray which is usually regarded as the principal source of accretion.

The characteristics of vessel-generated spray are difficult to prescribe in terms of the usual parameters: droplet temperature, size distribution and water content. This is largely because the spray is produced by a complex and highly-variable process. It is known, for example, that these parameters depend upon the state of the sea, the wind speed and the heading of the vessel, but it is not yet possible to use these indicators quantitatively in the determination of droplet spectra and trajectories. Consequently, it is not possible to follow the precise thermal history of a droplet from the moment of generation to the moment of impact, although it is possible to estimate the attendant water flux (Zakrzewski, 1986). As Table 5.4 indicates, vessel-generated spray possesses a very high water content and is composed of very large droplets, typical diameters being in the range of 1–3 mm. The second of these characteristics, coupled with the limited exposure of a droplet to subzero air temperatures, implies that shipboard accretion is frequently wet. The time of the droplet flight is not long enough nor the diameter small enough[15] to permit significant supercooling; consequently, the Stefan number upon impact is usually very small. Because of this, and the high water content, vessel-generated spray is often modelled as a continuous flux arising from a monodisperse cloud of droplets having the median volume diameter (i.e., the diameter at which half the water is contained in droplets larger than this) of the estimated or assumed distribution. Indeed, the use of this continuum model, in contrast to the approach to aircraft icing discussed in the previous section, is usually extended to most sprays formed over large bodies of water.

Although sprays originating in the marine environment may often be treated as a uniform water flux, the existence of discrete droplets may still be acknowledged through the collision efficiency concept applied to the median volume diameter. The absence of such a refinement implies that the local collision efficiency β would be determined completely by the geometry of the substrate or accretion: i.e., the spray would not then be affected by air flow around the surface on which impingement takes place. It is clear that the incorporation of the collision efficiency thus improves the continuum spray model, but analysis is usually limited to monodisperse characteristics.

The temperature of the spray flux may be estimated from the thermal behaviour of a single droplet (diameter d) in flight from water at temperature T_w through air at temperature T_a. When the droplet Biot number $Bi_d = \bar{h}d/k \ll 1$ ($\bar{h}$ is the overall transfer coefficient and includes both sensible and latent heat contributions), a simple heat balance requires that

$$\left(\frac{\rho_w c_{pw} d}{6}\right) \frac{d\theta}{dt} = -\bar{h}\theta$$

where $\theta = T - T_a$. Hence the droplet temperature is given by

$$\theta = \theta_w \exp\left(-6\bar{h}t/\rho_w c_{pw} d\right) \qquad (5.39)$$

where $\theta_w = T_w - T_a$. The principal difficulty in using this equation arises from current lack of knowledge of sea spray generating mechanisms; it is these which determine the droplet trajectories and the distribution of water flux (droplet size and concentration) with elevation. They also control the time of flight, the droplet velocity relative to the air, and hence the heat transfer coefficient. It is therefore difficult to make a general statement about the droplet temperatures determined through equation (5.39) except perhaps to note that sufficiently low air temperatures permit the generation of shipboard icing from seas which are well above their freezing point: for example, sea temperatures up to 6°C (Borisenkov and Pchelko, 1972).

To the extent that the collision efficiency distribution is considered important, it may be accommodated in the manner discussed in Sections 5.2 and 5.4. The energy balance in the accretion zone is always important because it controls the rate at which latent heat may be removed. For marine icing in general, and spray icing in particular, the energy balance is not unlike that for the atmospheric ice discussed in Section 5.3. Aerodynamic heating and droplet kinetic energy are not likely to be significant, and the principal contributions are therefore the sensible and latent heat losses to the air. The subcooled substrate (especially those mast, stay and deck areas not protected from the wind), which is obviously absent in naturally occurring atmospheric ice, modifies the balance by introducing a conductive flux through the ice.

In further contrast to aircraft icing, marine icing is characterized by the effect of gravity on the unfrozen water film and by the intermittency of the spray action. Although accretion is commonly modelled as a steady state process for purposes of comparison with controlled laboratory results (Horjen and Vefsnmo, 1983; Stallabrass, 1980; Lozowski and Gates, 1984), it would be more accurately described in general field terms as a cyclic process in which periods of quasi-steady spraying are alternated with intervals during which either:

(1) continued action of the wind freezes entrained water and creates subcooling of the accretion, especially near the surface, where it has the greatest effect on subsequent impingement; or

(2) the accretion is bathed in copious amounts of relatively warm water which, in the absence of the wind tends to ablate the ice surface, but which soon recedes to leave a thin film that may be quickly cooled and frozen.

Such spray-free processes are clearly different from accretion but they are not too difficult to describe in terms of heat and mass transfer. Their overall effect is to change the 'initial' conditions for each new burst of spray. In particular, they control the state and amount of runoff water and thereby influence the rate of accretion (Gates, 1985).

No mention has yet been made of the distinction between fresh water ice and salt water ice, and indeed most experimental work is done using fresh water, which is less corrosive and easier to provide. Fresh water experimental results, and the models used to interpret them, are directly applicable to vessels operating on the Great Lakes, for example, but in the marine context they must be modified to account for salt rejection during the freezing process. As noted in earlier chapters, the interfacial rejection of dissolved salts under equilibrium conditions will lower the freezing temperature of the remaining solution, thereby implying a reduced

freezing rate for given thermal conditions. Under typical spray icing conditions, the time scale of the freezing process, including post-impact
freezing of the water film, is small enough to prevent the attainment of
chemical equilibrium, but this fact alone is not enough to justify the neglect
of salinity effects. At the very least, saline solution will be trapped in
intergranular spaces. More significantly, salt water may be trapped in the
ice matrix which is nourished by a supercooled saline spray. This growth is
similar to the growth of spongy fresh water ice (Gates, Narten, Lozowski
and Makkonen, 1986; Makkonen 1986a) except that the retained unfrozen
water is no longer salt-free. As with pure water, this unfrozen fraction
increases the (spongy) ice growth rate but, unlike pure water, it lowers the
latent heat. On the other hand, the equilibrium freezing temperature of the
salt-enriched, retained water is lowered. These features evidently account
for the observed differences between fresh water and salt water accretions
(Stallabrass, 1980; Horjen, 1981; Horjen and Vefsnmo, 1983).

The discussion thus far has been focussed on shipboard icing generated
by vessel-generated spray; a similar discussion would apply to shipboard
icing of almost any origin provided the spray characteristics were specified.
A similar comment might be applied to icing on offshore structures but
these do exhibit some features worthy of additional discussion. The first,
and perhaps most obvious, point is the fixed location and the implications
that this may have in meteorological terms. Broadly speaking, the offshore
climate will reflect the presence of neighbouring land, especially if it forms
the edge of a continent. Coastal topography thus exerts an effect on the
prevailing winds, their temperature, speed and gustiness, and on the size
and shape of waves; all of these influence the character of fog and spray. In
particular, impact-generated spray on a fixed structure will be different
from that generated by a moving, deep sea vessel, for which heading is an
additional factor which not only alters the relative direction of the wind but
changes the character of the spray. This directionality tends to be fixed on
offshore structures, so long as wind shifts are limited, and may be completely absent in conditions where a structure such as a drilling rig is capable
of interacting with waves in an almost omni-directional manner. Spray
characteristics appropriate to such fixed structures have been investigated
by Itagaki (1984) who suggests the simple droplet distribution function

$$n(r) = A(U_\infty)/r^2$$

where r is the droplet radius and $A(U_\infty)$ is a coefficient dependent on wind
speed U_∞.

Another characteristic of offshore structures is their vertical slenderness.

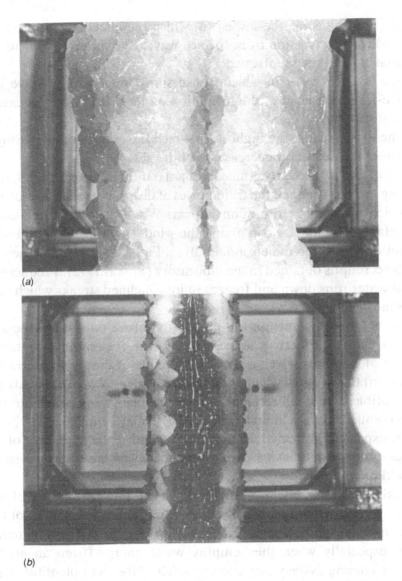

(a)

(b)

Fig. 5.25. Frontal view of icing on a vertical cylinder after 30 min, $T_a = -10\,°C$, $U_\infty = 22\,m\,s^{-1}$: (a) $r_d = 75\,\mu m$ (b) $r_d = 30\,\mu m$ (following Sroka (1972)).

This has two main implications: firstly, they typically extend to heights well above the water surface; secondly, with the exception of deck surfaces, the majority of the structural elements are either vertical or inclined. The variation of spray characteristics with elevation is thus an important factor in determining the overall distribution of ice. Three zones have been identified:

(1) the *ice-free zone*, the lowest, in which the effect of sensible heat and water advection from surface waves and the sea body ensure that no ice may be formed;

(2) the *wet ice zone*, in which dense spray immediately above the ice-free zone ensures the formation of dense ice bathed in excess water;

(3) the *dry ice zone*, the highest, in which the density of increasingly rime-like accretion decreases with height.

The effect of gravity is clearly noticeable, not only in influencing the spray-generating mechanisms and the trajectories of drops, but in the flow of any water which does not freeze upon impact. When the wind direction is fixed, a plane of symmetry containing the wind vector forms and divides the accretion into three-dimensional halves. Fig. 5.25 is a frontal view of two such accretions obtained in the laboratory (Sroka, 1972). Frequently, unfrozen water runs down and freezes in long inclined streaks which may develop into icicles.

Finally, a few comments on the prevention of ice on waterborne and offshore structures. Clearly, the best strategy for prevention is rooted in the design of elements which are not susceptible to drop collision in the first place but this is a goal which is difficult to achieve. Large elements are less susceptible than small elements, all other things being equal, but this guide evidently has two limitations: firstly, it may unduly restrict[16] the designer, especially for small vessels; and secondly, the *final* mass of accretion does not always have a simple dependency on the object's original size (Makkonen, 1984c).

Generally speaking, chemical methods, which rely on coatings of ice-phobic materials or fluid films of freezing point depressants, are not too effective, except in short term use. Thermal methods perhaps offer greater promise, especially when they employ waste energy (from an engine exhaust or cooling system), or energy which is free and plentiful, e.g., sensible heat from the ocean: in both instances the energy is low grade and must therefore be sought more for its sheer thermal capacity than for its temperature excess. Various arrangements of thermosyphons and heat pipes have been considered (Lock, 1972; Larkin and Dubuc, 1976; Okihara *et al.*, 1980).

5.6   Icing of land-based structures and equipment

The physics of icing on land-based structures and equipment is, of course, the same as that governing icing on airborne and waterborne structures; the discussions in Sections 5.1 and 5.2 are therefore germane to

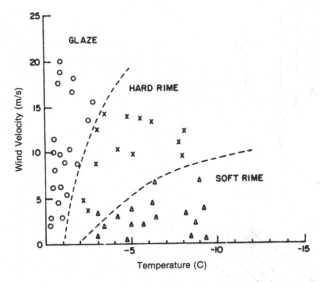

Fig. 5.26. Categories of in-cloud icing (following Kuroiwa (1965)).

this section also. Typically, icing on land-based structures originates in winds which are comparable with those at sea, or offshore, but not as strong as those encountered by aircraft. The water contents of air streams on land may vary considerably (at least in the range $10^{-1} - 10\,\mathrm{g\,m^{-3}}$) but, they are less likely to represent a serious threat when they are in the range typical of aircraft icing. Much the same comment applies to the droplet size distribution, although it must always be borne in mind that even small droplets moving with low velocities will eventually deposit substantial amounts of ice if the exposure time is long enough e.g., several days. Despite these similarities, it is worth mentioning certain differences of circumstance on land, where icing ranges from light frost on air-conditioning equipment to heavy accretion on overhead cables.

A principal difference arises from the fact that land-based structures are not often subjected to wave-generated spray: winds may be gusty but the process is not likely to be cyclic. Hazardous conditions are usually associated with air flows carrying thick fogs and dense clouds, although the effect of freezing rain, and its similarity with heavy sea spray, must not be overlooked. Shoreline structures such as piers, breakwaters and lighthouses, are perhaps better grouped with the offshore structures discussed in Section 5.5. Another significant aspect of land-based accretion is its occasional origin from clouds or fogs produced by other land-based structures and equipment: i.e., by man. Emissions from power plants, chemical plants, manufacturing plants and automobile exhausts, for example, may

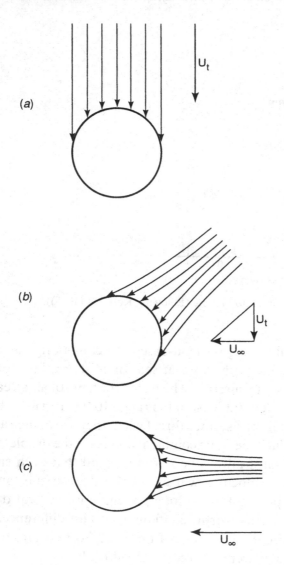

Fig. 5.27. Effect of gravity and wind on particle trajectories: (*a*) ignoring wind; (*b*) combining wind and gravity; (*c*) ignoring gravity.

produce an immediate local hazard from condensed water vapour directly, or they may become widespread sources of nucleants which seed an otherwise benign atmosphere elsewhere. Urban emissions, especially in metropolitan[17] areas, must therefore be considered as a potential source of icing.

The wide range of environmental conditions which give rise to ice on land-based structures is reflected in three different, but overlapping, categories of ice (Kuroiwa, 1965). Fig. 5.26 shows the somewhat arbitrary dividing lines between:

(1) *glaze ice*, which is hard, highly adhesive and very dense
   ($\rho_i \approx 900 \, kg \, m^{-3}$);
(2) *hard rime*, which is opaque, less adhesive and less dense
   ($\rho_i \approx 600 \, kg \, m^{-3}$);
(3) *soft rime*, which is even less adhesive and less dense
   ($\rho_i < 600 \, kg \, m^{-3}$).

In essence, these data confirm the expectation that smaller, cooler droplets tend to give a more porous and weaker type of accretion. They also reveal that the land icing hazard is limited to a temperature range of about[18] 0°C to −15°C, and is likely to be greatest in the range 0°C to −5°C when higher winds and larger droplets combine to produce glaze ice. This range of conditions also embraces accretion from sleet, wet snow and freezing rain but it is important to keep in mind that phenomenological details vary greatly. For example, the open, frozen structure of an immobilized snowflake would not be expected to lead to the same form, or rate, of accretion as the relatively dense, unfrozen water of a supercooled, but fluid, raindrop.

Freezing precipitation creates accretion processes which have yet to be fully modelled successfully. In particular, the effect of gravitation produces marked differences between vertical and horizontal surfaces; the relative icing characteristics depend not only upon the mean wind speed but upon gustiness. Chainé and Skeates (1974), for example, in studying the relationship between ice accumulation on a vertical surface δ_V, the wind speed U_∞, and ice accumulation on a horizontal surface δ_H, used the simple form[19] $\delta_V \propto \delta_H U_\infty$ to analyse 15 icing storms in which they found that δ_H is typically less than, but comparable with, δ_V. Such an approach, although somewhat arbitrary, does focus on precipitation as the primary source of accretion, treating the effect of the wind on icing over vertical surfaces as secondary. In Fig. 5.26, the two effects are combined while in Fig. 5.27 they are separated. Fig. 5.27(*a*) shows the trajectories of particles falling at their terminal velocity in the absence of a wind (the local collision efficiency is then fixed by the cylinder geometry as noted in Section 5.2), while Fig. 5.27(*c*) shows the trajectories when the effect of gravity is ignored. In reality these two effects are combined, as indicated in Fig. 5.27(*b*), to pose a collision efficiency problem and resulting wet icing problem which is a more general form of the accretion problem discussed in previous sections. Moreover, capillary effects may have to be considered in wet snow accretion.

In the absence of precipitation, accretion on land-based structures may be understood in terms equally applicable to aircraft and marine icing.

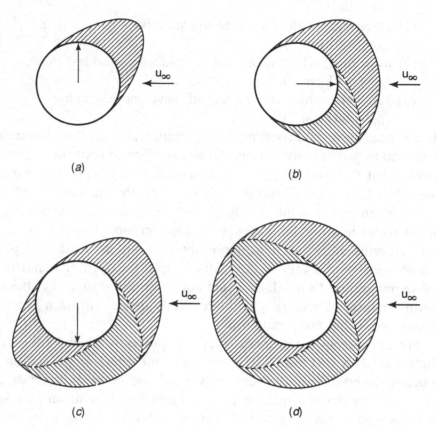

Fig. 5.28. Schematic representation of accretion build-up on a wire having torsional flexibility, showing four hypothetical stages of development from early (*a*) through intermediate (*b*) and (*c*) to final (*d*).

Kinetic and compressibility effects are, of course, much smaller, but the basic analysis still consists in predicting the catch of supercooled droplets and then estimating the rate at which they freeze in the presence of convective cooling. Situations in which the icing occurs, and the adverse effects which it may produce, are many and varied. Major categories include the damage to trees and crops, the collapse of electrical transmission and communication towers, and the loss of service on roads, railroads and runways. A very special problem, and one which has received much attention over the years, is the icing of overhead cables.

The overhead cable bears an obvious resemblance to the horizontal cylinder and therefore much of the earlier discussion of cylinder icing applies. Even so, the initial catch-freeze mechanism must be regarded merely as a starting point in the wide range of circumstances which prevail.

Gravity, in particular may play a crucial role. For example, over and above the effect of gravity in determining the catenary hanging profile, and thus the variation of cable slope between the end supports, there will also be its effect when precipitation is combined with a wind. In the absence of a wind, freezing rain produces a tailed form of accretion, symmetric about the vertical plane through the cable. As noted above, however, the action of the wind is to destroy this symmetry by skewing the bulk of the accretion towards the wind. Gravity then acts on this skewed distribution and generates a twisting moment along the cable, thus altering the angle of attack of the accretion as a result of its own weight and distribution along the cable. The aerothermodynamic coupling of the accretion is also coupled to the elastic and aeroelastic characteristics of the cable-support system (Davenport, 1986).

Fig. 5.28 provides a schematic illustration of the gradual progress of ice growth on a cable with torsional flexibility. The early distribution (Fig. 5.28(*a*)) arising from a given combination of wind and precipitation causes clockwise rotation of the cable thereby exposing an altered surface to the impacting drops. If the cable is flexible enough, this process would continue (Fig. 5.28(*b*)–(*d*)) until, theoretically, the cable had rotated once and thus experienced direct icing over its entire surface. However, the gradual evening out of the ice distribution circumferentially reduces the torsion and therefore tends to correct the situation which gave rise to it in the first place (the eccentric accretion also creates an aerodynamic moment). A variation of this rotational theme may occur without torsion of the cable when the adhesive strength of the accretion is particularly low; i.e., when the ambient temperature approaches 0 °C. If the accretion were to extend less than half way round the cable an adhesive break would simply allow the ice to fall off, but a more extensive accretion, such as those illustrated in Fig. 5.28(*b*) and (*c*), would slip round the wire, following the tendency to bring the centre of gravity of the ice beneath the cable centre. One-sided build up and rotation thus combine to alter the accretion shape dynamically, and often erratically.

The icing of moving structures on land, and of vehicles in particular, presents no general problem which cannot be approached along the lines suggested elsewhere in this chapter. Perhaps the only problems worthy of a separate discussion are those associated with prime movers, and with their induction systems in particular. Carburettor icing, for example, has been recognized as a potential hazard since the early decades of this century, while icing damage to gas turbines has been noted since the Second World War. Coles, Rollin and Mulholland (1950) distinguished three types of icing in reciprocating engine induction systems:

(1) *impact icing*, which is attributed to the accretion of ingested super-cooled drops;
(2) *throttle icing*, which results from vapour condensation and freezing caused by expansion of air through passage constrictions;
(3) *evaporative icing*, which results from vapour condensation and freezing caused by evaporation of volatile fuel.

(The freezing of water drops in fuel lines may also be important but is not a form of atmospheric icing.) Of these, the second and third, and especially the latter, are considered to be the principal causes of carburettor icing. Their severity may be reduced substantially through careful attention to pressure variations, the addition of surface coatings and the use of fuel additives (Gardner and Moon, 1970). On the other hand, gas turbines tend to be susceptible to the first and second of the above three, and especially to the former (Chappell, 1972; Lacey, 1973). Fine mesh inlet screens having high collection efficiencies provide some protection, but they also produce accretion which wind loading and machine vibration may detach, thus injecting into the turbine a variety of solid fragments which may create mechanical damage. Further downstream, accretion on inlet guide vanes and first stage compressor blades not only alters the performance of the flow-sensitive axial compressor, but provides yet another potential source of damaging ice fragments.

Whether engine induction systems should be treated as structures or as pieces of equipment is a moot point. However, there is no doubt in the classification of many other pieces of equipment which have a perform-ance either determined by, or compromised by, the formation of atmos-pheric icing. The high-pressure water jet issuing into a subzero environ-ment provides examples in both categories. As a high volume spray it may be used to cover horizontal surfaces, ranging from a skating rink to a drilling rig pad, with drops supercooled in flight; the purpose is to form an ice slab. As a fire hose, on the other hand, the supercooling may be an insignificant aspect of the main function, but it inevitably aggravates the problem of subsequent freezing. Indeed the operation of fire-fighting equipment under severe winter conditions appears to be one of the least studied despite, or perhaps because of, its complexity. Potential icing problems encountered include: frazil generation during pipe filling, spray icing and the freezing of water films.

By way of contrast, frost and ice formation in air-conditioning and cryogenic equipment have received a great deal of attention (Smith, Edmonds, Brentari and Richards, 1964; O'Neal and Tree, 1984). Most engineering studies have been concerned with frost-induced changes in thermal resistance and have therefore attempted to develop empirical correlations for the thickness and density of the frost; not much attention has been paid to the details of deposition, although these often provide useful insights. Vapour enters the frost layer at its outer boundary which it shares with the environment, either depositing directly on crystal tips located at the boundary or penetrating deeper into the frost matrix and depositing there. From the discussion in Section 5.1 it appears that deposition within the matrix will be much slower than that taking place on the tips. Therefore, once the period of initial growth has passed, and the frost has become an established matrix, the overall deposition process may be represented approximately by an idealized surface process very similar to that which accompanies the stable freezing of water, i.e., the situation may be described as a classical Stefan problem.

The energy balance at a frost surface growing in the X-direction is given by

$$k_i \left(\frac{\partial T_i}{\partial X}\right)_I = \bar{\rho}_i \lambda_{iv} \frac{d\hat{X}}{dt} + k_a \left(\frac{\partial T_a}{\partial X}\right)_I$$

in which $\hat{X}$ is the frost thickness, $\bar{\rho}_i = \rho_i (1 - \mathcal{N}_I)$ and $\mathcal{N}_I$ is the matrix porosity at the frost surface (the role of air in the pores is neglected). Using the normalized variables

$$\tau = \frac{t}{t_c}, \; x = \frac{X}{X_c}, \; \xi = \frac{\hat{X}}{X_c}, \; \phi_i = \frac{T_i - T_s}{\theta_i}, \; \phi_a = \frac{T_a - T_\infty}{\theta_a}$$

where T_s is the substrate temperature, the interface equation may be written

$$\left(\frac{\partial \phi_i}{\partial x}\right)_I = \frac{d\xi}{d\tau} + \frac{k_a \theta_a}{k_i \theta_i} \left(\frac{\partial \phi_a}{\partial x}\right)_I \tag{5.40}$$

by pre-supposing that the latent heat is removed by conduction through the frost matrix to the substrate i.e., $t_c = O(\bar{\rho}\lambda_{iv}X_c^2/k_i\theta_i)$. Under this condition, the relative importance of heat transfer from the environment is evidently given by $k_a\theta_a/k_i\theta_i$ compared with unity. Given that $k_a \ll k_i$, it follows that thermal convection is not too important unless $\theta_a \gg \theta_i$, a feature which is likely to be restricted to a short, initial period when the matrix model fails. Otherwise, the fact that $k_a\theta_a/k_i\theta_i \ll 1$ implies that the frost growth rate will not be very sensitive to external conditions in the air stream, in which case the approximate solution to equation (5.40) is the simple Stefan solution[20] for which $\hat{X} \propto t^{\frac{1}{2}}$.

This relation has been supported by experiment (Schneider, 1978; Cremers and Mehra, 1982). Re-writing the convective heat flux as $k_a(\partial T_a/\partial X)_I = h_Q\theta_a$, the second term on the right hand side of equation (5.40) becomes $(h_Q X_c/k_i)(\theta_a/\theta_i) = Bi\,\theta_a/\theta_i$ in which the Biot number in turn may be written as $Bi = Nu_D(\delta/D)(k_a/k_i)$, where D is a representative substrate dimension and δ replaces X_c as a measure of frost thickness. Thus, convection may be ignored when

$$Nu_D \frac{\delta}{D} \frac{k_a}{k_i} \frac{\theta_a}{\theta_i} \ll 1$$

which makes clear that the effect of Rayleigh number (for free convection) or Reynolds number (for forced convection) will usually be small even though they may vary over a wide range.

As noted in Section 5.1, the continued growth of a frost layer slowly increases its thermal resistance and thus tends to increase the surface temperature T_I as the growth rate decreases, quasi-steadily. While not an invariable occurrence, this trend may lead to substantial changes. The relative importance of latent heat will gradually diminish until, eventually, equation (5.40) reduces to a balance between convection and conduction, the latter by then being very much smaller. This limiting condition yields the maximum frost thickness δ_m as

$$\delta_m = \frac{\bar{k}(T_{If} - T_s)}{h_Q(T_\infty - T_{If})}$$

where $\bar{k}$ is the mean thermal conductivity and T_{If} is the final surface temperature.

Should T_{If} reach as high as the triple point temperature, a further complication arises. Condensation then begins to take place over the frost surface and the water thus produced enters the frost matrix under the influence of capillary action. As mentioned in Section 5.1, this will increase the effective thermal conductivity $\bar{k}$, thereby reducing T_I temporarily. Equally important is the effect of uneven convection over the frost surface. In boundary layer flow over a flat surface, for example, the heat transfer coefficient decreases with distance from the leading edge which is therefore the point where T_I would reach the triple point temperature first. Water formed upstream may then be blown over the frost surface in the form of a wave which re-freezes further downstream (Chung and Bywater, 1984).

6

Ice and Earth

Ice in and on the Earth takes a variety of forms ranging from a light, shortlived snow cover to a dense belt of permafrost, from pingos to patterned ground and from a small naled to an immense ice sheet. The origins of these forms, and their variations over short or long periods of time, depend to a large extent on the details of thermal history. In turn, these details are the consequence of microclimate, in the short term, and global (or regional) climate in the long term; and of course climate also embraces the movement of moisture and hence determines the distribution and form of precipitation.

This chapter examines the thermal energy balance at the Earth's surface and explores the consequences of disturbing that balance. To begin with, events are treated on a large scale, both spatially and temporally, in order to discuss the large bodies of ice which exist nearer the poles and at higher altitudes. Attention is then turned to the microphysics of water and ice as occupants of the pores of soil and rock. Based on this outline of fundamentals is the final section which deals with the processes of freezing in soils treated as continua. The context is typically geophysical and geotechnical, and the entire study may therefore be considered as an aspect of *geoglaciology*.

6.1   Surface energy exchange

It was noted in Chapter 1 that glaciation was closely associated with periodicities on an astronomical time scale. In particular, precession of the Earth's axis and changes in the eccentricity of the Earth's orbit were identified as major factors controlling the solar energy flux reaching the polar regions. A decrease of less than 2% in the solar input would cause the global temperature to drop well below 0 °C (Schneider and Gal-Chen, 1973), largely because of the positive feedback produced by the albedo of the increased snow cover. Once the snow cover begins to extend with decreasing global temperature (and presumably corresponding alterations in precipitation patterns), its high albedo causes tropospheric cooling which tends to discourage sublimation and melting while promoting further snowfall (Flohn, 1974; Williams, 1975): the effect of the snow cover is thus reinforced by more snow and a longer period of coverage.

The details of this mechanism are not fully understood, and perhaps this is not surprising, given the complexity of heat and mass transfer processes in the atmosphere, particularly on a global scale. None the less, it is clear that the mechanism is triggered and driven by astronomical events and is therefore likely to be of long duration. Even a small effect may ultimately account for global causes of glaciation. Near the poles themselves, and particularly in the Antarctic, the process evidently continues in a reduced fashion during interglacial periods and is therefore the principal cause of the ice sheets, covers and shelves which currently exist in those regions.

Seasonal changes in climate on a regional or local scale also affect the occurrence and growth of ice, though in a much more direct way: i.e. through the effect of subzero temperatures on water already present on or in the ground. With a mean global surface temperature near $280\,K$, freezing temperatures are the result of local alterations in the surface energy exchange, specifically in the thermal energy flux balance. These fluxes fall into two broad groups: radiation and convection, each of which may be either upward or downward. The magnitude and sign of the net thermal flux is dependent on many factors which are perhaps understood following a brief discussion of topography and vegetation.

Topography influences the exchange in two ways: through the effect of elevation and of slope. As a general rule, an increase in elevation implies a lower mean air temperature. However, at any given altitude there may be significant temperature variations from latitude to latitude and from season to season; there may also be substantial differences between night and day. On the slopes of Mount Kenya, for example, the air temperature fluctuates daily in the range $-10\,^{\circ}C$ to $+10\,^{\circ}C$, thus causing the alpine plant *Lobelia teleki* to freeze and thaw diurnally (Krog, Zachariassen, Larsen and Smidsrod, 1979). The effect of slope is felt mainly during the day, when a south-facing slope receives greater insolation than a north-facing slope (in the northern hemisphere; the opposite is true in the southern hemisphere); there is, of cause, a gradation of effect with latitude. The inclination of the slope in relation to the elevation of the Sun determines the magnitude of the effect which therefore alters not only with latitude, but with season and time of day. Topography also has an indirect effect on the surface heat balance by influencing the type and extent of vegetation, and by creating or altering local wind patterns. Local depressions of a few metres, for example, trap cold night air, thus producing frost hollows in which temperature gradients may be significant.

The type of vegetation, or lack of it, depends largely upon the local microclimate which in turn depends upon the regional and global climates. For reasons discussed in Chapter 7, the scale and density of vegetation generally decrease with a decreasing mean annual air temperature. The trend continues thus with increasing latitude[1], leading eventually to the tree line beyond which the vegetative cover declines from brush to lichen to organisms which barely function as vegetation, at least from a micro-climatological point of view. Here again, variations may be substantial, largely because of the presence and role of water: in the bulk form of an ocean or lake; in the flowing form of an aquifer, river or estuary; or in the disperse atmospheric form of vapour or precipitation. Water is obviously essential to the survival of plants which therefore grow in response not only to its abundance but its mobility (inside and outside of the organism) in the presence of subzero atmospheric temperatures. Vegetation and the ice regime are thus interconnected through the existence of unfrozen water which varies greatly in both amount and form between a coastal delta and the landlocked tundra.

More generally, vegetation is an active element in the surface heat balance. It affects the radiation exchange through the processes of absorp-tion, reflection, transmission and emission; and it affects the convective exchange through both the process of evapotranspiration and alterations in the turbulent wind profile. Many types of northern vegetation tend to collect and rely upon moisture in the form of snow. The snow cover itself, as it changes in ability to transmit heat and moisture while undergoing metamorphosis, becomes an additional active element affecting the sur-face heat balance (Berry, 1981).

Overall, the heat flux $\dot{Q}$ at the Earth's surface may be described in its simplest terms by

$$\dot{Q} = \dot{Q}_R + \dot{Q}_C \tag{6.1}$$

where $\dot{Q}_R$ is the net radiative flux and $\dot{Q}_C$ is the net convective flux, both measured positively downward (in the negative Z-direction). This expres-sion may be used to undertake an elementary study of the freezing regime created in subsurface materials. For example, a downward heat flux rep-resents a thermal energy gain, the rate of which is given by

$$\dot{q} = k(\partial T / \partial \bar{Z})_s \tag{6.2}$$

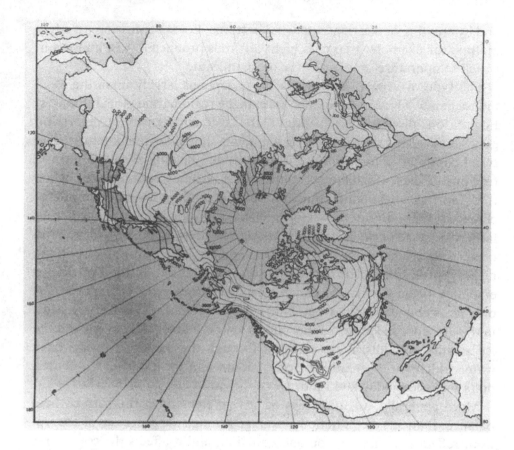

Fig. 6.1. Freezing indices (in °F-days) in the northern hemisphere (following Washburn (1973)).

where $\dot{q}$, the heat flux density, is $\dot{Q}$ per unit surface area, k is the Earth's thermal conductivity and $(\partial T/\partial Z)_s$ is the vertical temperature gradient in the Earth at its surface. Thus, when the surface temperature falls *below* the freezing point the corresponding heat *loss* causes subsurface freezing, or 'frost penetration' of depth Z, at a rate given approximately by

$$\frac{\mathrm{d}\hat{z}}{\mathrm{d}t} = -\frac{k}{m_i \lambda_{iw}} \left(\frac{\partial T}{\partial Z} \right)_{\hat{z}}$$

Integrating, the 'frost' depth may be written as

$$\hat{Z}(t) = \frac{k}{m_i \lambda_{iw}} \int_0^t -\left(\frac{\partial T}{\partial Z} \right)_{\hat{z}} \mathrm{d}t \tag{6.3}$$

in which the subsurface ice content m_i (the mass of ice per unit volume), the latent heat of fusion λ_{iw} and the thermal conductivity are all taken as constants, and $(\partial T/\partial Z)_{\hat{z}} \approx (\partial T/\partial Z)_s$.

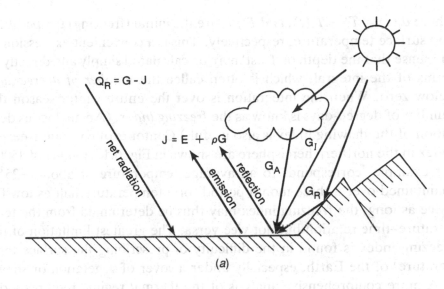

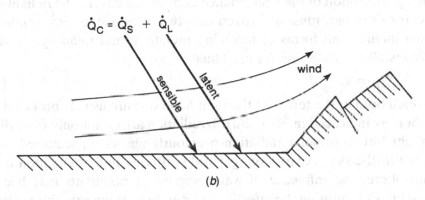

Fig. 6.2. Surface energy fluxes: (*a*) radiation components; (*b*) convective components.

For simplicity, the initial subsurface temperature is often taken to be uniform at the equilibrium freezing point. Under such a condition it is evident from equations (6.1)–(6.3) that a constant rate of heat loss corresponds to a 'frost' depth which is proportional to the time elapsed. More generally, the depth will be some function of time, as determined by the heat balance in equation (6.1). It will be recalled from Chapter 3 that the Stefan solution, in which the temperature gradient is taken to be $-\theta_s/\hat{Z}$, is represented by

$$\hat{Z}^2 = \frac{2k}{m_i \lambda_{iw}} \int_0^t \theta_s(t) \, dt \qquad (6.4)$$

where $\theta_s(t) = T^i - T_s(t)$, and T^i, T_s are the initial (freezing) temperature and surface temperature, respectively. This is a convenient expression in the sense that the depth of 'frost' may be calculated simply and directly in terms of the integral, which is often called the *number of degree-days* below zero. When the integration is over the entire winter season the number of degree-days is know as the *freezing index*. (An analogous definition of the thawing index is also useful.) Contours of constant freezing index in the northern hemisphere are shown in Fig. 6.1. A value of 4500^2, for example, corresponds to an average temperature of about $-25\,°C$ maintained over a three month period, or a temperature half as low for twice as long; the freezing index may thus be determined from the temperature–time relation but not vice versa. The greatest limitation of the freezing index is found in the difficulty of prescribing the surface temperature[3] of the Earth, especially under a cover of vegetation or snow.

A more comprehensive analysis of the thermal regime requires a detailed prescription of the heat balance and surface cover. The radiant heat flux, for example, must be broken into two components: *irradiation*, G, which includes all forms of incoming radiation; and *radiosity*, J, which includes all the outgoing forms. Thus

$$\dot{Q}_R = G - J \tag{6.5}$$

in which both of the terms on the right hand side are usually broken down further, as indicated in Fig 6.2(*a*). Irradiation arises not only from direct sunlight but from solar radiation previously absorbed, scattered or reflected in the sky. Apart from general scattering and attenuation in the atmosphere, the influence of water vapour or pollutants may become crucial in determining the effective temperature of the sky; this is especially true in the presence of fog or cloud cover which can virtually eliminate the direct irradiative flux. Such effects are also important when the sun is low in the sky during winter or, at other times of the year, near the beginning or end of the day. Topography may also alter radiation in a number of ways. Significant variations in elevation in some areas may place neighbouring areas in shade during sunrise, for example, and thus eliminate all but indirect irradiation. However, this early morning deficit may be partly offset later in the morning when direct insolation is augmented with radiation reflected from the elevated slopes.

In more precise terms, the irradiation may be written:

$$G = G_I + G_A + G_R \tag{6.6}$$

where G_I, the insolation, is that part of the sun's incident[4] radiation which is transmitted to the Earth's surface directly[5] or diffusively: it is neither

Table 6.1. *Albedo of various surfaces reflecting solar radiation (from Geiger (1965) and Berry (1981))*

Surface	Albedo (%)
Fresh snow cover	75–95
Dense cloud cover	60–90
Old snow cover	40–70
Clean firm snow	50–65
Clean glacier ice/sea ice	30–46
Desert	24–30
Bare fields/tundra	12–25
Forests	5–20
Dark, cultivated soil	7–10
Water[a]	3–10

[a] Excluding large angles of incidence.

absorbed nor reflected; G_A is irradiation coming from the atmosphere by virtue of the absorption and re-emission of that fraction of the Sun's incident radiation which is neither reflected nor transmitted; G_R is the irradiative flux which has been reflected off other surfaces.

The components on the right hand side of equation (6.6) begin as part of the Sun's incident radiation but they are significantly different when they arrive at the Earth's surface (Geiger, 1965). G_I is from the solar[6] spectrum of 'short' wavelengths grouped around the visible spectrum i.e., between $0.3 \mu m$ and $3 \mu m$, and typically about $0.5 \mu m$; whether this radiation is transmitted directly, or is scattered, it remains in this wavelength band. G_A, on the other hand, has been first absorbed and then emitted by the various gases, vapours and particles of the atmosphere which, having a temperature of the order of 273 K, radiate energy with 'long' wavelengths i.e., between $5 \mu m$ and $100 \mu m$, and typically of the order of $10 \mu m$. G_R will, in general, be a mixture of short and long wavelengths, depending on the characteristics of the reflecting surfaces. Almost all of G_A is absorbed at the Earth's surface, while G_I and G_R are reflected in varying degrees. Table 6.1 lists typical albedo (reflectance) ranges for various surfaces, clearly revealing the significant distinction between frozen and unfrozen water covers: such data are at the root of the glaciation feedback mechanism mentioned earlier. It is also worth noting that shorter wavelengths have a higher penetrating power, and this is another distinction which is important to the understanding of the radiation balance because it implies that short wave irradiation may have a volumetric heating effect while the longer waves are limited to surface heating.

The radiosity term in equation (6.5) subdivides into two components: emission and reflection, as illustrated in Fig. 6.2(*a*). Thus

$$J = E + \rho G \tag{6.7}$$

in which $E = eA$, is the emission from a surface of area A, and ρ is the reflectance. Long wave emission from the Earth's surface, like that from the atmosphere, is governed by the Stefan–Boltzmann expression for diffuse radiation:

$$e = \epsilon e_b = \epsilon \sigma T^4 \tag{6.8}$$

where e_b is the emission rate (per unit area) for a black body (i.e., a perfectly absorbing body), σ is the Stefan–Boltzmann constant ($5.669 \times 10^{-8} \ \mathrm{W\,m^{-2}\,K^{-4}}$), and T is the absolute temperature of the emitting material. The quantity ϵ, the emittance or emissivity is a surface property with magnitudes lying between 0 and 1.0. For the Earth as a whole, $\epsilon \approx 1$, as may be demonstrated by combining the reflectance ρ, the absorptance α and the transmittance τ in the conservation expression

$$\rho + \alpha + \tau = 1$$

to which Kirchoff's law $\alpha = \epsilon$ (strictly, $\alpha_\lambda = \epsilon_\lambda$ for each wavelength) may be applied to yield

$$\epsilon = 1 - \tau - \rho.$$

Since the Earth as a whole is opaque ($\tau = 0$), and ρ is typically less than 0.05 (Geiger, 1965), it is evident that the Earth radiates much like a black, diffuse surface, as implied in the simple calculation performed in Chapter 1. The individual emittances of water and snow are both about 0.97 (Ashton, 1986), thus leading to the general approximation that $J = E_b$ when equations (6.7) and (6.8) are applied to long wave radiation at the Earth's surface; however, this is not generally true for short wave radiation, as Table 6.1 reveals.

A representative month-by-month radiation balance, measured at Hamburg (Fleischer, 1953/4), is shown in Fig. 6.3 for conditions in which $G_R = 0$. From equation (6.5), the net radiation gain at the Earth's surface is given by

$$Q_R = (G_I + G_A) - (E_b + \rho_I G_I)$$

where ρ_I is the short wave reflectance and ρ_A, the long wave reflectance, is taken to be very small, thus implying that $E \approx E_b$. More precisely, the long wave contribution to the radiosity is $\epsilon E_b + \rho_A G_A$ but, since $\epsilon = 1 - \rho_A$, it is also given by $E_b + \rho_A(G_A - E_b) \approx E_b$ because $G_A \approx E_b$ and $\rho_A \ll 1$. If this balance is re-written as

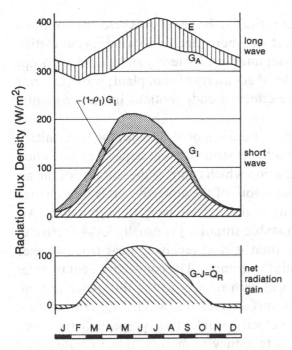

Fig. 6.3. Radiation balance measured at Hamburg, West Germany (following Fleischer (1953/4)).

$$Q_R = (G_A - E_b) + (1 - \rho_I)G_I \qquad (6.9)$$

it is seen to subdivide into a long wave balance $(G_A - E_b)$ and a short wave balance $(1 - \rho_I)G_I$, as illustrated in Fig. 6.3. At this particular latitude, the net short wave gain more than compensates for the net long wave loss. The net radiation is therefore positive except during the winter months.

Before leaving the radiation balance, two important observations must be made on the data presented in Fig. 6.3. Firstly, the measurements are monthly averages and therefore mask daily and weekly fluctuations. At night, for example, $G_I \rightarrow 0$, and hence the instantaneous balance is tilted in favour of a substantial net radiation loss. Secondly, it is clear that if the albedo ρ_I increases significantly, the net gain may once more be turned into a loss, as noted earlier.

The make up of the convective flux $\dot{Q}_C$ in equation (6.1) is also complex but it too may conveniently be divided into two separate though inter-related effects: convective *heat* transfer and convective *mass* transfer, both of which are represented schematically in Fig. 6.2(b). The former corresponds to a sensible heat flux $\dot{Q}_S$ and the latter entails a latent heat flux $\dot{Q}_L$. Convective heat transfer is the result of air movement close to the ground and therefore depends upon the temperature and velocity distribution in that region. Theoretically, the heat flux is given by Fourier's law applied at the air/earth interface where there is no wind speed relative to the interface itself, but such an interface is merely an idealized concept which takes no account of the local geometry of soil, plant, wave, etc; nor does it include the penetrating effect of eddy motions in the frequently turbulent air.

The wind may have arisen out of the regional climate, the local microclimate, or a combination of both. Its structure will depend in part upon the vertical temperature distribution which controls the stability of the air mass and thereby influences the vigour of the advective eddy interchange through alterations in both the scale and intensity of turbulence. Air flowing over cooler ground (a stable situation) is usually less effective at transferring heat to the earth than it is at receiving heat from warmer ground when the flow is potentially unstable. The convective heat transfer rate is therefore not symmetrical with respect to the difference in temperature between the Earth's surface and the wind. The temperature profile is often coupled to the velocity profile; this is especially true over inclined surfaces, which have a tendency to modify, if not create, local winds. For example, the interplay of slope and valley winds throughout the day (Defant, 1951) illustrates the complexity of mountain winds; the interaction of a cold and stable glacier wind (which may be only a few tens of metres deep) with a summer upvalley wind provides another example (Hoinkes, 1954).

Vegetation and snow cover may also play a significant role in the determination of convective heat and mass transfer rates. This is partly because of the highly variable surface roughness they present: the roughness parameter[7] varies from 0.01 cm, for snow over short grass, to 10 cm for a forest (Geiger, 1965); representative values of Z_0 for snow covers may be expected to fall in the range 0.1–1 cm. The role of moisture in the air is crucial where the temperature profile induces a humidity profile which causes moisture to flow towards the region of lowest temperature. Moisture flow thus enters into the advective exchange process and becomes especially important when a phase change occurs. Deposition on the surface of (or within) the snowpack, evaporation from leaves and condensation on grass, all represent moisture transfer processes in which latent heat is either liberated or absorbed, and since this latent heat is usually much larger[8] than the accompanying sensible heat it follows that the contribution of $\dot{Q}_L$ to the overall convective flux will then be significant. Precipitation also makes a contribution through the provision of new material at the Earth's surface, but such an advective flux of sensible heat may be

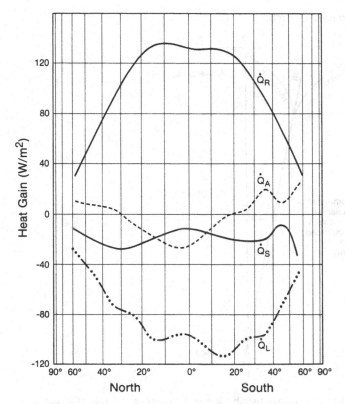

Fig. 6.4. Global heat balance (following Budyko (1958)).

much less important than the subsequent release or absorption of latent heat when the precipitation undergoes a phase change: a freezing raindrop tends to produce a warming effect while an evaporating raindrop produces a chilling effect.

Fig. 6.4 shows the sensible and latent heat flux contributions to the overall heat balance of the Earth on a latitudinal basis (Budyko, 1958). Also shown is the net radiation flux which may only be made to balance the net convective flux if an advective flux $\dot{Q}_A$ towards the poles is taken into account[9]; otherwise, the excess near the equator and the deficiency near each pole would cause the temperature difference between the equatorial and polar regions to continue to increase, the poles getting colder and the equator getting hotter. It is evident from Fig. 6.4 that $\dot{Q}_R$ and $\dot{Q}_C$ are in balance at 30°N and 20°S, and are roughly equal and opposite elsewhere. As a rule of thumb, the principal radiation gain, the short wave irradiation $(1 - \rho_1)G_1$, is roughly balanced by the principal convective loss, the latent heat flux $\dot{Q}_L$. These two major components are each much greater than the net overall heat flux $\dot{Q}$, which must have an average value close to zero. It

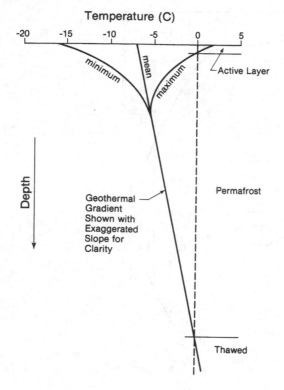

Fig. 6.5. Ground temperature distribution in a northern latitude (following Brown (1970)).

therefore follows that a small percentage change in either of the major components will have a much larger effect on $\dot{Q}$ which, as noted earlier, determines the thermal regime near the surface of the Earth. Alterations to the vegetation cover, for example, cause changes to evapotranspiration and advective exchange patterns, thus altering $\dot{Q}_L$. Such alterations may also modify ρ_I, as would the appearance or disappearance of ice or snow. Airborne pollutants and water vapour, especially in the form of fog or cloud cover, change the transmissive and scattering properties of the atmosphere thereby affecting $\dot{Q}_R$ by altering G_I. In each of these modifying events, changes in the radiation or convective fluxes have a leveraged effect on the thermal regime.

Over the long term, the temperature distribution below the Earth's surface in higher latitudes may be described by the familiar 'trumpet' curve shown in Fig. 6.5 (Brown, 1970). At great depth, the mean temperature is a fixed curve reflecting the geothermal gradient, but near the surface the precise shape varies from month to month, always lying somewhere within the limiting envelope shown. The mean annual surface temperature de-

Table 6.2. *Sensitivity of ground temperature T_G at depth to various parametric changes (following Gilpin and Wong (1973)).*

Input parameter	Influence coefficient	Input range $\pm(\%)$	T_G $(\pm\,^\circ C)$
$\Delta G_I/G_I$	0.030	30	0.9
$\Delta\rho$	−0.016	10	0.2
$\Delta\epsilon$	−0.029	5	0.2
ΔRH	0.032	10	0.3
ΔCL	0.023	10	0.2
$\Delta U/U$	−0.042	30	1.3
$\Delta K_S/K_S$	0.036	100	3.6
$\Delta K_W/K_S$	−0.056	100	5.6

RH – relative humidity CL – cloud cover U – wind velocity
K_S – surface cover conductance (summer)
K_W – surface cover conductance (winter)

creases with increasing latitude. The flare of the trumpet is determined from the microclimatic fluctuations discussed above, and typically decays to zero within a depth of ten or so metres.

From the analysis of Chapter 3 it is evident that, for a given *surface* temperature excursion above and below the freezing point, freezing will penetrate much further than thawing because the thermal properties of water are quite different from those of ice. This could lead to the continuing aggradation of permafrost in latitudes where the mean surface temperature was near 0 °C. However, the insulating effect of the snow cover reduces the impact of subfreezing *air* temperatures by introducing a heat valve effect in which winter cold is less effective than summer heat in altering the surface temperature; the net result of this heat valve is a mean ground temperature often 3–4 °C higher than the corresponding mean air temperature (Brown, 1970), though still decreasing with latitude.

Precise predictions of the effect of surface heat exchange on ground temperatures, either at the surface or at depth, require precise specification of all of the factors entering into the heat balance. Since these vary greatly from location to location, it often becomes necessary to adopt either of two strategies. The surface heat flux data, compiled from meteorological records, may be incorporated into a numerical model of the heat balance which may then be computed at will (Outcalt and Carlson, 1975). The net heat flux thus obtained provides the necessary input for a numerical analysis of ground freezing, as discussed at length in Chapter 3. Alternatively, it may be simpler to use the results of a

Table 6.3. *Densities of snow and ice (following Paterson (1981))*

	Density (kg m⁻³)
New snow (immediately after falling in calm)	50–70
Damp new snow	100–200
Settled snow	200–300
Depth hoar	100–300
Wind packed snow	350–400
Firn	400–830
Very wet snow and firn	700–800
Glacier ice	830–910

sensitivity analysis. Gilpin and Wong (1973), for example, have obtained a representative set of sensitivity estimates for Norman Wells, Canada. These are shown in Table 6.2 which clearly reveals the dramatic effect of alterations in the vegetation or snow cover. A sensitivity analysis not only provides insight and perspective but often permits the development of realistic, though simplified, models which may then be used to analyse the thermal regime in detail. Such an analysis may, for example, be applied to the heat valve effect (Gilpin and Wong, 1976).

6.2   Freezing on the Earth

This planet, unlike its neighbour Mars and several of the moons of Jupiter, holds most of its ice above, rather than below, the surface. During the repeated periods of glaciation which have taken place over the past three million years, land ice has covered up to 30% of the Earth's land surface. In the current interglacial period the figure is closer to 10%. None the less, this mass of ice (which is shared largely between Antarctica and Greenland in the rough proportions of 90% and 10%, respectively) contains about 70% of the Earth's fresh water. It consists mostly of ice sheets, ice caps, ice shelves and glaciers; added to these are minor components such as naleds and needle ice. (Naleds and needle ice will be discussed in detail later. The former results from various types of winter flooding, while the latter may occur when water exudes from a porous surface in winter.) All of the major land ice forms are associated in some way with snow, and in typical circumstances it is falling and blowing snow which provides the raw material from which they are built. This section therefore begins with a discussion of snow, from the moment it gently alights on a surface to the time, perhaps millennia later, when it has become an integral part of a massive layer of solid ice.

Table 6.3 contains a list of typical densities for snow at various stages of transformation into ice. It is worth mentioning that the term *snow* is usually reserved for falling, blowing or newly-deposited snow crystals; *ice* is used when the density is greater than about $830\,kg\,m^{-3}$. Between snow and ice the material is usually described as *firn* (from a German word meaning last year's snow), but it must always be borne in mind that this term applies not to a single type of material but to one which may be at any point in a complex and gradual transformation. As Table 6.3 indicates, this transformation occurs over a density range of $400–830\,kg\,m^{-3}$.

The density of newly-deposited snow depends upon the circumstances of deposition. In a cold calm, for example, freshly fallen snow will tend to be more porous than at temperatures nearer the melting point or when driven by the wind into drifts. Similarly, the opportunity to settle may increase the packing density. These factors alter the porosity and thereby influence the permeability; the ability of snow to transmit water or water vapour through its pores is therefore highly variable. Vapour fluxes are created whenever there is a temperature gradient in the snow: this gradient generates a vapour pressure gradient in the same direction, and thus leads to a vapour flux which is in the same direction as the heat flux. For example, late autumn or early winter cooling near the surface of a snow cover may produce a vertically upward vapour flux which is manifest in two ways: it may flow out of the snow and thereby dehydrate the substrate below (Santeford, 1976); or it may re-deposit higher up in the cover as depth hoar (Akitaya, 1975; Paterson, 1981). On the other hand, water fluxes are independent of the temperature gradient and may occur as the result of rain, flooding from beneath or melting on top; clearly, this water may subsequently freeze, thaw and re-freeze.

The formation of ice from snow may occur in a number of ways, any one of which is complex and variable. Shumskii (1964) has organized the various possibilities into hypothetical zones within the cryosphere (i.e., the collection of those regions on Earth where freezing temperatures exist). In essence, these zones define specific mechanisms and processes of ice formation, and are therefore only loosely associated with geographical regions. In general, they represent variations on a basic theme: namely, the formation of ice from snow taking into account both the mean annual temperature and excursions above and below the freezing point. It is a theme of metamorphosis controlled by climates ranging from the continental to the maritime.

The metamorphosis of dry snow begins immediately after it has been deposited. For temperatures consistently below the triple point in arid

Table 6.4. *Illustrative surface heat fluxes over ice (a) Winter in Antarctica (following Paterson (1981))*

Station	Elevation (m)	Surface heat flux (W m^{-2})			
		$\dot{Q}_R$	$\dot{Q}_S$	$\dot{Q}_L$	Total
Vostok	3400	−17	15	0	−2
S. Pole	2800	−28	25	1	−2
Byrd	1530	−17	12	0	−5
Mawson	150	−43	39	−6	−10
Little America	44	−16	6	8	−2
Maudheim	37	−22	15	4	−3
Mirny	36	−25	47	−24	−2
Wilkes	12	−10	5	0	−5

regions, such as central Greenland or Antarctica, the snow crystals sublimate, minimizing their free energy (see Chapter 2) by reducing their surface to volume ratio. This has two main effects: the dendritic snow crystals discussed in the previous chapter tend to become less needle-like, thus reducing their effective diameter; and larger crystals tend to grow at the expense of smaller crystals. Both of these effects, and any fracturing which took place through collision during deposition, help produce a collection of rounded, isometric crystals. It is the settlement of these crystals, and compaction under their collective weight, which gradually creates a density increase from perhaps less than 100 kg m^{-3} up to 550 kg m^{-3}; at the end of this sublimation–consolidation process the porosity will have been reduced to the order of 40% from an initial value perhaps in excess of 90%.

At this stage, the snow has become firn, and densification attributable to re-packing of fine grains has gone about as far as it can. None the less, sublimation, diffusion and re-deposition of vapour continue in the neighbourhood of crystal contacts in a slow sintering process (Hobbs and Mason, 1964) during which the contact points develop growing necks. As the necks thicken, the sublimating surface area, together with the porosity (and permeability), gradually decrease; metamorphosis is then more accurately described as re-crystallization of the grains, which at that stage occupy most of the space. Mobility of water vapour slowly becomes less important while the significance of volume and surface diffusion, through and over grains, becomes correspondingly greater. Eventually, all the pore air will either be expelled or trapped in small pockets; the material will then be a form of polycrystalline ice[10] in which re-crystallization

Table 6.4 (*cont.*) (*b*) *Summer on Chogo-Lungma Glacier (following Untersteiner (1957))*

	Surface heat fluxes ($W\,m^{-2}$)			
	Firn at 4000 m in June		Ice at 4300 m in July	
	Day	Night	Day	Night
$\dot{Q}_R$	279	−73	450	−71
Q_S	35	41	23	17
Q_L	−10	6	−8	−2
Total	304	−26	465	−56

adjusts itself to intergranular stresses. In the absence of a bulk flow, the stress field, and the ice itself, ultimately become isotropic.

The climate of central Antarctica is representative of the conditions which engender this type of metamorphosis. With an annual water equivalent precipitation of 30–200 mm, this region may be described as a desert (Lorius, 1973; Dolgushin, Yevteyev and Kotlyakov, 1962). A positive ice accumulation rate is only possible because of the absence of ablation. Table 6.4 shows the principal components of the heat balance under winter conditions when the radiation balance consists essentially of long wave radiation, thus creating a net overall heat loss. During the summer, the short wave radiation gain, limited by the high albedo, increases both the long wave and latent heat losses; the surface temperature therefore changes little and usually remains below 0 °C. Also shown in Table 6.4 is some data from Chogo–Lungma Glacier in Pakistan. At both of the altitudes indicated, the data clearly illustrate the effect of incoming solar radiation on the net heating rate; the surplus thermal energy adds both sensible and latent heat to the cover. The data indicate a net heat loss during the night but a daily gain which more than offsets it. The large difference between the two is a measure of the latent heat consumed during melting.

It is thus to be expected that metamorphosis of snow on the polar ice would be quite different from that on a temperate glacier where temperatures closer to the melting point tend to accelerate the 'rounding', settling, sintering and re-crystallization processes. The typically higher snow falls on a temperate glacier tend to warm the firn and insulate it from subzero air temperatures. But the most significant distinction is a positive heat balance too strong to be absorbed merely as sensible heat in the snow, firn and underlying ice. The meltwater which results may be partly

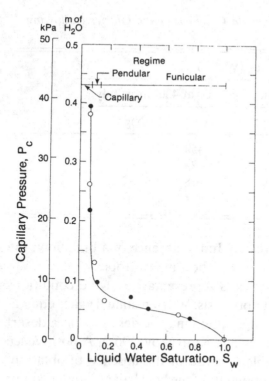

Fig. 6.6. Capillary pressure in wet snow as a function of water saturation. The two saturation regimes are shown. Density: filled circles, $550\,kg\,m^{-3}$; open circles, $590\,kg\,m^{-3}$ (following Colbeck (1973)).

evaporated but much of it percolates downward where it produces a warming effect as it re-freezes.

Meltwater displaces air and assists the consolidation process by lubricating the grains and creating cohesive capillary forces. The extent of the infiltration depends upon precipitation rates and upon the precise details of the diurnal and seasonal heat balance. In the extreme, when the sensible heat deficit in snow and firn is much less than the latent heat absorbed during the surface melting process (i.e., the Stefan number is much less than 1.0), only a small amount of re-freezing is needed to raise the underlying temperature uniformly to the melting point. Any unfrozen melt then becomes runoff water or, if the local topography prevents the water from leaving, creates a slush which, having a lower albedo, melts rapidly.

Metamorphosis in wet snow shares some features with metamorphosis in dry snow: for example, larger crystals grow at the expense of smaller ones. But the degree of wetness (or saturation), i.e., the fraction of the pore volume occupied by water, exerts a strong influence on behaviour, as indicated in Fig. 6.6. In very small amounts, e.g., less than 7% of the pore

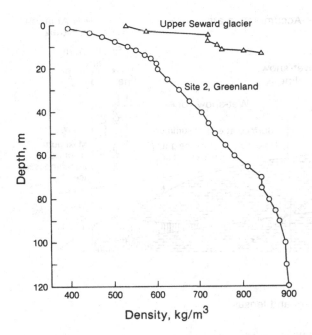

Fig. 6.7. Variation of firn density with depth in a temperate glacier and in the Greenland ice sheet (following Paterson (1981)).

volume, the water is held in place by a large capillary pressure (i.e., atmospheric pressure minus water pressure: positive values indicate suction in the water). As the degree of saturation increases to about 15%, the water pockets thicken, spread and merge to become a continuous film in what is known as the pendular regime. Air and water vapour continue to have access through interconnecting pores. Beyond this point, is the funicular regime, gases are trapped in air bubbles and thus no longer form a continuous phase. In the pendular and funicular regimes the equilibrium freezing point of the air/water/ice mixture may be calculated using the Gibbs–Duhem equation developed in Chapter 2 (Colbeck, 1973, 1974). It is found that a fall in the degree of saturation, creating an increase in capillary pressure, produces a fall in the equilibrium freezing temperature below 0°C. (This relationship is explored more fully in Section 6.4.) The degree of saturation also determines the mobility of the water. In the capillary regime the 'corner' water is essentially immobile, while in the pendular regime the snow will tend to drain under the influence of gravity. In general, the flow is Darcian (Colbeck and Davidson, 1973).

Climatic influence on the release and re-freezing of water is also reflected in the densification process. In Greenland or Antarctica, for example, where meltwater is often negligible, the settlement process is

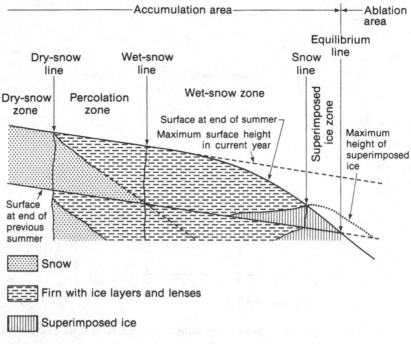

Fig. 6.8. Glacier zones (following Paterson (1981)).

much slower but continues year after year under the increasing weight of accumulated snow. It may take hundreds, and perhaps thousands, of years to convert snow crystals to polycrystalline ice. Fig. 6.7 provides a comparison of firn densities in an ice sheet and a temperate glacier. Three distinct, but variable, zones are evident. On the Greenland ice sheet, the transition from snow to firn occurs at a density near $600 \, \text{kg m}^{-3}$ ($\mathcal{N} \approx 40\%$) and requires over 100 years to complete. On the Seward Glacier, a temperate glacier, the same transition lies closer to $720 \, \text{kg m}^{-3}$ and occurs within 3–5 years (Paterson, 1981). The second transition – from firn to ice – is not evident in the glacier data but is marked by a slope change near $830 \, \text{kg m}^{-3}$ for the ice sheet. The upper layer of the ice sheet is a particularly porous zone where the temperature gradient which accompanies heat loss often gives rise to the appearance of depth hoar; this may be used as a mark of annual growth. On temperate glaciers, however, not only are the temperature gradients typically less, but the reduced quantity of hoar is frequently destroyed by meltwater.

In general, glacial behaviour reflects a wide range of climates: from the continental to the maritime, and including substantial variations in latitude and altitude. Glaciers must therefore be regarded as complex, variable, and above all dynamic, systems. Fig. 6.8, following Paterson (1981), is a schematic illustration of the various zones found in a hypothetical glacier as it flows downward through a substantial change in elevation. The *dry snow* zone is located near the head where atmospheric conditions prevent melting and thus imply a limited metamorphosis in a given year. The *percolation* zone is found where atmospheric conditions allow some meltwater to form and percolate downwards thus raising the temperature in the upper layers of firn. Where this warming effect has reached down to the limit of the entire year's snow cover marks the beginning of the soaked or *wet-snow* zone, in the lower reaches of which the ice of previous years becomes noticeable. This ice extends into the *superimposed ice* zone at the point where it emerges from beneath the firn i.e., at the snow line. Superimposed ice may grow further down slope into the *ablation* zone but melts back to the equilibrium line by the end of summer. Although highly idealized, this description is based on the physical processes discussed earlier. During a typical northern winter, some of these processes may be seen in miniature on a house with a low-sloping roof.

The thermal regime in a glacier or ice sheet is determined from the energy equation which may be written in the form

$$\frac{\partial T}{\partial t} + U\frac{\partial T}{\partial X} + V\frac{\partial T}{\partial Y} + W\frac{\partial T}{\partial Z}$$
$$= \varkappa\left(\frac{\partial^2 T}{\partial X^2} + \frac{\partial^2 T}{\partial Y^2} + \frac{\partial^2 T}{\partial Z^2}\right) + \dot{s}_Q/\rho c_p \qquad (6.10)$$

where X is taken in the direction of bulk motion (downslope), Y vertically (upward) and Z transversely; $\dot{s}_Q$ is a volumetric heat source consisting principally of strain energy and latent heat. In the upper layers of a wide snow/ice layer, transverse effects may be ignored along with the corresponding strain energy. Therefore, since the maximum temperature penetration is felt in the absence of a phase change[11], the heat source will be neglected. In essence, this implies that $\dot{s}_Q \ll k_i\theta_c/\delta^2$ where δ is the penetration depth of the thermal regime which extends over a temperature difference θ_c. The corresponding form of the continuity equation

$$\frac{\partial U}{\partial X} + \frac{\partial V}{\partial Y} = 0 \qquad (6.11)$$

indicates that the vertical velocity scale is given by $V_c = O(\delta U_c/L)$, where L is the reach of the layer. Thus, since $\delta \ll L$ in a typical situation, it follows that equation (6.10) reduces to

$$\frac{\partial T}{\partial t} = \varkappa_i\frac{\partial^2 T}{\partial Y^2} \qquad (6.12)$$

if also $U_c \ll L/t_c$, where t_c is the period of the surface temperature change.

The solution to equation (6.12) provides a definition of the thermal regime, the extent of which may be estimated simply by using the implicit scales of length and time. Thus

$$\delta^2 = O(\varkappa_i t_c)$$

which for ice[12] of $\varkappa_i = 36 \text{ m}^2 \text{ yr}^{-1}$ leads to $\delta \approx 6 \text{ m}$ in a year. However, this is only an order of magnitude. A more precise value requires the complete solution of equation (6.12) which, for a periodic surface temperature $\theta_0 \sin \omega t$, is given by

$$\theta = \theta_0 \exp(-y) \sin(\tau - y) \tag{6.13}$$

where θ is measured from the mean value, θ_0 is the amplitude, $y = Y(\omega/2\varkappa_i)^{\frac{1}{2}}$ and $\tau = \omega t$. Since $|\sin(\tau - y)| \leqslant 1$, the magnitude of the temperature fluctuation at depth depends principally upon the exponential function. If the penetration depth is defined arbitrarily by the point at which temperature variations are equal to 3% of the surface variations i.e., $\theta/\theta_0 = 0.03$, then $y = 3.5$. Given this tolerance, it is found that diurnal penetration is about 0.6 m whereas the annual penetration is about 12 m. Longer term climatic variations penetrate deeper again: e.g., over a century the penetration depth is about 120 m.

Generally speaking, the thermal regime in the upper layers of a glacier controls the extent of ablation and infiltration, the effects of which may be felt to some depth beneath. In an old ice sheet, however, the temperature distribution reflects a quasi-steady balance that extends right down to the base, a distance h which may be hundreds, and perhaps thousands, of metres. To estimate the temperature field in such an ice sheet, it is possible to use the two-dimensional form of equation (6.10) with the expectation that the transient term would be unimportant. Upon examination of the equation it is evident that a steady state solution requires $h^2 \ll \varkappa_i t_c$: for a period of the order of 10 000 yr, for example, this implies $h \ll 600 \text{ m}$; thicker ice sheets would require correspondingly longer periods of equilibration. The longitudinal advection term may be neglected when $U_c \ll \varkappa_i L/h^2$, where U_c is the longitudinal velocity scale of the ice sheet; the longitudinal conduction term is unimportant if $h \ll L$; and internal dissipation may be ignored if $\dot{s}_Q \ll k_i \theta_c/h^2$. The lateral advection term, on the other hand, provides the only means by which the basal (geothermal) heat supply rate may be reconciled with the withdrawal of sensible heat into the atmosphere. This withdrawal is achieved through the addition of colder snow, thus implying that

$$k_i (\partial T/\partial Y)_b = \rho_i c_{pi} V_s (T_s - T_b) \qquad (6.14)$$

where V_s is the equivalent ice accumulation velocity at a surface temperature T_s; T_b is the basal temperature. Hence, for a steady state,

$$V_s h/\varkappa_i = O(1).$$

When all of the above conditions are satisfied, equation (6.10) reduces to

$$V \, \mathrm{d}T/\mathrm{d}Y = \varkappa_i \, \mathrm{d}^2 T/\mathrm{d}Y^2 \qquad (6.15)$$

in which $V(Y)$ is unknown *a priori*. Substituting the simple form $V = -V_s Y/h$, which satisfies the requirement that $V(0) = 0$ and $V(h) = -V_s$, equation (6.15) yields the closed form solution

$$T - T_s = \theta = A \, \mathrm{erf} \, y + B$$

where $y = Y/(2\varkappa_i h/v_s)^{\frac{1}{2}}$. Satisfying the boundary conditions,

$$Y = h; \, \theta = 0$$

$$Y = 0; \, -k_i (\partial T/\partial Y)_b = G, \text{ the geothermal heat flux,}$$

we obtain

$$\theta = \frac{\pi^{\frac{1}{2}}}{2} \frac{G}{k_i} \left(\frac{2\varkappa_i h}{V_s} \right)^{\frac{1}{2}} \left[\mathrm{erf} \, y_s - \mathrm{erf} \, y \right] \qquad (6.16)$$

a result first obtained by Robin (1955).

This result implicitly embodies the energy balance stated in equation (6.15) and is therefore restricted by the consequences of that balance. For example, equation (6.15) does not allow for transient[13] effects, which have been studied separately (Wexler, 1958), ignoring the lateral advection term. However, the Robin equation (6.16) does provide a valuable base from which to investigate various refinements. Weertman (1968), for example, has used the steady state model to determine the importance of strain energy dissipation and longitudinal advection, concluding that together they have little effect. Philberth and Federer (1971) used the incompressible form of the continuity equation (6.11) to refine the lateral advection velocity as a function of depth and obtained better agreement with field data in the lower half of the sheet. Alterations in climate may offer the best prospect for reconciliation nearer the surface: in the short term, by superimposing a transient perturbation on the deep steady state solution or, in the longer term, by treating the Robin solution as a quasi-steady form. The order of magnitude analysis presented earlier implies that, in general, the coefficient of the transient term $(h^2/\varkappa_i t_c)$ may be significant, especially for depths greater than 1000 m. A full, unsteady solution is then required, almost certainly using the numerical techniques discussed in Chapter 3.

Glaciers and ice sheets present an interesting contrast to the *naled*[14] which is usually much smaller and correspondingly quicker to come and go. Naleds form whenever mobile water finds itself suddenly thrust into a freezing environment. Examples are encountered in a wide range of circumstances: runoff from a drain pipe; effluent from a sewer; overflow on river ice; and seepage of groundwater. Typically, a large and relatively warm source gives rise to a slow, film-like flow of water which freezes in the presence of sub-zero temperatures. A subcooled substrate (including ice itself) provides a sensible heat deficit capable of removing both sensible and latent heat from the water, while subzero air temperatures, coupled with evaporative and long wave radiation losses, induce supercooling and thereby promote heterogeneous nucleation and *a*-axis crystalline growth. The details vary enormously and the naleds which form give rise to effects ranging from the nuisance of a blocked drain to the hazards of spring flooding. A particularly widespread form results from the constriction of groundwater (Carey, 1970, 1973).

The appearance of naleds in natural circumstances is frequently associated with a number of characteristic conditions:

(1) A large, and perhaps swollen, source of moving surface water which is frozen over annually.

(2) Proximity of an impermeable surface, such as the permafrost table or an ice cover, restricting vertical movement of water within or above the ground.

(3) Abnormally high winter heat losses in the vicinity: this reduces the cross sectional area through which groundwater or river water may flow.

(4) Abnormally low spring heat gains, which impede spring thawback.

In general, such conditions tend to create a constricted flow of water, thereby altering the local hydraulic conditions, the pressure head in particular. This often leads to a forced 'leakage' which is the source of the naled. Leakages vary from 1–100% of the main flow (Carey, 1973; Kane and Slaughter, 1972; Froehlich and Slupik, 1982), and have produced naleds up to 15 m thick extending as far as several kilometres in their downstream direction (Gavrilova, 1972; Van Everdingen, 1982; Slaughter, 1982; Froehlich and Slupik, 1982).

In view of the complex and variable interaction between the initial water flow system and the local heat balance, it is to be expected that the thermal conditions immediately adjacent to the forming naled would vary considerably. The leakage water itself usually has a fairly narrow range of temperatures: typically 0–1°C (Gavrilova, 1972; Akerman, 1982; Van

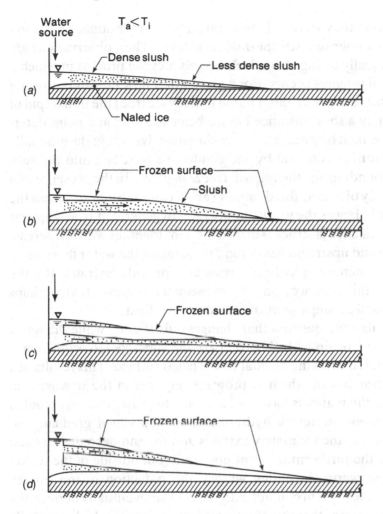

Fig. 6.9. Growth of a two-dimensional naled (following Schohl and Ettema (1986)).

Everdingen, 1982). On the other hand, Gavrilova, (1972) reports surface soil temperatures of −5°C and −20°C recorded when the neighbouring air temperatures were less than −25°C and −37°C, respectively; at depth (5–10 m) the temperature was of the order of −1°C. It thus appears that the relative importance of soil subcooling and air-induced water supercooling may vary widely, thereby influencing both the mechanism and extent of the ice formation.

The freezing mechanism of a thin film of water spilling over a cold substrate in subzero air temperatures has been studied recently under controlled laboratory conditions by Schohl and Ettema (1986). Using a refrigerated flume they were able to create a two-dimensional simulation of naled growth under realistic thermal conditions. Their observations are shown schematically in Fig. 6.9 which reveals a cyclic process in which a layering of naled tongues occurs. For a given water flow rate and substrate geometry (in this case a horizontal flume initially ice free) the initial spill of water travels only a short distance before being arrested at a point determined by the toe heat balance: i.e., once the advective flux in the gradually cooling toe water is overcome by the conductive heat flux into the substrate and surrounding air, the toe will freeze in place. In the presence of a continuing supply of water, this damming effect cannot continue unless the rate of freezing balances the water supply rate. In general, it would appear that a dynamic balance is struck: surface tension, together with an increasing naled slope and upstream head, tend to maintain the water flow, while a frozen toe and increasing hydraulic resistance provide restraint. It is the sudden loss of this balance, and its subsequent redress, that perhaps explains the peculiar steps seen on naleds in the field.

Fig. 6.9 reveals a toe position that changes only slightly while the water supply freezes. To begin with (Fig. 6.9(a)), the supercooled water film is invaded by platelets growing normal to the naled surface. This results in a slush, the ice fraction of which is progressively less in the downstream direction where the platelets have had less time to grow. The supercooled water flow thus encounters a hydraulic resistance which gradually increases with time as the ice matrix extends upward and becomes less and less porous. At the furthermost point upstream (but outside of the region occupied by the warm, incoming water), the reduction in the surface velocity of the water film in contact with the cold air eventually permits the surface to freeze over, thereby creating a thin ice cover which gradually extends downstream, as shown in Fig. 6.9(b). However, the consequent narrowing of the water flow path beneath this cover increases the hydraulic resistance further until the water supply is forced to spill over the newly-formed ice cover and thus create a new film: this is illustrated in Fig. 6.9(c). The process is evidently capable of repeating itself, each time with a foreshortened tongue. As the slush channels eventually freeze shut they produce a layering effect which is commonly observed in the field.

A much rarer event, which takes place on an even smaller scale, is the appearance of needle ice (also known as mush frost, stalk ice, hair frost, ice fibres and pipkrake (Geiger, 1965)). This may occur above the surface of soil, or any other porous substrate with a plentiful supply of water beneath, when the air temperature falls below 0°C. However, the conditions must be such that the water does not simply freeze *in situ*, but is drawn up to the surface before freezing. This is a good example of primary heaving, which will be discussed in more detail later. Suffice it to say that if the water temperature is initially around 0 °C the lowering of the surface temperature may generate a lower pressure in the unfrozen water immediately beneath; this is capable of creating an upward flow from depth. Upon reaching the surface and emerging from the substrate, the water freezes into a segregated form of ice consisting of closely packed needles of about 1.0 mm diameter (Fukuda, 1936).

The essential requirement for the growth of needle ice appears to be a sufficiently mobile source of cold water in a suitable soil e.g., sandy loam (needle ice has also been observed on pine needles and decaying wood). Growth occurs parallel to the temperature gradient causing the water flow, and is therefore not restricted to the vertical direction, which is most common. As discussed later, the segregation is accompanied by a thrust which is capable of lifting surface objects such as soil particles and rocks, though seldom more than a few centimetres above the surface. Needle ice has also been known to grow beneath a snow cover.

6.3 The pore system: its architecture, hydrodynamics and thermodynamics

An important distinction between freezing *on* the Earth and freezing *in* the Earth is the fact that the latter takes place inside the pores of earth material. Therefore, before considering ice formation processes in soils generally, it is advisable to develop a basic understanding of the circumstances and events which precede and accompany the freezing process in a single pore. Accordingly, this section contains a brief morphological review of soil and rock, an introduction to heat and mass transfer processes therein, and a description of the dynamics and thermodynamics of ice formation in small cavities.

6.3.1 Soil and rock

The morphology of natural porous media found on and immediately beneath the Earth's surface reflects the processes involved in their origin; these may be physical, biological, or a combination of both. Physically, such media are formed initially in two essentially different ways: by the freezing of liquids and by the aggregation of solid fragments. In the process, nucleation and growth lead to the juxtaposition of crystals or grains whose size and orientation depend upon the prevailing conditions: in particular, they vary with the rate of cooling and the rate at which the melt is mixed. Typically, the intergranular spaces thus formed are very

Table 6.5. *Soil particle fractions (following Jumikis (1962))*

Designation of soil fractions	Diameter of soil particle (mm)		
	US Department of Agriculture, Bureau of Soils (1951)	American Society for Testing Materials (1958)	International Society of Soil Science
Fine gravel	2–1	>2.00	
Coarse sand	1–0.5	2.0–0.42	2.0–0.2
Medium sand	0.5–0.25		
Fine sand	0.25–0.10	0.42–0.074	0.2–0.02
Very fine sand	0.10–0.05		
Silt	0.05–0.002	0.074–0.005	0.02–0.002
Clay	<0.002	<0.005	<0.002
Colloids		<0.001	

slender and the polycrystalline solid is not especially porous or permeable[15], at least not on that account. However, if the melt also includes gas bubbles, the final solid may be very porous if the inclusions are large enough and numerous enough to form a continuous labyrinth. Again, the rate of cooling and the melt mixing rate will influence the final result.

The aggregation process is obviously very different and can only begin after continuous solids, inorganic or organic, have been reduced to fragments by glaciological processes, weathering or biological decay. At one extreme, the fragments may be subjected to very high pressures which consolidate, deform and fuse them. At the other extreme they merely rest upon each other with no more force than the weight of the material above, which may, however, be considerable.

The structure of porous media in the lithosphere, hydrosphere and biosphere is thus a function of the formative process[16], and this is particularly evident in attendant properties such as the porosity and permeability. Degasification in molten rock, for example, a fairly evenly distributed phenomenon generating spherical bubbles, tends to produce a porous solid which is more or less homogeneous in porosity and permeability, but subsequent extrusion of the melt tends to elongate the cavities in the direction of motion and thus creates anisotropy.

Notwithstanding the history of the material, it is always possible to draw conclusions about its porous characteristics from the microscopic details. Obviously, the porosity depends upon the size and frequency of the cavities. It also depends upon the cavity geometry or, for aggregations, on

Table 6.6. *Relative values of soil hydraulic conductivities (following Jumikis (1962))*

Degree of permeability	Range of hydraulic conductivity K(cm s^{-1})	Approximate textural soil fraction
High	$>10^{-1}$	Medium and coarse gravel.
Medium	10^{-1}–10^{-3}	Fine gravel; coarse, medium and fine sand; dune sand.
Low	10^{-3}–120^{-5}	Very fine sand; silty sand; loose silt; loess; rock flour.
Very low	10^{-5}–10^{-7}	Dense silt; dense loess; clayey silt; clay.
Impervious	$<10^{-7}$	Homogeneous clays.

the fragment geometry: for example, a cubic arrangement of spherical particles exhibits a higher porosity ($\mathcal{N} = 0.476$) than a rhombohedral arrangement ($\mathcal{N} = 0.26$). It is thus important to know both the geometry of the pore (or particle) and the geometrical arrangement. The same comment applies to the permeability which depends not only upon the size and frequency of the pores but on their interconnections, both in the direction of flow and normal to it.

Given the importance of the aggregation process and the consequent size and shape of the particles, it is useful to divide soils into two different but overlapping categories: sediments, which are usually mineral or inorganic soils; and aggregations of decayed or decaying organic fragments known as organic soils. The former group is the result of various fragmentation processes, largely attributable to weathering and chemical reactions, followed by a variety of depositional processes involving gravity and displacement by air, water and ice. When depositional processes dominate, it is usual to speak of *transported* soil; fragmentation *in situ* produces a *residual* soil. Without dwelling on the details, it is evident that they will determine the precise size distribution of the fragments, the geometry of the fragments and their subsequent geometrical arrangement. The depositional history is therefore important in controlling the size, shape and interconnections of the pores and thus in determining porosity and permeability. History may also influence the degree of anisotropy. In clay, for

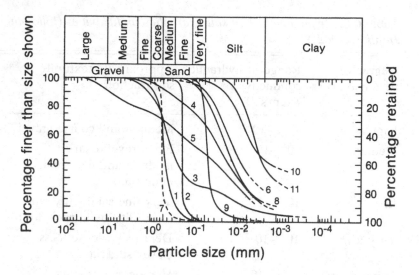

Fig. 6.10. Various soil particle size distribution curves: 1 Cape May, NJ, sand; 2 Daytona Beach, Florida, sand; 3 Dunellen, NJ, silty sand; 4 Montalto, NJ, sandy silt; 5 Pennsylvania Pike, Del., silty, gravelly sand; 6 Delaware Memorial Bridge, Del., sand from bridge approach; 7 Lewis Beach, Del., sand; 8 gneissic rock flour, NP, NL; 9 fine sand NJ; 10 raritan clay, NJ; 11 varved clay NJ (following Jumikis (1962)).

example, the ratio of horizontal to vertical permeability may exceed 10:1, though only rarely.

A common way to classify mineral soils is by the dominant particle size. Table 6.5 shows the size range of the principal soil types, while Table 6.6 illustrates the strong correlation between hydraulic conductivity, which is proportional to permeability, and soil type. Fig. 6.10 reveals how particle size may vary within any given soil. At the clay end of the spectrum soil behaviour is modified by surface electrical forces which become more and more important as the total surface area increases with decreasing particle size. The colloidal nature of small clay particles, and the role these particles play in the retention and transport of soil moisture, will be discussed more fully later. Useful though particle size may be as an indication of soil properties, it neglects other important factors such as the degree of compaction and the influence this has on the size and interconnections of the pores.

The morphology of fibrous media in general, and organic soils in particular, is usually quite different from that of the granular aggregates discussed above, especially when the fibre fragments are much longer than their diameter. Randomly orientated fibres tend to produce a medium which is isotropic, and in that respect fibrous soils are not too different from many mineral soils, except in the shape and interconnections of the pores. However, partly or completely aligned fibres may produce very different properties. Firstly, they tend to form a less permeable barrier if their axes are aligned and the fibres are in intimate contact with one another. Secondly, they tend to align any fluid flow parallel to the fibre axes. The distribution of axis orientation thus exerts an influence on the magnitude and direction of the permeability: anisotropy is a consequence of a dominant orientation, the principal axis being aligned with it.

The density of organic soil is usually much less than that of mineral soils and the porosity is therefore correspondingly greater. A study of sphagnum peat moss, for example, revealed a porosity of 90% at depth, while the porosity in the uppermost layer, where it was still growing, was almost 100% (Kay, Hons and Coit, 1975). More typically, the values range from 60% to 85%. The skeletal surface area within organic soils is not very large but yet, like the colloidal clays, they expand when wetted, though not to the same extent. Organic surfaces are often markedly hygroscopic. This is particularly noticeable at low moisture contents, e.g., less than 10%, when the heat of wetting is significant enough to alter the effective latent heat of the soil (latent heat of water minus the heat of wetting, the latter being negligible in mineral soils, except clays (Anderson, 1963)). The effect may be ignored when the moisture content exceeds 50% (Kay *et al.*, 1975). Presumably, the role of water in a freezing organic soil is similar to that in the wet snow discussed earlier, bearing in mind the differences in skeletal surface behaviour.

From the standpoint of thermal and hydrodynamic processes, the size, shape and interconnections of the pores constitute the most important morphological characteristics. There are, however, other characteristics which may become extremely important under certain special circumstances, notably freezing or thawing. Principal among these is bonding, by which is meant the ability of the material matrix to resist deformation and distortion. Porous media formed through a polycrystalline growth process will exhibit high bonding, and the resulting solid, which is macroscopically homogeneous, will exhibit correspondingly high tensile and shear strengths, and a low compressibilty. Much the same may be said of sintered and indurated materials, although the bonding strength will clearly depend upon the intergranular fusion process and the size of the contact areas. As might be expected, the most poorly bonded media are the loose aggregates formed from fragments which exhibit no natural adhesive forces at their surface and whose shape is not conducive to interlocking or intertwining. However, interparticulate cohesion may also be induced by small water films (also described as pellicular water (De Wiest, 1965)) which wet the surface of the particles in the vicinity of their contact points, thus introducing menisci and their accompanying surface tension forces which may be of considerable magnitude when integrated over a large number of small particles. Such cohesion is a function of the moisture content, being zero in clay-free soils when they are either saturated or completely dry.

6.3.2   *Transport processes*

The retention and mobility of water, in either or both of its fluid phases, play an important part in the freezing and thawing processes inside pores. This is true not only in a granular soil, for example, where ice may occur interstitially within a mineral matrix, but in snow or frost, where the matrix is ice itself. In any given instance, water alone may fill the pores (bearing in mind that air will usually be present too), or it may share the space with liquid water which, as noted earlier, appears first as a thin surface film spread throughout the matrix but may grow to fill the pores completely. For water in particular, the volumetric fraction $v_w = V_w/V$ both defines the amount of water present and influences its ability to flow.

The maximum value of the volumetric water fraction in any medium is the porosity[17] $\mathcal{N}$, defined as the ratio of the pore volume to the total volume V. The porosity thus provides a measure of the fluid storage capacity of the medium but it does not provide a measure of mean pore diameter or describe pore geometry and interconnections, all of which influence the hydrodynamics and thermodynamics surrounding ice formation. Fluid flow is usually described by the Darcian form[18]

$$\mathbf{V} = -K\nabla H \tag{6.17}$$

in which $\mathbf{V}$ is the local mean fluid velocity, and the piezometric head H is derived from the pore pressure $H\rho g$ given by $H\rho g = P + Z\rho g$, where P is the excess pore pressure and $Z\rho g$ is the hydrostatic pressure. The quantity K is known as the hydraulic conductivity and may be written (De Wiest, 1965)

$$K = cgd^2/\nu$$

where d is the mean pore diameter, ν is the kinematic viscosity of the fluid and c is a non-dimensional coefficient determined either empirically or from the details of a flow model: e.g., in a coarse medium it may be written in terms of the porosity using the form $c = b \mathcal{N}^2/(1 - \mathcal{N})$ where b is a geometrical coefficient (Rumer, 1969). The intrinsic permeability of the medium[19] is then defined by $k = cd^2$: hence $K = kg/v$.

The continuity equation for a single fluid saturating a porous medium is

$$\partial m/\partial t + \nabla \cdot (\rho \mathbf{V}) = 0 \qquad (6.18)$$

where ρ is the fluid density at a point and $m = \mathcal{N}\rho$ is the fluid content, i.e., the mass of fluid per unit medium volume. It is important to note that $\mathbf{V}$ is the Darcian velocity defined by equation (6.17) which enables the second term in equation (6.18) to be written as $-\rho K \nabla^2 H$ if the medium is homogeneous and the fluid is incompressible. the first term may then be written

$$\partial m/\partial t = \rho \alpha \, \partial P/\partial t = \rho^2 g \alpha \, \partial H/\partial t$$

where α is the matrix compressibility coefficient, and hence

$$\partial H/\partial t = C \nabla^2 H \qquad (6.19)$$

where $C = K/\rho g \alpha$, sometimes called the coefficient of consolidation (Jumikis, 1962). More generally, the piezometric head equation must be written

$$\rho g (\alpha + \mathcal{N}\beta) \, \partial H/\partial t = \nabla \cdot (K \nabla H) + \rho \beta g K (\nabla H)^2 \qquad (6.20)$$

to accommodate the two effects of fluid compressibility β: an added storage capacity (the left hand side of the equation); and the introduction of a non-linear term (on the right hand side). Spatial variations of K are also included.

The transfer of heat within a pore is very complex since, in general, it consists of conduction (through the matrix and pore contents), radiation (in absorbing media) and convection accompanied by phase change processes such as evaporation, condensation, sublimation and, of course, freezing and melting. The treatment given here, which broadly speaking applies to both heat and mass transfer, will therefore be limited to water-saturated media in which radiation may be neglected. To begin with, it is important to recognize two separate, though related, phenomena: diffusion and dispersion. The first of these often accompanies, but does not require, fluid motion; it occurs whenever the medium is subjected to a temperature or concentration gradient. Because it is a molecular process, constrained by pore geometry and interconnection, its macroscopic effect must somehow accommodate the tortuous path which the diffusive flux actually follows. The effective molecular diffusion coefficient $\mathscr{D}^m$ is therefore written[20]

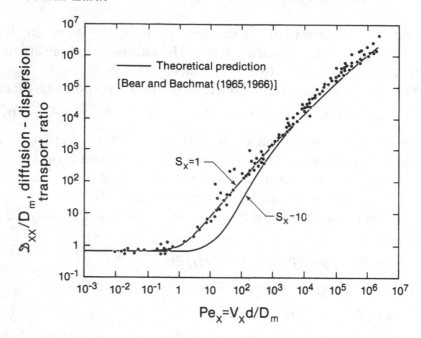

Fig. 6.11. Effect of unidirectional flow velocity on diffusion-dispersion transport (following Bear (1969)).

$$\mathscr{D}^m = D_m \tau \tag{6.21}$$

where D_m is the molecular[21] diffusivity in the fluid and τ is the tortuosity of the medium.

If the molecular diffusivity were zero, transport in a pore would not cease unless the fluid were at rest; otherwise, the fluid continues to carry and mix while passing through the pores. This is the process of dispersion which is heavily dependent upon the local (fluid) velocity and the interconnection of the pores, both laterally and longitudinally. The dispersion coefficient $\mathscr{D}^d$ is usually written in the form[22]

$$\mathscr{D}^d = \alpha V f \tag{6.22}$$

to reflect the effect of fluid velocity V: α is defined as the dispersivity of the medium and f is a function of local conditions, notably geometry. The two transport coefficients given by equations (6.21) and (6.22) are combined to yield the overall diffusion–dispersion transport coefficient $\mathscr{D}$. Thus

$$\mathscr{D} = \mathscr{D}^m + \mathscr{D}^d \tag{6.23}$$

This enables the heat flux density in the X-direction, for example, to be written

$$j_X^Q = -\rho c_p \mathscr{D}^Q \frac{\partial T}{\partial X} \tag{6.24}$$

in which $\mathcal{D}^Q$ is the overall heat transport coefficient, including both diffusion and dispersion in the pores.

The relative importance of pore diffusion and dispersion depends upon both the architecture of the pores and the piezometric head distribution (which determines the velocity field[23]). In the absence of fluid motion, $\mathcal{D} = \mathcal{D}^m$ which, for a loose sand, is $2D_m/3$: i.e., $\mathcal{D}/D_m = \tau = \frac{2}{3}$. Fig. 6.11 shows this asymptotic value in a general plot of $\mathcal{D}/D_m$ against the pore Peclet number $Pe_X = Re_X Pr$: the results refer to unidirectional flow in the X-direction. The figure also compares experimental data with theoretical curves based on two different slenderness ratios measured in the direction of fluid flow. It is interesting to note that for cold water $(Pr = O(10))$ infiltrating fine sand $(d \le O(10^{-1}) \text{mm})$ with a velocity of 1mm s^{-1} the Peclet number is $O(1)$. Under such circumstances it is evident that dispersion is unimportant, as it would be for lower water velocities in finer soils. However, it is equally clear from Fig. 6.10 that the effect of dispersion could be much more important in coarse sand or gravel, and is capable of increasing the overall heat transport coefficient by an order of magnitude or more. Fig. 6.11 indicates that as a rule of thumb, $\mathcal{D}_{XX}/D_m \approx Pe_X$ for $Pe_X \ge 1$ while $\mathcal{D}_{XX}/D_m = \frac{2}{3}$ for $Pe_X < 1$. This suggests important differences in the role of water transport during freezing or thawing in different media. In snow, for example, a velocity range of $0.1-1.0 \mu\text{m s}^{-1}$ (Colbeck and Davidson, 1973) makes it possible to neglect dispersion.

It is worth noting that when $Pe < 1$, $Re = Pe/Pr \ll 1$ if $Pr \gg 1$, which is true of cold water. It therefore follows that the absence of the dispersive mechanism then coincides with the absence of inertial effects in the equation of motion: i.e., diffusion will be the dominant pore transport mechanism in water if Darcy's law is valid. In such common circumstances, $\mathcal{D}$ is a scalar if the medium is isotropic. This greatly simplifies the energy equation which is then usually written with the properties averaged[24] between the solid matrix and the pore fluid. The average specific enthalpy $\bar{h}$, for example, may be written

$$\bar{h} = \frac{h_f \rho_f \mathcal{N} + \rho_s (1 - \mathcal{N}) h_s}{\rho_f \mathcal{N} + (1 - \mathcal{N}) \rho_s} = \frac{h_f \rho_f \mathcal{N} + \rho_s (1 - \mathcal{N}) h_s}{\bar{\rho}} \qquad (6.25)$$

in which the numerator is the volumetric average. Similarly, the heat flux density (equation 6.24) becomes

$$\mathbf{j}^Q = -[\mathcal{N} k_f + (1 - \mathcal{N}) k_s] \nabla T = -k^Q \nabla T \qquad (6.26)$$

in which k^Q is a volumetric average. The energy equation, which is discussed extensively in Chapter 3, may thus be written

$$\bar{\rho} \partial \bar{h}/\partial t + \rho_f (\mathbf{V} \cdot \nabla) h_f = \nabla \cdot (k^Q \nabla T) + \dot{s}_Q \qquad (6.27)$$

where $\dot{s}_Q$ is a volumetric heat source, examples of which include distributed latent heat, chemical or nuclear reactions, electric resistive heating and viscous dissipation.

The specific enthalpy of any component is a function of both temperature and pressure but, as noted in Section 3.3, there are many practical situations in which it may be approximated by the simple form

$$\mathrm{d}h = c_p\,\mathrm{d}T \tag{6.28}$$

where c_p is the specific heat. The terms on the left hand side of equation (6.27) may therefore be re-written with temperature as the principal dependent variable, but this leaves two complications:

(1) c_p is often a function of temperature;
(2) the masses of individual components change as a result of phase transformation, in which case the enthalpy changes must include latent heat.

The first of these merely implies that $\mathrm{d}h = c_p(T)\,\mathrm{d}T$, which is a minor point. The second requires that the continuity equation be re-written to incorporate a phase change. For example, when water is being converted to immobile ice at a rate $\partial m_i/\partial t$ per unit volume, where m_i is the volumetric ice content, the continuity equation must be written

$$\partial m_w/\partial t + \nabla \cdot \rho_w \mathbf{V} = -\partial m_i/\partial t \tag{6.29}$$

which is an extension of equation (6.18). This immediately requires an extension of equation (6.28) which, when applied volumetrically to the ice and water, becomes

$$\mathrm{d}h_v = m_w c_{pw}\,\mathrm{d}T + m_i c_{pi}\,\mathrm{d}T - \lambda_{iw}\,\mathrm{d}m_i \tag{6.30}$$

which includes sensible and latent heat.

Finally, it is worth recalling that much of the above discussion assumes that the pores are filled with water alone. The presence of other pore occupants not only complicates the calculation of average properties but may introduce several phase changes simultaneously, not to mention several modes of heat transfer. Currently, the behaviour of such systems is not well understood. It is always important to remember that freezing and thawing in porous media may be significantly affected by the mobility of the water present. This mobility may be strongly influenced by the phase change itself, notably through alterations in the effective size and shape of the spaces which control the permeability.

6.3.3   Growth of ice in a pore

Nucleation and growth of ice in a pore are, in general, complex

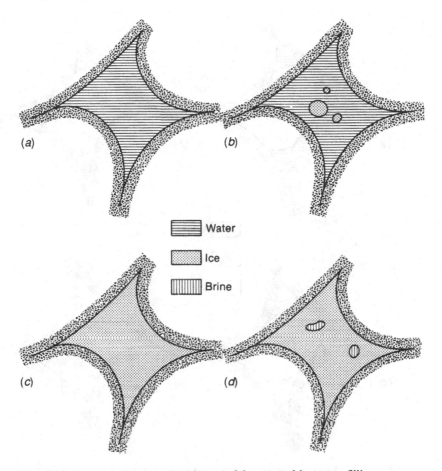

Fig. 6.12. Freezing in a saturated pore: (*a*) metastable water filling pore, $T \leqslant T^i$; (*b*) stable ice in pore water, $T = T^i$; (*c*) stable ice filling pore, $T \leqslant T^i$; (*d*) stable ice with brine inclusions, $T < T^i$.

processes, the details of which depend upon several key factors: the pore constituents and their relative amounts; the area, geometry and material of the pore surface; and the thermodynamic state of the occupants. Perhaps the simplest situation occurs when the pore is cooled while filled only with moist air. Recalling the discussion in Section 5.1, if the water vapour pressure is lower than the triple point value, the vapour will deposit on the cavity surface. Leaving aside hygroscopy, the large latent heat of sublimation tends to warm the solid matrix containing the pore, thus delaying a further drop in temperature until the pore is filled with deposited ice, or at least until the ice fraction is large enough to inhibit vapour access.

If the pore were completely filled with water, the situation would be very different, as suggested in Fig. 6.12 which describes ice growth in a water-

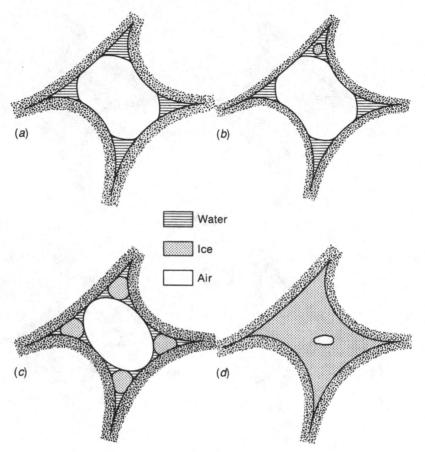

Fig. 6.13. Freezing in an unsaturated pore: (*a*) $T > T^i$ (stable) or $T < T^i$ (metastable); (*b*) $T = T^i > T_{cr}$ (stable); (*c*) $T \lesssim T_{cr}$ (unstable); (*d*) $T < T_{cr}$ (stable).

saturated medium. As the temperature is lowered uniformly in the absence of any superimposed water flow, it may pass beneath the equilibrium freezing temperature T^i and create the metastable state indicated in Fig. 6.12(*a*). Metastability is more likely to occur under the carefully controlled conditions of the laboratory than in the field where there are usually enough foreign particles to ensure heterogeneous nucleation close to T^i. In general, however, the freezing point is slightly depressed i.e., $T^i < T^0$.

Fig. 6.12(*b*) represents the attainment of stable, thermodynamic equilibrium at T^i. Small crystals of ice have grown from their nuclei, and may have merged with others. Further growth requires removal of the latent heat of fusion, while decay requires a corresponding addition in processes which, being a succession of stable states, are reversible. Should growth continue, the ice will eventually fill the pore, as indicated in Fig. 6.12(c): subsequent removal of (sensible) heat will lead to temperatures less than T^i. The presence of solutes does not change the freezing process in open pores too much, at least for low cooling rates: the only essential differences are further depression of the freezing temperature, through freeze concentration, and the possible appearance of eutectics. These were discussed in Chapter 2. If the freezing is rapid enough to prevent stable equilibrium, concentrated pockets of solution (e.g., brine) may form and remain as inclusions. This situation is illustrated schematically in Fig. 6.12(d).

When the pore is not completely filled with water, nucleation and growth are complicated further by various interfaces which exist between substrate, ice, water and air (including water vapour). Fig. 6.13 illustrates the freezing process in an unsaturated pore, again assuming a unique equilibrium freezing temperature T^i. To begin with, ice-free water is cornered beneath the menisci, as depicted in Fig. 6.13(a): the water may be in either stable or metastable equilibrium. After the onset of nucleation, further cooling produces stable, and reversible, crystal growth, as depicted in Fig. 6.13(b). Up to this point, there is no essential difference from behaviour in a saturated pore, except for the influence of the water film dimensions on T^i. However, in unsaturated pores the crystal growth process may be abruptly interrupted (as in a Haines jump (Haines, 1930)) at a critical temperature T_{cr} which marks the division between stable, reversible ice growth and unstable, irreversible ice growth (Miller, 1973). The situation is represented by Fig. 6.13(c). This critical temperature, which is typically only a small fraction of a degree below T^i, evidently arises from the sudden loss of mechanical equilibrium at the ice–water interface: the ice then spontaneously increases to fill the entire space available, thus expelling air and drawing water from neighbouring pores or, if this water is limited, creating a suction. The sudden growth of ice fed by pore water is characteristic of ice segregation phenomena, and will be discussed in detail in the next section. It is the cause of the needle ice discussed in the previous section.

Thus far, it has been assumed that the substrate plays a passive role in the growth of pore ice. Nucleation seldom takes place on the substrate surface, except in the early stages of deposition from the vapour phase. Freezing begins in the interior of the pore and might be expected to continue until the pore is completely blocked. However, it has been found that the temperature required to freeze the water left between the ice and the substrate decreases progressively as the film thickness decreases. That

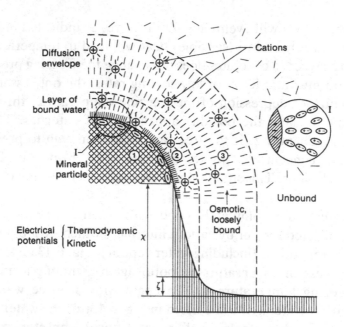

Fig. 6.14. The effect of a mineral substrate on ions and water molecules (following Tsytovich (1975).

is, T^i is not a unique value under these conditions, but extends over a range dictated by the unfrozen water content w_w. The explanation of this lies partly in the size of the pore and partly in the surface chemistry of the substrate.

The importance of surface effects was noted in Chapter 4 during the discussion of particle engulfment where ionic forces are reflected in the chemical potential and pressure distribution of water near the substrate. This continuum model has its counterpart in the double-layer molecular model depicted in Fig. 6.14 (Tsytovich, 1975). In essence, this model expresses the idea that a negatively charged substrate, being electrically balanced by a neighbouring cloud of cations, creates an electric field which orientates the polar water molecules. Very near the substrate, the force of mutual attraction is strong and therefore 'binds' molecules; further from the substrate the molecules are only loosely, or osmotically, bound. It is interesting to note that some lamellar crystals with this surface property (e.g., the trimorphic forms present in montmorillonite clay) naturally tend to arrange themselves into multi-layered but thin platelets, in the inter-lamellar spaces of which the osmotic pressure may be much higher than in surrounding water. Such platelets are naturally hygroscopic and create soils which swell on wetting and shrink on drying. This behaviour greatly influences their freezing and thawing characteristics (Kohnke, 1968).

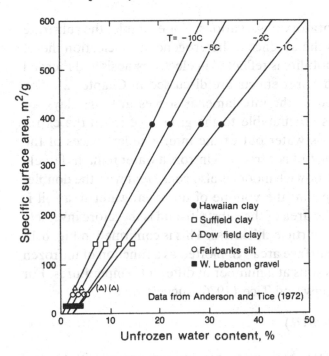

Fig. 6.15. Relation between unfrozen water content (by dry weight), specific surface area and temperature (from McGaw and Tice (1976).

In recent years it has become apparent that the double-layer model is merely a point of departure in our understanding of pore freezing (Anderson and Morgenstern, 1973). The region between the pore ice and the substrate is not simply a bound film of normal water (e.g., its viscosity is many times greater (Graham, *et al.*, 1964), thus impeding its mobility further). More accurately, it must be described as a liquid-like sandwich with two outer zones of disordered molecules separated by a partially ordered zone. The molecules of the central zone may be regarded as clusters or cages which are continually being formed and broken (the flickering cluster model is discussed in Section 7.2): to a certain extent the water has enhanced order. The disordered zone adjacent to the substrate is characterized by monomers which are not strongly hydrogen bonded, but which are engaged in dipole interactions facilitating proton mobility. Much the same appears to be true of the disordered zone contacting the ice. The adjacent ice has itself been subdivided into zones (Drost-Hansen, 1967) which extend from the disordered water zone to normal crystalline ice: between these are several layers with order in varying degrees suggested by observed dipole and proton behaviour.

The sandwich model provides a more substantial basis on which to account for field and laboratory observations. For example, the substrate disordered zone explains the absence of heterogeneous nucleation there, and the high molecular mobility is reflected in electroosmosis and thermal osmosis (the Dufour and Soret effects are discussed in Chapter 2). The model is readily extended to the interlamellar spaces in those clays for which freeze-shrinkage is attributable to the growth of ice in the extra-lamellar spaces; this draws water out of the interlamellar spaces of the platelet causing it to shrink but not freeze. Unfrozen water remains largely within the platelet. The sandwich model is also consistent with the double-layer model in predicting that the volume of unfrozen water in a soil is directly proportional to the area of the substrate and is therefore inversely proportional to the mean particle thickness. This is confirmed in Fig. 6.15 which shows the specific surface area[25] S_s plotted as a function of unfrozen water content for various soils at a number of different temperatures. For the data presented, McGaw and Tice (1976) suggest that

$$w_w = a(\theta) + \frac{S_s}{600} \, b(\theta) \tag{6.31}$$

where $\theta = T^0 - T$ and $a(\theta)$, $b(\theta)$ were determined empirically, to be

$$\left.\begin{aligned} a(\theta) &= 0.0210\,(1 - \log\theta) \\ b(\theta) &= 0.478 - 0.194 \log\theta \end{aligned}\right\} \tag{6.32}$$

Since the unfrozen water film may be only a few tens of angstroms thick, its volume only becomes significant in fine soils e.g., $r < 2\,\mu$m. It is therefore to be expected that unfrozen water would be most important in silts and clays, especially the latter.

Most of the above discussion applies to particulate soils consisting of grains or platelets separated by pores filled with varying amounts of ice, water and air. In organic soils, the 'particles' are formed from the frag-ments of organs and tissue in varying stages of decomposition. Chapter 7 treats the structure of living material in some detail; here it is sufficient to make a few general observations. Most plants and animals abound in fibrous and tubular structures, and it is therefore not surprising to find organic debris with a similar structure. In the early stages of decomposi-tion the skeletal structure tends to remain while water and other consti-tuents are lost. In the face of the subsequent forces of weathering, this fragile structure is fragmented and decomposed further until, eventually, it reaches a stage where it is described as *humus*. It no longer decomposes, except through the possible action of bacteria. The structure of an organic soil thus lies somewhere between that of humus and that of last year's deposit of leaves, stems and roots.

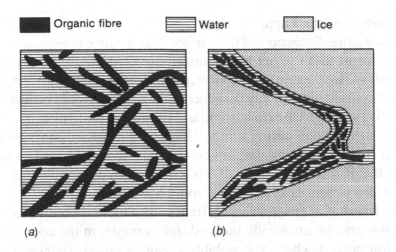

Fig. 6.16. Freezing of a saturated organic soil (*a*) before freezing; (*b*) after freezing (following Kay *et al.* (1975)).

Humus itself may consist of very small fragments and it may be present in organic soils in significant amounts (though usually less than 50%). It is also found in mineral soils in small amounts. Humus is colloidal matter (Hillel, 1980a) and plays a role somewhat analogous to that of fine clay particles. A particle of humus, which is typically amorphous (cf. clay which is typically crystalline), is often negatively charged and its behaviour may therefore be described in terms of the double-layer model discussed above. The similarity with mineral soils ends there, however, because the bulk of the organic material is fibrous and may be hydrophilic or hydrophobic, depending on the original structure and the state of decay. Despite the low surface area associated with the large pores found in organic soils, surface effects are clearly noticeable (Kay *et al.*, 1975), particularly at lower values of water content.

When the temperature of an organic soil decreases, nucleation begins in the interior of the larger pores which then attract more water from the smaller pores and may partly dehydrate the organic fibres. As indicated in Fig. 6.16, the organic matter is increasingly concentrated into slender regions as a result of freezing; ice crystals grow to fill and then enlarge the pores. Such texturing of the fibrous matrix clearly produces changes in its properties, notably in the heat capacity and the permeability. The freeze-shrinkage produced is thought to be analogous to that which takes place in clay suspensions (Norrish and Rausell-Colom, 1962).

6.4   Freezing in the Earth

During the discussion of the surface heat balance in Section 6.1 it became apparent that the net heat transfer rate $\dot{Q}$ at the Earth's surface may be positive (i.e., a gain) or negative, depending upon the location, the season and even the time of day. In general, the balance would be increasingly negative with increases in latitude and altitude; in particular, it would be negative at night and during winter. Given the presence of water in the regolith, it is not surprising that environmentally-induced heat loss may cause much of it to freeze. Some of this ice has existed for millennia and some of it comes and goes annually; some has perhaps appeared overnight. The freezing of groundwater is frequently a natural occurrence but freezing may also be artificially-induced: for example, in the course of a construction project where the stability of soil is crucial (Kinosita and Fukuda, 1985).

During freezing, the 'cold wave' sweeps through the earth thereby generating a cooling rate $\partial T/\partial t$ which is a function of both time and position. Typically, the drop in temperature and the cooling rate both decrease with distance from the cold source, and hence the thermodynamic state in general, and phase composition in particular, will also vary with distance from the cold source. During the artificial freezing of soil, the cooling rates are often particularly high and may prevent the local state from reaching stable equilibrium; any data or behaviour based upon stable conditions may not then apply. For example, solutes may not be able to diffuse ahead of the interface fast enough, and water mobility may be restricted. Under natural circumstances, however, when cooling rates are typically much lower, stable conditions are more likely to occur: phase behaviour determined in the laboratory may then be used directly, provided it is recognized that field data are seldom subject to the same high degree of experimental control.

Natural heat loss from the Earth's surface leads to a freezing problem which is nominally one-dimensional, at least on a horizontal scale of tens to thousands of metres and a corresponding vertical scale less than one-tenth of that. In a simple, Stefanian description the freezing front is a (horizontal) surface of fixed temperature moving away from the cold source above. This situation was discussed at length in Chapter 3. Whenever the depth of freezing is systematically greater than the depth of thaw, which is usually called the *active layer*, the seasonal cycles will gradually create permanently frozen soil at depth, as noted in Section 6.1. This *permafrost* is a widespread indication of climate, from the recent to the remote past.

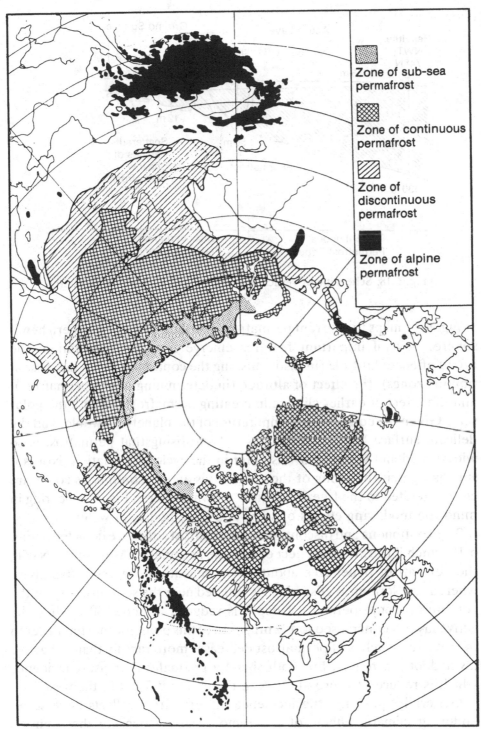

Fig. 6.17. Distribution of permafrost in the northern hemisphere (following Péwé (1983)).

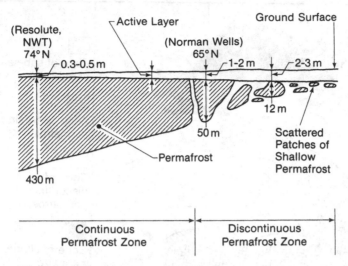

Fig. 6.18. Schematic of vertical distribution and thickness of permafrost.

Fig. 6.17 maps the current permafrost distribution in the northern hemisphere. Several important features emerge from this illustration: the broad effect of latitude (in distinguishing the continuous from the discontinuous zones); the effect of altitude (in determining alpine permafrost); and the effect of earlier climate in creating permafrost beneath the polar seas. Given the equilibrium temperature of the planet as a whole, and the delicate surface heat balance, it is not surprising that permafrost is so widespread and variable at this point in the cycle of glaciation. Nor is it surprising that the limits of the continuous and discontinuous zones are closely related to the freezing indices shown earlier in Fig. 6.1, bearing in mind the modifying effects of snow and large bodies of water.

The continuous and discontinuous permafrost zones are depicted in Fig. 6.18 which illustrates the effect of latitude in Canada. At Resolute NWT, the permafrost penetrates about 430 m beneath a shallow active layer, whereas at Norman Wells, which is situated near the southern edge of the continuous zone, permafrost depth is reduced to about 50 m while the active layer has increased to 1–2 m. South of this point the maximum depth of scattered patches of permafrost decreases more rapidly than the active layer depth increases, but both signal a gradually decreasing incidence which is reduced to zero at 50° N in the east and 57° N in the west.

Generally speaking, the incidence of permafrost reflects large scale, enduring climatic effects of the remote past, especially those effects accompanying periods of glaciation. But within particular permafrost zones, the effects of more recent climate are manifest in a great many other

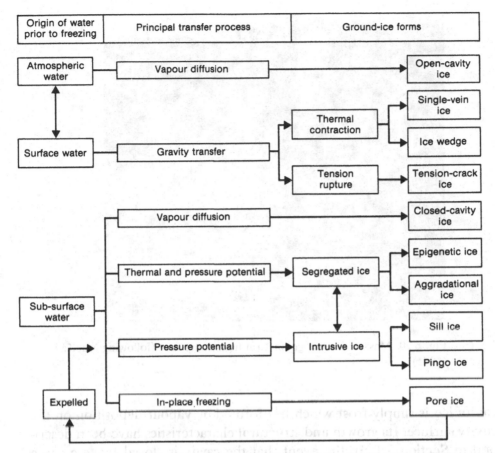

Fig. 6.19. Classification of ground-ice forms (following MacKay (1972)).

forms of ground ice. These have been described in a series of papers by MacKay (1971, 1972, 1973) who has divided them into the various categories displayed in Fig. 6.19. The overall organization is guided by two principal factors: the source of the water from which the ice was formed, and the transport processes by which the water was moved to the point where freezing eventually took place. The first factor may be divided further into three main sources: these are atmospheric water vapour, surface water and ground water. The second factor includes the multiplicity of heat, mass and momentum transfer processes which have been discussed elsewhere in the book; notably in Chapters 2 and 3 and in the previous section.

In the simplest situation, water vapour from the atmosphere may enter an open cavity the size of which may range from that of a single pore to the macroscopic dimensions of cracks formed through tensile failure. Open

Fig. 6.20. Massive ice wedge on 10 m high slump face (following MacKay (1971)).

cavity ice is simply frost which has formed by vapour deposition on the cavity surface. Its growth and structural characteristics have been described in Section 5.1. In the event that the cavity is closed (as, e.g., in a pressurized gas pocket found in frozen sediment (MacKay, 1965)), vapour deposition may produce geometrically-regular crystals which are reminiscent of depth hoar. This has been described as closed cavity ice. Although the depositional process is very similar to that of open cavity ice, the source is groundwater.

Cavity ice may be greatly modified by the subsequent presence of water in the liquid phase; this is especially true when surface water, under the influence of gravity, drains into open cavities formed at the Earth's surface. As Fig. 6.19 indicates, cracks may be flooded with water which, being close to the freezing point, soon freezes. Capillarity and hydrostatic head both encourage the water to fill the pores of the cavity ice. Prior subcooling of the ice matrix may begin the freezing process but is unlikely to complete it unless the Stefan number is 1.0: i.e., unless the cavity ice is at a temperature[26] not greater than $-160\,°C$. Any unfrozen water remaining is frozen through the subcooled material flanking the cavity and by heat loss to the atmosphere through the cavity mouth. Cracks in the permafrost are frequently filled by surface water: if the cracks are thermal in origin, the ice is called vein ice, whereas if the crack origin is mechanical rupture caused, for example, by the growth of adjacent segregated ice, it is called tension-crack ice. Vein ice invariably runs vertically but tension-crack ice is orientated according to the ice-induced stress field which caused it.

Mechanical rupture is usually caused by an adjacent displacement which is progressive; i.e., as time progresses the crack would gradually widen whether filled with ice or not. On the other hand, thermal contraction accompanying falling temperatures is a pervasive, seasonal event which occurs at numerous locations over large areas of the Earth. Each year the isotropic (in the horizontal plane) tensile stress field may induce a pattern of cracks which, in the spring, fill with meltwater from snow. Initially, this produces a pattern of thin, vertical ice veinlets which, being relatively incompressible, cause the surrounding soil to be compressed slightly as it warms during the summer and tries to expand. Thermal contraction the following year causes the cracks to re-open, thus repeating the process. Veins may grow in this cyclical manner for many years until they occupy as much as 50% of the ground surface and perhaps extend down 3 m (Brown, 1966). This form of massive ground ice, called an ice wedge, is illustrated in Fig. 6.20.

In contrast to surface water, groundwater does not generally play a passive role in the formation of ground ice, except perhaps in providing the water vapour from which closed cavity ice grows. At the very least, groundwater filling or partially filling soil pores will freeze *in situ*, following the dictates of thermodynamics discussed in the previous section. This produces pore ice occupying a greater volume than the water from which it was formed, and hence, in saturated sand or gravel, creates either a build up of pressure or an expulsion of water ahead of the freezing front. Provided the surplus water is free to move easily, as it often is in extensive coarse-grained soils, the pore ice thus formed will be capable of filling the entire space originally occupied by the water and may become a very effective supporting skeleton compromised only by any solutes released and trapped in concentrated pockets during the freezing process.

The pore pressure of groundwater may, on occasion, become high enough to give rise to an intrusive flow which splits the earth before freezing. Analogous to igneous intrusions, such aqueous intrusions are most common in tabular form (the equivalent of a sill) and domed form (equivalent to a laccolith). Both forms are engendered by the freezing of a body of pore water confined by boundaries impermeable enough to create a substantial build-up of pressure which finds release by splitting the

Fig. 6.21. Sill ice (following MacKay (1971)).

confining material at its weakest point. The intrusive thrust thereby created is then subject to a heat loss: to the subcooled flanking material or the subzero atmosphere. The intrusion thus freezes.

In tabular form, such intrusions are know as sill ice. They may occur at the base of the active layer which contains much of the summer thaw water and perhaps snowmelt. Upon re-freezing from the top, this water is sealed between the descending ice front and the top of the permafrost, both of which are essentially impermeable. The consequent rise in pore pressure at the base of the active layer may then give rise to the intrusion. Fig. 6.21 shows sill ice over a metre thick found in the Brock River Delta, NWT, Canada. The principal domed form of intrusive ice is the pingo (an Innuit word for hill) which originates in a surface eruption of surplus pore water constrained by the permafrost table. The source of this water is typically a former lake which, deep enough not to freeze to the bottom, has created an extensive depression in the permafrost table through summer heat absorption. When such a lake drains, the earth immediately beneath begins to form a frozen crust, thus enclosing a bowl of saturated, unfrozen soil. Frequently, this is a coarse soil in which the pore pressure continues to rise as the freezing front travels downwards and inwards. The water may not escape through the permafrost beneath and immediately surrounding it, and therefore seeks release in surface weaknesses. These are often

Fig. 6.22. Ibyuk Pingo, 49 m high, located just to the southwest of
Tuktoyaktuk, NWT (following MacKay (1987)).

provided by the thinnest parts of the crust underneath any remaining
ponds of surface water which, because of radiation-trapping ability, have
maintained a local depression in the crust. The final eruption may be quite
spectacular, as Fig. 6.22 reveals.

The growth of pingo ice is illustrated schematically in Fig. 6.23 which
shows how the depressed surface at the bottom of a pond rises in three
successive stages. At first, the uplift is simply the local bulging of the frozen
crust under the influence of a rising pore pressure in the unfrozen water
beneath. The onset of uplift may therefore signal the completion of the crust
beneath the pond during the period immediately after the original lake has
drained; this is when the rate of crust formation and thickening is at its
highest. The pond is thus spilled over the surrounding terrain while a lens-
like core of ice appears and grows intrusively at the new permafrost base, as
indicated schematically in Fig. 6.23(b). This is a period of rapid growth
attributable to large pore pressures, and during which the tensile stresses in
the domed 'overburden' frequently exceed the tensile strength of the soil
which consequently fractures, producing a pattern of radial cracks. These
cracks, which are evident in Fig. 6.22 are soon filled with tension-crack ice
which may extend down into the pingo ice core. Eventually, the base of the
ice core is sealed over by the continuing aggradation of the surrounding
permafrost, by which time the rate of freezing of pore water, and the rate of
uplift, will be much reduced. This is depicted in Fig. 6.23(c).

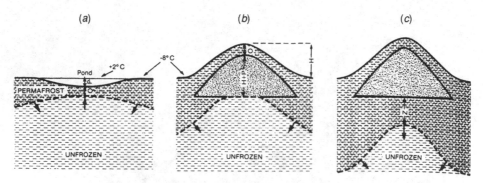

Fig. 6.23. Diagram (*a*) shows a residual pond of depth *P* and permafrost beneath it of thickness *O* at the moment that pingo growth commences. As the ice grows, the freezing front remains stationary, and permafrost of thickness *O* becomes uplifted to form the 'overburden' thickness. Diagram (*b*) shows a pingo with the freezing front at the bottom of the ice-core. The overburden thickness is *O*; the height of the pingo above the lake bottom is *H*; and the maximum height of the ice-core is *H* + *P*. Diagram (*c*) shows a pingo with the freezing front at a depth *Z* below the ice-core. The pingo can continue to grow from ice segregation and the freezing of pore water in a confined system (following MacKay (1971)).

Once the ice core has formed it is fed by pore water under pressure from beneath. In an open pingo system, this water is supplied from the general pattern of groundwater flow in the vicinity, and may therefore be provided indefinitely. In the closed pingo system described above, however, the supply is not only finite but will become disconnected from the ice core. Even before this happens the pore pressure immediately beneath the ice may be insufficient to maintain an intrusive thrust, in which case growth must then be attributable to some other mechanism. MacKay (1973) suggests that the explanation may be found in the process of primary heaving which is exemplified in the growth of needle ice, discussed in Section 6.2. It will be recalled that this phenomenon was attributed to the growth of segregated ice in the presence of a plentiful supply of subsurface water. The segregated ice grows only if the pore water pressure immediately beneath it is in a restricted range: if it is too high it will not create a water flow from depth towards the ice, and if it is too low the ice will invade the pores. Thus, as intrusion in a closed pingo progresses and the pore water pressure falls, it may eventually become low enough to permit a primary heaving process to begin and thus sustain growth, though at a lower rate. However, if the pore water pressure P_w subsequently falls to the point where

$$P_i - P_w = 2\sigma_{iw}/r_p \tag{6.33}$$

in which P_i is the ice pressure in the base of the core, and σ_{iw} is the surface tension of water over ice in a pore of radius r_p, the water will no longer be able to prevent the ice from entering the pores beneath; the permafrost then grows to seal off the base of the intrusion.

The point at which intrusive ice becomes segregated ice is unclear, and the two forms are often difficult to distinguish in the field. Moreover, the primary heaving mechanism does not seem to apply to most forms of segregated ice. It is found, e.g., that the freezing of fine silts often produces a banded ice structure with lens-like segregations lying normal to the direction of heat loss i.e., horizontally. On the other hand, some clays (e.g., montmorillonite) exhibit a reticulated ice pattern under similar thermal conditions: superimposed on horizontal lenses are vertical ice bands evidently made possible by interlamellar water. These are particular manifestations of secondary heaving. Rooted in the thermodynamics of pore ice, this is a more complex mechanism which is perhaps best viewed in the context of frost penetration or, more specifically, in light of the temperature and pore pressure fields produced by freezing. Therefore, before attempting a discussion of secondary heaving it is worthwhile reviewing these field equations, the regions in which their forms are valid and the variables they contain.

The simple Stefanian description of freezing is usually adequate for the determination of general trends as they relate to the prevailing climate. More often than not, however, local predictions require a more precise knowledge of local conditions: details of the microclimate; variations in thermophysical properties; and the role of water, mobile or quiescent. The classical Stefanian description is restricted to two regions separated by a surface over which the temperature is independent of position and time. More generally, it is important to recognize the existence of a freezing temperature range attributable to the unfrozen water. This may be done by dividing the field into three zones:

(1) an *unfrozen zone* in which groundwater may or may not flow, and where temperatures are above the initial equilibrium freezing temperature T^i;

(2) a *frozen zone* which, for practical purposes, may be assumed to contain only immobile, unfrozen water, and in which temperatures are less than some lower freezing temperature $T_f \ll T^i$;

(3) a *freezing fringe* in which unfrozen water is mobile and the temperature ranges between T^i and T_f.

The characteristics of these zones are generally very different.

The principal thermophysical and transport properties also vary greatly,

Table 6.7. *Comparison of thermal properties (following Hillel (1980a)).*

Constituent	Density		Volumetric heat capacity		Thermal conductivity	
	$(gm\,cm^{-3})$	$(kg\,m^{-3})$	$(cal\,cm^{-3}K^{-1})$	$(Jm^{-3}K^{-1})$	$(mcal\,cm^{-1}s^{-1}K^{-1})$	$(Wm^{-1}K^{-1})$
Quartz	2.66	2.66×10^3	0.48	2.0×10^6	2.1	8.8
Other Materials						
(average)	2.65	2.65×10^3	0.48	2.0×10^6	7	2.9
Organic	1.3	1.30×10^3	0.6	2.5×10^6	0.6	0.25
Water (liquid)	1.0	1.00×10^3	1.0	4.2×10^6	1.37	0.57
Ice	0.92	0.92×10^3	0.45	1.9×10^6	5.2	2.2
Air	0.00125	1.25	0.003	1.25×10^3	0.06	0.025

not only from zone to zone but from soil to soil. Where these properties are not available in empirical form they may be calculated as the weighted average[27] of the component values. Thus, the mean density is given by

$$\bar{\rho} = M/V = m_w + m_i + m_a + m_s \qquad (6.34)$$

where m represents the volumetric content of the component, and the subscripts w, i, a and s signify water, ice, air and soil, respectively. It is important to distinguish between the component material density ρ, its volumetric content m and its specific content w. Thus, for the ice component,

$$\rho_i = M_i/V_i, \quad m_i = M_i/V \quad \text{and} \quad w_i = M_i/M_s$$

in which M_s is the mass of the soil particles in the volume $V = V_w + V_i + V_a + V_s$. Hence

$$m_i = \frac{M_i}{M_s}\frac{M_s}{V} = w_i m_s = w_i \rho_s (1 - \mathcal{N}) = \frac{w_i \rho_s}{e + 1}$$

relates the volumetric and specific contents through particle volumetric content, particle density, porosity and void ratio.

In similar manner, the mean intrinsic specific heat capacity may be calculated from

$$\bar{c} = \frac{m_w c_w + m_i c_i + m_a c_a + m_s c_s}{\bar{\rho}} \qquad (6.35)$$

the numerator of which is simply the average volumetric heat capacity $c_v = \bar{c}\bar{\rho}$. A representative set of component densities and volumetric heat capacities is included in Table 6.7 from which it is evident that air plays only a minor role. The intrinsic volumetric heat capacities of some soils are given in Table 6.8. Given the range of data in Table 6.7, it is to be expected that soil heat capacities would depend upon both the porosity and the

Table 6.8. *Average thermal properties of soils and snow (following De Vries (1963))*

Material	Porosity	Volumetric wetness	Thermal conductivity (10^{-3}cal cm^{-1} s^{-1} K^{-1})	Volumetric heat capacity (cal cm^{-3} K^{-1})
Sand	0.4	0.0	0.7	0.3
	0.4	0.2	4.2	0.5
	0.4	0.4	5.2	0.7
Clay	0.4	0.0	0.6	0.3
	0.4	0.2	2.8	0.5
	0.4	0.4	3.8	0.7
Peat	0.8	0.0	0.14	0.35
	0.8	0.4	0.7	0.75
	0.8	0.8	1.2	1.15
Snow	0.95	0.05	0.15	0.05
	0.8	0.2	0.32	0.2
	0.5	0.5	1.7	0.5

contents of the pores. It is worth noting that an increase in the volume fraction of water v_w (sometimes called the wetness) causes a substantial increase in the heat capacity; this is especially true of organic soils.

The data in Table 6.8 are intrinsic and do not incorporate the heats of fusion or wetting. To accommodate the latent heat effect, for example, it is necessary to add the term introduced earlier in equations (6.29) and (6.30). Thus, the apparent specific heat capacity $\bar{c}'$ is given by

$$\bar{c}' = \bar{c} - \frac{\lambda_{iw}}{\bar{\rho}} \frac{\partial m_i}{\partial T}$$

and hence the apparent, or effective, volumetric heat capacity is no longer c_v but

$$c_{ve} = \bar{c}'\bar{\rho} = c_v - \lambda_{iw} \partial m_i/\partial T \qquad (6.36)$$

Alternatively, the heat capacity may be based on the mass of dry soil rather than on the total mass of all the components. Thus

$$\bar{c}'' = w_w c_w + w_i c_i + w_a c_a + c_s - \lambda_{iw} \partial w_i/\partial T$$

When air and vapour are neglected and the total water content $w = w_i + w_w$ remains constant, this reduces to the more convenient form

$$\bar{c}'' = w c_i + c_s + w_w(c_w - c_i) + \lambda_{iw} \partial w_w/\partial T \qquad (6.37)$$

The mean intrinsic thermal conductivity may also be represented as a volumetric average (a more thorough discussion is given by Farouki (1982)). Thus

$$\bar{k} = v_w k_w + v_i k_i + v_a k_a + v_s k_s \qquad (6.38)$$

which implies that pore geometry and interconnectedness are unimportant, and components conduct in parallel, each being subject to the same temperature gradient. Table 6.7 indicates that the mineral materials, followed by ice, are the most effective heat conductors. At first glance it appears that the effect of air on thermal conductivity could be ignored, but Table 6.8 reveals that air is such a poor conductor that when it is displaced from the pores by water, which conducts at more than 20 times the rate, the effect is very significant. The general transport of heat was discussed in the previous section from which it will be recalled that 'fluid thermal conductivity' must, in general, be represented by an expression of the form

$$k_f^Q = \rho_f c_{pf} \mathscr{D}^Q = \rho_f c_{pf} (\mathscr{D}^m + \mathscr{D}^d)$$

When the Reynolds number of the pore water flow is small, this reduces to

$$k_f^Q = \rho_f c_{pf} D_m \tau = k_f \tau. \qquad (6.39)$$

On the other hand, the effective overall thermal conductivity used in equation (6.26) is given by

$$k^Q = \mathcal{N} k_f + (1 - \mathcal{N}) k_s \qquad (6.40)$$

where, in general, $\mathcal{N}$ is not the porosity but the volume fraction filled by fluids (here by air and water) and $(1 - \mathcal{N})$ is therefore the complementary volume fraction occupied by solids (ice and particulates). By inspection, it is evident that the fluid contributions to thermal conductivity in these expressions are different and essentially unrelated: in equation (6.40) the systemic effect of tortuosity has been ignored. However, this difference will be unimportant if $(1 - \mathcal{N}) k_s \gg \mathcal{N} k_f$. Table 6.7 indicates that $k_s \gg k_f$ in mineral soils, for which the neglect of tortuosity will therefore be permissible provided that the porosity is not too great e.g., $\mathcal{N} < 50\%$. The ice and mineral components dominate the thermal conductivity in mineral soils, but this is less true for organic soils or snow.

Returning to the zonal model of freezing, it is now possible to substitute the thermophysical properties which appear in the energy equation (6.27). In the one-dimensional (vertical) form suitable for the study of frost penetration, this may be written

$$c_{ve} \frac{\partial T}{\partial t} + c_{vw} V_Z \frac{\partial T}{\partial Z} = \frac{\partial}{\partial Z}\left(\bar{k}\frac{\partial T}{\partial Z}\right) \tag{6.41}$$

where c_{ve} is given by equation (6.36), $\bar{k}$ by equation (6.38), and c_{vw} is simply the volumetric heat capacity of pure water (neglecting the presence of air and water vapour (see Table 6.7)). The (latent) heat source $\dot{s}_Q$ in equation (6.27) has been absorbed in the effective volumetric heat capacity c_{ve}. The vertical water velocity V_Z must be determined from the continuity equation (6.29) which becomes

$$C_m \frac{\partial H}{\partial t} + \frac{1}{\rho_w}\frac{\partial m_i}{\partial T}\frac{\partial T}{\partial t} = \frac{\partial}{\partial Z}\left[K(H)\frac{\partial H}{\partial Z}\right] \tag{6.42}$$

thus extending equation (6.19) to include the water sink inherent in the formation of ice: $C_m = \rho_w g \alpha$, the specific storage,[28] includes any effect of ice fraction on the matrix compressibility α.

Equations (6.41) and (6.42) together pose the heat and fluid flow problem for each of the three zones described earlier. In the unfrozen zone ($T > T^i$), $\partial m_i / \partial t = 0$; hence c_{ve} reduces to its intrinsic form and the left hand side of equation (6.42) simplifies accordingly. In the frozen zone ($T < T_f$), the entire continuity equation reduces to the simple balance
$$\partial m_w / \partial t = - \partial m_i / \partial t$$
if the unfrozen water is assumed to be immobile, in which circumstance the advective term is eliminated from the energy equation, but the apparent specific heat must be retained. The freezing fringe is the most complex zone to treat. This is partly because none of the terms vanishes or reduces to a simpler form, and partly because the thermophysical and transport properties are difficult to prescribe, even empirically (Horiguchi and Miller, 1983).

It is now clear that thermophysical and transport properties of soils, even averaged values, are not simple constants; frequently, they are functions of the dependent variables. This is true of the water content and hydraulic conductivity in particular. For an unsaturated soil, Gardner (1970) has suggested that

$$v_w(H) = a(v_w)_{sat}/H^m \tag{6.43}$$

where a and m are empirical constants. Gardner (1960) has also suggested that

$$K(H) = b K_{sat}/H^n \tag{6.44}$$

where b and n are empirical constants. These expressions ensure that equation (6.42) is non-linear. A similar complication arises in the energy equation (6.41) because temperature-related changes in ice or water content have an effect on both c_{ve} and k. The role of water content on c_{ve}, for

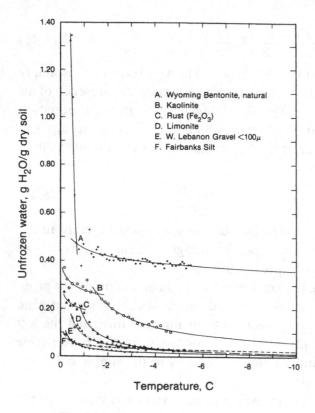

Fig. 6.24. Phase composition curves for six representative soils and soil constituents (following Anderson and Morgenstern (1973)).

example, is implicit in equation (6.36) but may be seen more clearly when the pore volume is constant and the equation is re-written in the form

$$c_{ve} = c_v + \rho_w \lambda_{iw} \left(\partial v_w / \partial T \right) \tag{6.45}$$

The unfrozen water plays an important role in each of the zones: in the frozen zone where it is immobile, or least mobile, and where v_w is lowest; in the freezing fringe where it is more mobile and v_w is almost as low as in the frozen zone; and in the unfrozen zone where both v_w and mobility are greatest. The equilibrium thermodynamic state of a partially-frozen soil is not defined unless the unfrozen water content is given. In general, an equilibrium state may be specified in terms of the work variables: the chemical work variables (the chemical potentials μ and the mass or mole fractions n); and the mechanical work variables (the pressures P and volumes V of each phase, together with the surface tensions σ and surface area S of the interfaces). It is therefore difficult completely to specify the state of a soil in easily measurable terms. However, for a given soil in which the pores contain only water and ice, the description simplifies and

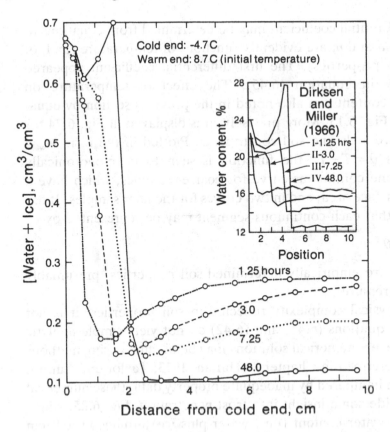

Fig. 6.25. Numerical simulation of soil freezing (following Taylor and Luthin (1976)).

the state may often be specified in terms of only two variables: the pressure (or piezometric head) and volume of the unfrozen water. Any other thermodynamic variable may then be expressed in terms of this pair. Thus, e.g.,

$$T = T(H, v_w)$$

and

$$\bar{h} = \bar{h}(h, v_w)$$

determine the system temperature and enthalpy,[29] respectively. The first of these may be re-written

$$v_w = v_w(T, H) \tag{6.46}$$

from which it follows that

$$dv_w = \left(\frac{\partial v_w}{\partial T}\right)_H dT + \left(\frac{\partial v_w}{\partial H}\right)_T dH$$

The second differential coefficient may be determined from equations of the type (6.43): a and m are evidently functions of temperature (and, of course, the soil properties). The first differential coefficient appeared earlier, notably in equation (6.45). The effect of temperature on unfrozen water content was also noted in the previous section by equation (6.31) and Fig. 6.15. More precisely, it is displayed in Fig. 6.24 for six representative soils and soil constituents. Plotted in the form $w_w = v_w \rho_w / m_s$ versus $\theta = T - T^0$ the function is seen to be monotonically decreasing: in one continuous curve for coarser samples, which have a lower specific surface area; and in two curves for the finer samples. It has been observed that each continuous segment may be represented by

$$w_w = \alpha \theta^\beta$$

where α and β are empirically determined soil properties, presumably dependent on pressure.

Given the physical complexity reflected in soil properties, it is not surprising that equations (6.41) and (6.42) do not yield simple analytic solutions. However, numerical solutions may be obtained using methods such as those described in Chapter 3 (Harlan, 1973; Taylor and Luthin, 1976). Although hampered by inadequate property data, these numerical models do provide some insight into frost penetration. Fig. 6.25, which shows the total water content (i.e., water plus ice) profiles in a 10 cm sample, compares the numerical predictions of Taylor and Luthin (1976) with the measurements of Dirksen and Miller (1966) obtained under similar conditions. The agreement, although qualitative, lends support to the physico-mathematical model. The effect of water suction developed at the colder (left) end of the sample is clearly evident. After an hour or two the soil near the warmer end is progressively dried by the freezing action. Likewise, the role of the freezing fringe, in which $T_f < T < T^i$, is also evident. It will be recalled that when $T = T_{cr} \lesssim T^i$, ice spontaneously fills the pores, if pore water is available nearby; this tends to create higher ice (plus water) contents, as shown. Initially, the freezing front at T^i moves very rapidly, leaving behind pore ice in soil which then has a very low permeability (assumed to be zero in the zonal model). At the same time, if the permeability of the unfrozen soil and its initial water content are low enough, complete ice saturation may be prevented, and the early water content profiles within the frozen and freezing zones would then tend to decrease with distance from the colder end of the sample. However, as the freezing front velocity decreases relative to the pore water velocity, the water content in the freezing fringe may be able to rise above earlier values; this evidently creates the U-shaped profile in the frozen zone.

Finally, we return to the question of secondary heaving as a manifestation of ice segregation during frost penetration. It will be recalled from Section 6.3 that nucleation in a pore begins not at the substrate but somewhere in the pore interior. For a saturated soil this occurs when $T = T^{i} \lesssim T^{0}$. For an unsaturated soil, nucleation occurs in the 'corners' where the inhibiting effect of surface forces tends to reduce the nucleation temperature yet further by an amount which is dependent on the volume of the corner water: i.e., on the initial water content, the grain size and the grain composition (or water surface tension). Freezing of a saturated soil is thus a limiting case in which crystal growth continues, reversibly and isothermally, until no unfrozen water remains, except for that retained in interlamellar spaces or as a thin film surrounding each grain. Experience with macroscopic cavities containing slightly supercooled water suggests that initial growth will be dendritic followed by the slower solidification of interdendritic water and the gradual development of glaze ice.[30] In general, pore ice growth cannot continue unless sufficient water is made available; at the freezing front this may create a jump to ice saturation, as noted earlier.

This type of discontinuity is perhaps better seen through the stable equilibrium force balance which, for the water–ice interface, is given by

$$P_{\text{w}} - P_{\text{i}} = 2\sigma_{\text{iw}}/r_{\text{iw}} \qquad (6.47)$$

where the interfacial radius r_{iw} is positive when the water contains the centre of curvature. For an ice crystal growing in water, r_{iw} is negative and hence the equilibrium ice pressure is greater than the equilibrium water pressure. However, should $P_{\text{i}} - P_{\text{w}}$ exceed $2\sigma_{\text{iw}}/|r_{\text{iw}}|$, the interface would become unstable because the imbalance would permit ice growth to a greater radius, thus increasing the imbalance further. Hence, if the ice pressure were to become too great, or the water pressure too small, ice would spontaneously fill the pore, except for the region containing the unfrozen water film; nor would it penetrate interlamellar spaces which are much smaller.

The significance of this basic model of pore freezing is that it accommodates the effects of interfacial forces before and after nucleation, whether the pore is initially filled with water or not. In general terms, it is applicable to all particulate aggregates, although the details vary considerably with substrate material and geometry. It implies that the warm edge of the freezing fringe is actually a slender region in which $T_{\text{cr}} < T < T^{i}$. Behind

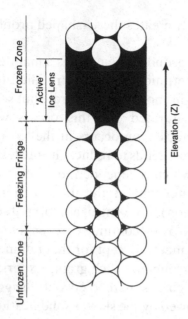

Fig. 6.26. Schematic representation of a frozen fringe in a soil consisting of spherical particles (following Gilpin (1982)).

this inner surface, i.e., where $T < T_{cr}$, the soil approaches very close to a state of complete ice saturation, assuming no serious impediment to the expulsion of any air initially present. In turn, this degree of ice saturation is required if segregation is to occur and thus provides a necessary, though not sufficient, condition. To understand the nature of segregation it is necessary to recall both the thermodynamics of pore ice and the equations governing heat and fluid transport in a porous medium; notably, in the freezing fringe.

Fig 6.26 is a schematic illustration of conditions beneath an active ice lens. This is taken from the work of Gilpin (1980, 1982) on which the following discussion is based. The pores of the freezing fringe are full of ice except for the unfrozen water content, which decreases as the segregating ice is approached. From the Clausius–Clapeyron equation (see Chapter 2), the stable equilibrium ice and water pressures at a temperature $\theta = T^0 - T$ are related through the expression

$$P_i = \frac{\rho_i}{\rho_w} P_w + \frac{\rho_i \lambda_{iw}}{T^0} \theta, \tag{6.48}$$

where $P_w = H \rho g$. This may be written

$$P_i = 0.91 P_w + 1.12 \times 10^3 \theta \tag{6.49}$$

when pressures are measured in kPa. Neglecting the density difference between water and ice, this equation reveals that the ice pressure is

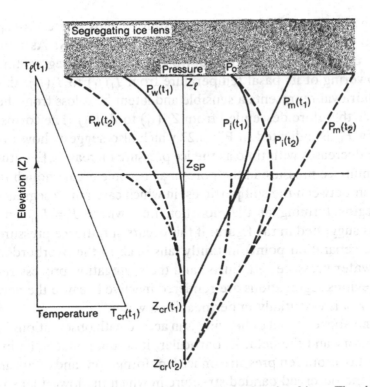

Fig. 6.27. Pressure development in the freezing fringe. The chain lines are ice and water pressure profiles shortly after new segregation forms at t_2.

simply the melting pressure $P_m = \rho_i \lambda_{iw} \theta / T^0$ plus the water pressure. Inside the thin freezing fringe below segregating ice the temperature profile may be approximated by a linear function of elevation, implying that P_m will also be linear. On the other hand, water flowing up towards the segregating ice demands that

$$V_{ff} = \frac{-K}{\rho_w g} \frac{\partial P_w}{\partial Z} \tag{6.50}$$

in which the water velocity V_{ff} will be independent of elevation only if ice and water do not accumulate in the fringe. As the segregating ice is approached from beneath it is to be expected that a decreasing temperature would engender an increasing ice fraction and a correspondingly lower value of K; this implies a water pressure gradient which steepens with elevation.

Taken together, the above features suggest the pressure profiles $P_m(t_1)$, $P_i(t_1)$ and $P_w(t_1)$ shown in Fig 6.27 for the time t_1. The water pressure has been shown as zero (it is, in fact, slightly negative as noted later) at the base of the fringe ($Z = Z_{cr}$), and the ice pressure has been equated to the overburden pressure P_0 at the base of the existing ice lens where $Z = Z_1$. The figure also shows the pressure profiles at a later time t_2. As time progresses, the continual removal of heat from the top of the segregating ice causes a lowering of its basal temperature from $T_1(t_1)$ to $T_1(t_2)$: this same heat withdrawal represents a sensible and latent heat loss from the fringe which has therefore descended from $Z_{cr}(t_1)$ to $Z_{cr}(t_2)$. The corresponding shift in P_m is indicated in Fig 6.27 which also suggests how the water pressure decreases with time as the ice pressure increases. Eventually, the maximum ice pressure in the pores may become great enough to cause separation between the soil particles, in which case a new segregation of ice begins forming at that location i.e., where $P_i = P_{SP}$ and $\partial P_i / \partial Z = 0$. As suggested in the figure, if this occurs at t_2 the ice pressure at $Z = Z_{SP}$ (the separation point) instantly falls back to the overburden pressure, the water pressure re-adjusts, and the segregation process repeats itself. Previous segregations are rendered inactive because the new segregation at Z_{SP} is essentially impermeable to water.

In general, the above model of heaving is in accord with observations in both the laboratory and the field. In particular, it accommodates the inhibiting effect of overburden pressure on ice lens formation, and offers an explanation of the observed banded structure in which the lowest lens is the thickest and the most recently active. It hinges on the assumption that the critical ice pressure at which particle separation occurs is determined by the force balance in the pore (following the analogous work of Haines (1925) and Fisher (1926)). In addition to incorporating the overburden pressure, this balance requires knowledge of the pore and ice–water interface geometries, together with the water and ice pressures, both assumed to be isotropic stresses. Gilpin used circular arcs to describe the geometry of the system and replaced the pressure of the pore occupants with a single parameter given by the ratio of the melting pressure P_m and a surface tension pressure $P_\sigma = 2\sigma_{iw}/r_p$, where r_p is the particle radius. This permits the ice pressure at separation to be written

$$P_{SP} = P_0 + \frac{2\sigma_{iw}}{r_p} f\left(\frac{P_m}{P_\sigma}\right) \tag{6.51}$$

in which $f(P_m/P_\sigma)$ must be found either from the assumed geometry or from experiment (Miller, 1978). Gilpin (1980) suggests that $f = 1$, and that r_p be determined from the fines (with diameters of about 10% of the nominal particle diameter).

Heat and mass flowing through the freezing fringe are bounded by two isothermal surfaces. On the colder surface, the temperature is given by

$\theta_1(t) = T^0 - T_1(t)$. Typically, $\theta(t)$ will be a monotonically increasing function of time, dependent on the rate of heat conduction through the solid ice above, and therefore upon the net rate of heat loss from the Earth's surface (or from the laboratory apparatus). On the warmer surface of the fringe, the temperature is given by $\theta_{cr} = T^0 - T_{cr}$; for higher temperatures (i.e., $T_{cr} < T < T^i$), ice may be present but will not fill the pores. Gilpin notes that the critical point corresponds to $P_m/P_\sigma \geqslant 2$, and hence $\theta_{cr} \geqslant 4\sigma_{iw} T^0/\rho_w \lambda_{iw} r_p$. The corresponding values of water pressure are: $P_w(t) = P_0 - P_{ml}(t)$ on the colder surface, where $P_{ml}(t) = \rho_w \lambda_{iw} \theta_1(t)/T^0$; and $P_w(t) = -\rho_w \lambda_{iw} \theta_{cr}/T^0 \approx -4\sigma_{iw}/r_p$ on the warmer surface. These boundary conditions are a necessary part of the problem prescription, but they do not complete it. It is also necessary to satisfy the requirements of continuity on the two surfaces of the fringe.

As Miller (1980) has noted, the freezing fringe is a heat engine. In fact it is a very inefficient heat engine (it has a Carnot efficiency of $\eta_{th} = 1 - T_1/T_{cr}$), accepting heat from the unfrozen soil at a rate

$$\dot{q}_{in} = -k_{uf}(\partial T/\partial Z)_{cr}$$

and rejecting it to the segregating ice at a rate

$$\dot{q}_{out} = -k_i(\partial T/\partial Z)_1$$

Ignoring the minor work transfer, the net rate of heat loss must be balanced by the rate of energy reduction within the material. Hence

$$\dot{q}_{out} - \dot{q}_{in} = \rho_i \lambda_{iw} V_1 - m_i \lambda_{iw} (dZ_{cr}/dt)$$

in which V_1 is the ice velocity at $Z = Z_1$, i.e., the heave rate. Both terms on the right hand side represent a latent heat withdrawal, the first term approximating the enthalpic reduction rate $\rho_i V_1(h_{cr} - h_1)$ of material flowing through the fringe;[31] the corresponding latent heat removal is assumed to occur at Z_1. The second term represents the latent heat removed from the fringe by the advance of the Stefanian freezing front, or frost front, at Z_{cr}. This equation may be split into two heat balances, one for each surface of the fringe. Thus

$$\left.\begin{aligned} -k_i\left(\frac{\partial T}{\partial Z}\right)_1 + k_{ff}\left(\frac{\partial T}{\partial Z}\right)_1 &= \rho_i \lambda_{iw} V_1 \\[2mm] -k_{ff}\left(\frac{\partial T}{\partial Z}\right)_{cr} + k_{uf}\left(\frac{\partial T}{\partial Z}\right)_{cr} &= -m_i \lambda_{iw} \frac{dZ_{cr}}{dt} \end{aligned}\right\} \tag{6.52}$$

in which

$$k_{ff}\left(\frac{\partial T}{\partial Z}\right)_{cr} \approx k_{ff}\left(\frac{\partial T}{\partial Z}\right)_1$$

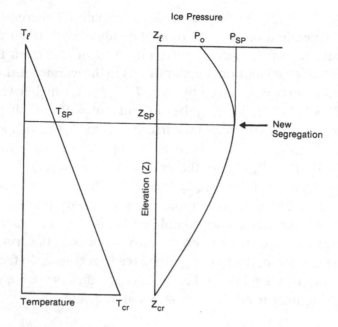

Fig. 6.28. Ice pressure and temperature fields immediately prior to the appearance of a new segregation.

is the heat flux through the fringe itself. The corresponding water flux balances also incorporate the assumption that water is being frozen at each interface. Thus, the material leaves the fringe at a rate

$$V_l = \frac{\rho_w}{\rho_i}(V_{ff})_l$$

and enters it at a rate

$$V_{uf} = (V_{ff})_{cr} + \frac{m_i(\rho_w - \rho_i)}{\rho_w^2}\frac{dZ_{cr}}{dt}$$

(6.53)

the second term on the right hand side representing water pushed back as a result of expansion during the ice formation. In these expressions,

$$(V_{ff})_l \approx (V_{ff})_{cr}$$

is the water flux through the fringe. This was defined earlier by equation (6.50), and may be re-written

$$V_{ff} = \frac{-K}{g}\frac{\partial}{\partial Z}\left(\frac{P_i}{\rho_i} - \frac{\lambda_{iw}\theta}{T^0}\right)$$

(6.54)

using equation (6.48).

This leads us quite naturally to the discussion of a segregation criterion. It is clear that ice segregation will not occur unless two conditions are

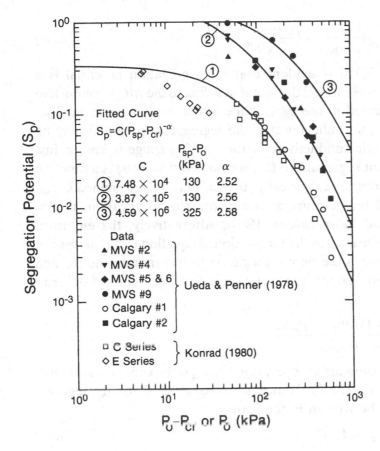

Fig. 6.29. Experimental data compared with a power law dependence of segregation potential (S_p) on overburden pressure (following Gilpin (1982)).

satisfied: the soil is ice-saturated; and the ice pressure at some point in the resulting freezing fringe is high enough to cause soil particles to separate. The situation is described in Fig. 6.28. As noted earlier, a new lens appears in the fringe when, $P_i = P_{SP}$ and $\partial P_i / \partial Z = 0$ at $Z = Z_{SP}$; equation (6.54) then reduces to

$$V_{ff} = \frac{K_{SP} \lambda_{iw}}{g T^0} \left(\frac{\partial \theta}{\partial Z} \right)_{SP} = \frac{\rho_i}{\rho_w} V_H \qquad (6.55)$$

using the first of equations (6.53) applied to the incipient growth rate V_H of a lens appearing at $Z = Z_{SP}$. This provides a means whereby the segregation pressure condition may be converted into a corresponding segregation temperature condition. Using this relation, the segregation potential[32] is now defined by

$$S_P = \frac{\rho_i \lambda_{iw} V_H}{k_{ff} (\partial \theta / \partial Z)_{SP}} = \frac{\rho_w \lambda_{iw}^2 K_{SP}}{k_{ff} g T^0} \qquad (6.56)$$

From equation (6.52) it is evident that the segregation potential is a thermal parameter, the ratio of the latent heat flux to the fringe conductive heat flux, both referred to the base of the incipient segregation at Z_{SP}.

Equation (6.56) also indicates that the segregation potential may be found if the hydraulic conductivity of the freezing fringe is known just beneath the incipient segregation. If it is assumed that $K(\theta)$ is a function which may be determined empirically, then S_P may be determined if θ_{SP}, the new lens basal temperature at the moment of segregation, is also known (Konrad and Morgenstern, 1980). Alternatively, the empirical function $K(\theta)$ may be used in the integration of equation (6.54) across the freezing fringe beneath the new segregation to incorporate the ice and water pressures. For example, if $K(\theta) = K_1 / \theta^\alpha$, the segregation temperature is

$$\theta_{SP} = \frac{(\alpha + 1)(P_{SP} - P_{cr}) T^0}{\alpha \rho_i \lambda_{iw}}$$

where P_{SP} is the ice pressure at separation and P_{cr} is the water pressure just below the freezing front at Z_{cr}. Hence, using equation (6.56), the segregation potential may be written in the form[33]

$$S_p = C / (P_{SP} - P_{cr})^\alpha \qquad (6.57)$$

which automatically includes the effect of overburden pressure,

$$P_o = P_{SP} - \frac{2\sigma_{iw}}{r_p} f\left(\frac{P_m}{P_\sigma}\right)$$

Fig. 6.29 shows equation (6.57) fitted to experimental data (Gilpin, 1982) for a variety of soils, and suggests that C may be regarded as a soil property: α is evidently slightly greater than 2.5 for the soils considered. When the overburden pressure exceeds 10^2 kPa, its effect on the segregation potential becomes increasingly significant.

Overburden pressure also affects the heave rate through the segregation potential. From equation (6.56) it is evident that the heave rate V_H of the new segregation may be predicted if both the segregation potential and the heat flux in the freezing fringe are known. The latter is not known *a priori* but may be written in terms of the heat flux in the segregating ice; in turn, the heat flux through the new ice segregation may be related directly to the upward soil heat loss causing the freezing to take place. Thus from the first of equations (6.52) applied to the new lens,

$$V_H = \frac{k_i \, (\partial \theta_i / \partial Z)_{SP}}{\rho_i \lambda_{iw} \, (1 + 1/S_p)} \tag{6.58}$$

gives the heave rate during an advancing frost front. This automatically incorporates the effect of overburden. As P_o approaches 10^3 kPa it is evident from Fig. 6.29 that $S_p \ll 1$, in which case the above expression reduces to

$$V_H = \frac{S_p \, k_i \, (\partial \theta_i / \partial Z)_{SP}}{\rho_i \lambda_{iw}}$$

This model of segregation emphasizes the importance of the heat fluxes in the new lens and the freezing fringe during incipient lensing. However, it assumes that the hydraulic conductivity in the fringe may be adequately modelled, at least as a function of temperature. The physical effect of K decreasing rapidly with temperature in a small temperature range beneath T_{cr} (which is often only slightly beneath T^0), is to increase the water pressure gradient: it is presupposed that little or no water freezes in its passage through the fringe. Once the ice pressure exceeds P_{SP}, it must then suddenly drop to the level of the overburden pressure, as suggested in Fig. 6.27. This evidently requires a sudden increase in water suction and suction gradient at the point of segregation if equation (6.48) is to be satisfied. The new segregation is apparently well nourished by an abundant supply of water during its early life.

Lastly, a word about the determination of the segregation potential. This was first done experimentally by Konrad (1980) who noted that when the frost front ($Z = Z_{cr}$) comes to rest the heat flux in the freezing fringe is given by

$$k_{ff} \left(\frac{\partial \theta}{\partial Z} \right)_{cr} = k_{uf} \left(\frac{\partial \theta}{\partial Z} \right)_{cr}$$

using the second of equations (6.52). Thus by using a one-dimensional test cell with fixed end temperatures, the final steady state condition may be used to measure the segregation potential, then given by

$$S_p = \frac{\rho_i \lambda_{iw} V_H}{k_{uf} \, (\partial \theta / \partial Z)_{cr}}$$

or, by using the first of equations (6.52) applied to the new lens,

$$S_p = \left[\frac{k_i \, (\partial \theta_i / \partial Z)_{SP}}{\rho_i \lambda_{iw} V_H} - 1 \right]^{-1}$$

This test procedure is very convenient and would appear to lend itself to the further investigation of the segregation potential: for example, in determining the influence of pressure on permeability, or of dissolved gases on attainable water suction.

7

Ice and life

It is well known that water is vital to human life; indeed it is essential to all forms of life on this planet. It is therefore not surprising that the freezing of water contained in, or immediately surrounding, living systems is potentially harmful to that life. The study of organisms at low temperatures – often referred to as *cryobiology* – may be regarded as a branch of biothermodynamics which includes, but places no special emphasis on, phase change. *Bioglaciology*, on the other hand, is perhaps a better term to describe the study of ice in organisms: its occurrence, its effect and its control. This chapter begins with an examination of water: its role as environment, structural component and, quite literally, as life blood. A description of the freezing characteristics of intracellular and extracellular fluids is then given followed by a brief review of the morphology and physiology of various cells. This leads quite naturally to a discussion of ice nucleation and growth in organisms of gradually increasing complexity: unicellular organisms, plants and animals. The chapter ends with an outline of freezing in organisms harvested as food.

7.1 Water in the biosphere

In Chapter 1 brief mention was made of the origin and abundance of water: in the Universe, the galaxy and in our solar system. Throughout its five billion year evolution the Earth has not always deserved to be called the water planet; in fact the hydrosphere and atmosphere as we know them today are of comparatively recent origin, appearing after the planet had acquired its now familiar core, mantle and crust structure. The long process of water formation is believed to be one in which previously formed iron oxide, dissolving at high pressure deep in a high-temperature iron core rich in hydrogen, made possible the formation of hydroxyl ions, OH^-, which were subsequently transferred into the Earth's mantle (Ringwood, 1978). Water thus bound in the mantle, and amounting to only about 0.1% of its

Table 7.1. *Distribution of water on the Earth (following Philip (1978))*

Component body of water	Water volume (as liquid) (10^6 km³)	Average depth (m)	Based on area of	Percentage of total water
Ocean	1338	3700	ocean	97.3
Ice caps and glaciers	29	80	ocean	2.1
Groundwater	8.4	56	land	0.61
Lakes and rivers	0.23			0.017
Atmosphere	0.013	0.025	planet	0.00094
Biological water	0.0006	0.004	land	0.000005
Planet	1376	2700	planet	100

Table 7.2. *Mean residence time of the Earth's waters (following Philip (1978))*

	Volume (10^6 km³)	Rate of turnover (10^6 km³ yr^{-1})	Mean residence time
Ocean	1338	0.5058	2600 yrs
Ice caps and glaciers	29	0.00255	1100 yrs
Groundwater	8.25	0.0119	70 yrs
Lakes	0.23	0.00173	13 yrs
Soil water	0.03	0.0706	155 d
Rivers	0.0017	0.0469	13 d
Atmosphere	0.013	0.5781	8.2 d
Biological water	0.0006	0.0651	3.4 d

total mass, is thought to be the primary source of the free water which eventually formed the oceans, rivers and lakes. Plate tectonic processes, in which the mantle rises at the mid-ocean ridge to produce a basaltic oceanic crust, continually generate water and water vapour, and it is a sobering thought that the life they support owes its origin almost entirely to these geological processes.

Table 7.1 shows the current distribution of water thus generated (Philip, 1978). As indicated, 97.3% is saline, and most of the remainder is frozen. Substantially less than 1% of the total is fresh, mobile water of which by far the greatest part is buried deep in the ground. All the water in the Earth's lakes and rivers amounts to less than two hundredths of 1%. Most significant of all, perhaps, is the fact that only five parts in 100 million are contained within the entire plant and animal kingdoms.

The time during which water dwells periodically within these bodies is also highly variable. Not surprisingly, this time varies in proportion to the size of the body. Table 7.2 gives the mean residence times for the groups and subgroups listed in Table 7.1. Since losses and gains take place at the body surfaces, these figures probably underestimate the residence time deep within the bulk; this is particularly true of oceans, ice caps and glaciers. The turnover of water in the soil and the atmosphere is among the principal determinants of the root and leaf systems of plants, and it is well known that in the absence of water many plants and animals experience rapid desiccation.

In general terms, the movement of water in the biosphere depends upon gradients in the hydraulic and thermal potentials: pressure gradients create bulk flow while concentration gradients create diffusive flow; at the same time, temperature gradients cause vapour pressure gradients and evaporation (or condensation). On land, heat exchange processes at the Earth's surface (see section 6.1) exert a considerable influence on plants, and vice versa. The albedo of the vegetative cover largely controls the radiant heat balance and the aerodynamic roughness controls the intensity of turbulent diffusion, thereby influencing both sensible and latent heat exchange in the foliage. Within the soil, capillarity, pore pressure distribution and temperature distribution all influence the general movement of water; in particular, rain (or irrigation) supplies water to the root zone from which it is lost either by drainage (to ground water) or by evaporation and transpiration (to the atmosphere). Evaporation and transpiration together are usually described as evapotranspiration. As noted in Section 6.4, freezing itself may produce a severe, additional form of desiccation in soil.

In higher latitudes and altitudes, the annual snow cover is often the dominant ecological feature (Pruitt, 1960, 1978; Laws, 1984), carrying with it major implications for the regulation of biological water. In much of the northern hemisphere, the retreat of the pleistocene ice sheets has left a continually changing landscape which is covered in snow for a significant portion of the year, and often underlain by permafrost. The functional characteristics of this snow have been described in detail elsewhere in this book, notably in Sections 6.1 and 6.2 from which it will be recalled that the principal physical effect of a snow cover is to dampen the annual variations in temperature and humidity felt at the surface of the ground. Beneath the snow, plants and animals both benefit from thermal insulation in an environment where moisture loss to a dry, cold wind is limited by upward diffusion. This slow diffusion is often accompanied by metamorphosis of the snow crystals near the bottom of the cover, eventually leading to a fragile, lattice-like structure, well-suited to the tunnelling practices of small mammals. At the same time, the permafrost beneath limits water drainage, should the local ground temperature exceed 0 °C. The ability of the snow to 'breathe' is crucially important if the moisture is not to re-deposit within the pack and thereby plug it. Exactly the same principle is observed in the traditional winter clothing of northern natives.

Sea ice provides further examples of biotic environments in which the solid and fluid phases of H_2O share an interdependency (Horner, 1985). Surface algae, for example, originate in cell-laden sea water flooding the snow-laden ice, or migrate up through brine channels leading to surface melt pools. Bottom ice biota, on the other hand, congregate largely in intergranular spaces, having arrived there as the result of scavenging by rising frazil, or through entrainment between the interfacial platelets so characteristic of ice growth in the presence of constitutional supercooling. The physical mechanisms associated with the creation of these habitats were described in Chapter 4, notably in the first three sections.

Many lower forms of animal life such as the amoeba, the jelly fish or the earthworm, are tied to local environmental conditions and are thus depen-dent on the suitability of those conditions for the take-up and turnover of essential water. In higher animals, such as the squirrel, the magpie or the reindeer, the existence of a tightly-regulated internal environment vir-tually eliminates such a dependency, at least for short periods. It is clear that the survival of an organism faced with freezing temperatures will depend upon its ability to respond appropriately, and it is equally clear that the nature of such a response will reflect the organism's freedom with respect to its environment. In part, this freedom will depend upon the degree of thermal isolation within its control, but it also depends upon the quality, quantity and distribution of water throughout the organism.

In the human body, water is distributed in uneven amounts (Bell, Davidson and Scarborough, 1961). In the lungs, for example, water con-tent is about 84% but the lungs contribute only 4% of the total mass. Striated muscle, which is about 32% of our mass, has a water content of 80%; adipose (fatty) tissue contributes 14% to the total mass with a water content of 50%. On the other hand, the skeleton contains only 32% water and adds 15% to our mass. The highest water contents are found in the organs, muscles, tissues and skin: values ranging from 50% to over 90% have a significant bearing on the art and science of cryopreservation.

This distributed mass of water is in continuous activity, physically and chemically. Generally speaking, it is transported between various regions of the body by mechanical effects such as heart pumping and peristalsis. Within a region, it is transported by additional non-mechanical effects such as diffusion (including the coupled electrical and thermal effects discussed in Section 2.4) and osmosis. The study of freezing may similarly be divided into two parts: the freezing of aqueous solutions and suspensions flowing, in a steady or pulsatile manner, within an array of branched, flexible-walled tubes; and the freezing of these solutions and suspensions, while quiescent, as their chemical components diffuse, passively or actively, through walls and membranes – specifically in and around cells.

In complex organisms, the distributed fluid may be divided into two categories: intracellular fluid and extracellular fluid (which includes, e.g., intravascular fluids such as blood plasma and lymph, together with interstitial fluid). In a typical human, with a water content of around 60%, 63% of the body fluid is intracellular; of the remainder, 28% is found in the interstices between cells while only 9% is intravascular. It is thus to be expected that the transport mechanisms which operate within and around cells will be of crucial importance, and it is therefore natural to suppose that the effects of freezing in biological systems will be manifest first and foremost at the cellular level. However, before discussing cellular characteristics it is worthwhile taking a closer look at the biofluids which comprise and perfuse the cell.

7.2   Freezing of biofluids

If water is not unique, by virtue of its characteristics and its very existence on this planet, then at the very least it must be regarded as a remarkable fluid. This is particularly apparent in a biological context. Water's high latent heat of evaporation, for example, permits relatively small quantities to be used in the evaporative cooling processes associated with sweating and breathing. Similarly, a high latent heat of fusion provides the first line of defence against freezing temperatures. Even between the extremes set by these changes of phase, the capacity of water to absorb heat is anomalously high (only ammonia has a higher specific heat), largely because its quasi-solid structure undergoes progressive 'melting' as the temperature is raised (Fletcher, 1978).

A flickering cluster model, in which several dozen molecules form a coherent, ice-like cluster, seems particularly appropriate to cold water. The clusters are dispersed in an ordered, bonded structure, but continually change as the result of molecular jumps: typically, a molecule will move to a new configuration every 10^{-11} s, having occupied the old configuration for 100 or so vibrations (Fletcher, 1978). It is the flickering which explains why such a highly-bonded liquid should possess a viscosity low enough to be comparable with that of other non-metallic liquids. On the other hand, the high surface tension of water may be attributed to the bonding.

Hydrogen bonding also permits molecular re-orientation which is then transmitted through the structure. Thus, even though the water molecule has a high dipole moment, the dielectric constant of water is notably higher than that of most common liquids, and this fact implies lower ionization energy levels. The mobility of ionization defects (OH^- and H_3O^+) was discussed in Section 2.6. It has been suggested (Fletcher, 1978) that these self-ions, in both ice and cold water, are analogous to electrons and holes in semi-conducting materials, thus providing some insight into the mechanism of electric charge generation during freezing.

The molecular structure of water enables it to behave in a very friendly way towards many other molecules. As a dipole, it can accommodate both positive and negative ions, even to the extent of relaxing strong crystalline bonds, as with sodium chloride. It may thus dissolve solid crystals, and is capable of shielding ions from each other through the process of hydration (Dyson, 1978). Being a good hydrogen bonder, it will bond with oxygen, carbon and nitrogen which, taken together with hydrogen, are perhaps the four most important elements of living systems on this planet, if not throughout the entire Universe (see Table 1.1). It will even accommodate hydrophobic substances by allowing them to form self-contained pockets or micelles consisting of several hundred molecules with a non-ionic interior surrounded by a charged soluble surface. Hydration is only one form of structural modification in water. The large molecules and membranous surfaces characteristic of biological systems may introduce other effects extending at least over a coherence length[1] of about 10 Å.

It is the great accommodating ability of water which accounts for the presence of nearly all of the solutes, molecular aggregates and suspended particles which may be found in the water of organisms. Dissolved in the body fluid of a human, for example, are the cations of sodium, potassium, calcium, magnesium, etc, along with the anions of chloride, bicarbonate, phosphate, sulphate, etc. Added to these are the atmospheric gases and the more complex molecules of carbohydrates, lipids, proteins, hormones, vitamins, etc. Distributed throughout the vascular system are mobile cells such as erythrocytes and leucocytes, while contained within the larger cells are sub-units such as the nucleus, mitochondria etc. The freezing of such biofluid complexes is correspondingly difficult to model. A completely

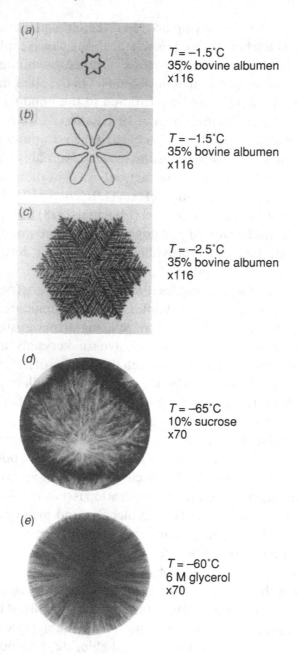

$T = -1.5°C$
35% bovine albumen
x116

$T = -1.5°C$
35% bovine albumen
x116

$T = -2.5°C$
35% bovine albumen
x116

$T = -65°C$
10% sucrose
x70

$T = -60°C$
6 M glycerol
x70

Fig. 7.1. Ice formation in aqueous solutions (following Luyet and Rapatz (1958)).

successful model must account for a multiplicity of solutes, wide variations in their concentration and reactivity, and the presence of cellular or sub-cellular units in varying numbers and with varying degrees of interaction.

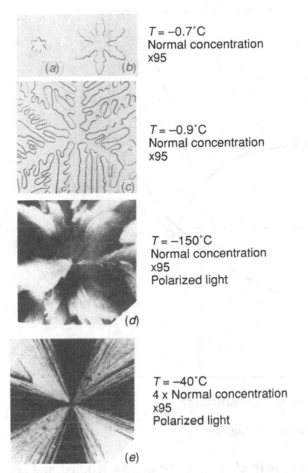

$T = -0.7°C$
Normal concentration
x95

$T = -0.9°C$
Normal concentration
x95

$T = -150°C$
Normal concentration
x95
Polarized light

$T = -40°C$
4 x Normal concentration
x95
Polarized light

Fig. 7.2. Ice formation in blood plasma (following Rapatz and Luyet (1960)).

A useful starting point is to investigate the effect of cooling rate. It is perhaps to be expected that the greater the rate the greater will be the excursion below the equilibrium freezing temperature; with a correspondingly greater chemical potential deficit, when nucleation does occur it will likely be more widespread. This expectation has been confirmed by many workers (Meryman, 1966), but confirmed predictions of nucleation temperature, nucleation density and subsequent growth rates in biological systems have been slower to appear. The variability observed in pure water is compounded in biofluids by the composition and concentration of solutes and by structural features, both biochemical and anatomical.

In general terms, crystal growth in most biofluids e.g., blood plasma, vitreous humour and muscle juice, closely resembles that in concentrated aqueous solutions (Luyet, 1966; Rapatz, Menz and Luyet, 1966). In

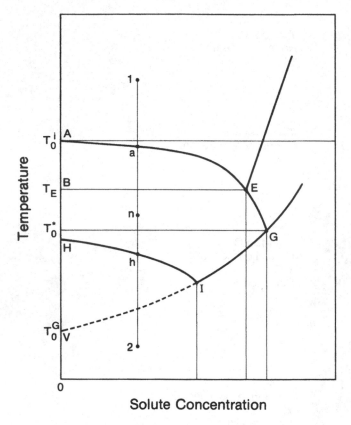

Fig. 7.3. Modified phase diagram.

particular, the *a*-axis growth characteristic of supercooled and constitutionally supercooled liquids is frequently observed: this behaviour was
discussed earlier, notably in Section 4.1. At sufficiently low cooling rates, a
regular, branching dendritic form is produced under conditions which
allow easy diffusion of heat and mass. As Figs 7.1 (*a*)–(*c*) and 7.2 (*a*)–(*c*)
indicate, the primary[2] crystallization process is highly ordered and
generates the familiar hexagonal structure reminiscent of ice crystals seen
elsewhere: a disc takes on a stellar form with the six principal branches
developing secondary and then tertiary branches. When the temperature
of the specimen is lowered, the probability of multiple nucleation increases and the opportunity for mono-crystalline growth is correspondingly limited. Figs 7.1(*d*) and 7.2(*d*), the latter in polarized light, illustrate
the irregular dendritic forms thus produced. On the other hand, increases
in solute concentration increasingly impede the diffusion of water towards
the ice–solution interface which becomes more spherical. Figs 7.1(*e*) and
7.2(*e*) reveal this spherulitic character.

These structural alterations are perhaps best understood in relation to the phase diagram shown in Fig. 7.3. Using the binary system for illustrative purposes, the first and most obvious effect is freezing point depression, as discussed in Chapter 2. Fig. 7.3 shows this as the curve AE on which A corresponds to the equilibrium freezing temperature of pure water T_0^i. Continued cooling below the eutectic temperature T_E theoretically produces no further change in composition but in a real situation, even with a low cooling rate, the process continues metastably along the curve EG. The glass transition point G represents a limit below which the solution has for all practical purposes become infinitely viscous.

The point H represents the spontaneous (homogeneous) freezing temperature of pure water T_0^*. As the solute concentration increases the magnitude of T^* decreases along the curve HI such that $\Delta T^* \approx 2\Delta T^i$. The point I marks another point on the vitrification curve VIG along which $T = T^G$ and below which the solution is an amorphous solid. Fig. 7.3 thus extends the stable equilibrium phase diagram (see, e.g., Fig. 2.4) to include the possibilities of supercooling and vitrification. As noted in Chapter 2, $T_0^i = 273\,\text{K}$, $T_0^* \approx 233\,\text{K}$ and $T_0^G \approx 140\,\text{K}$[3]. However, the figure gives no indication of the effects of molecular ordering attributed to micelles, macromolecules, membrane surfaces, etc, or of heterogeneous nucleation. As noted previously (see, e.g. Sections 4.1 and 6.3), surface force fields often have the effect of permitting some water to remain unfrozen under stable conditions beneath the bulk equilibrium freezing temperature; this effectively lowers the liquidus AE. On the other hand, impurities frequently raise the spontaneous freezing curve HI.

At extremely high cooling rates, a solution with the state indicated by point 1 would not nucleate at all as its temperature fell; it would be 'frozen' into a vitreous form at 2. In a natural biological situation, however, a much lower cooling rate would likely lead to nucleation at a point n below T^i (at a) but above T^* (at h). When this occurs, the release of latent heat immediately raises the temperature again, as discussed in Section 2.5 and illustrated in Fig. 2.12. Thus, under natural circumstances the temperature shortly after nucleation reaches the equilibrium freezing point, with the consequence that further cooling produces freeze-concentration of the solution; the temperature gradually moves along the liquidus towards the eutectic point and beyond along the metastable extension EG. The final concentration thus bears little relationship to the initial value, and is essentially determined by the final temperature. If this is less than T_E, the final mixture consists of ice crystals embedded in a supercooled solid solution, but the size distribution of the crystals is a function of the

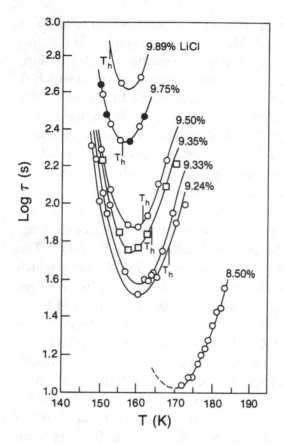

Fig. 7.4. Experimental determinations of curves for a series of emulsified LiCl + H_2O solutions. The temperatures marked T_h on each curve are the temperatures at which crystallization is observed to commence suddenly on continuous cooling at 10°C min^{-1}. The filled circles are data from an unemulsified solution proving that, for this composition range, the crystallization rates are dominated by a homogeneous process (following Angell and Senapati (1987)).

temperature–time curve of the mixture. As observed earlier, the higher the cooling rate, short of vitrification, the larger the number of crystals and the smaller their size; beyond this generalization the experimental confirmation of theoretical predictions is difficult to obtain, and is usually limited by the resolution of the light microscope. Warming of the frozen mixture will be discussed in Chapter 8.

The above thermodynamic description of freezing may be usefully complemented with another type of analysis. This consists of observing the time required to freeze[4] small droplets of solution held at a given temperature (MacFarlane, Kadiyala and Angell, 1983). Once nucleation takes place at a particular location in the drop, the rate of ice generation on the surface of the crystal is equal to $4\pi R^2 \mathscr{V}$, where R is the effective radius of the crystal and $\mathscr{V} = dR/dt$, assumed independent of time for that temperature. If the nucleation rate in the solution is given by the constant value $\mathscr{I}$, then the volumetric rate at which ice appears at time τ because of earlier nucleation during the time interval t to $t + dt$ is equal to

$$d\dot{v} = 4\pi \mathscr{I} \mathscr{V}^3 (\tau - t)^2 dt$$

bearing in mind that $R = \mathscr{V}(\tau - t)$. Thus, for all the $\mathscr{I}\tau$ nuclei appearing during the time period τ, and assuming none existed initially, it is evident that the rate of ice generation per unit volume is given by

$$\dot{v} = \frac{4\pi}{3} \mathscr{I} \mathscr{V}^3 \tau^3.$$

Therefore the total volume fraction of ice $v(\tau)$ is

$$v(\tau) = \frac{\pi}{3} \mathscr{I} \mathscr{V}^3 \tau^4$$

from which it follows that

$$\tau = \left(\frac{3v}{\pi \mathscr{I} \mathscr{V}^3} \right)^{\frac{1}{4}} \tag{7.1}$$

This freezing time is shown plotted against the droplet temperature in Fig. 7.4 for each of a series of concentrations using a lithium chloride solution. At higher temperatures, it is evident that a reduction in temperature produces a reduced freezing time. Equation (7.1), in which both $\mathscr{I}$ and $\mathscr{V}$ are variables, may be used to interpret this behaviour as a substantial increase in growth velocity while the nucleation rate increases only slightly. Freezing is limited by the ability to nucleate; a smaller number of crystals is created but these may eventually become quite large. On the other hand, if the temperature is reduced too far, the opposite behaviour occurs. This is attributed to a dramatic reduction in growth velocity $\mathscr{V}$ which more than offsets further increases in $\mathscr{I}$. Freezing is then limited by crystal growth velocity; this creates a larger number of crystals which remain quite small. The curves shown in Fig. 7.4 relate time, temperature and transformation; they are appropriately called *TTT* curves. To the right of the minima, dendritic growth may be expected; to the left, growth is altered through the limits imposed by molecular mobility and diffusion at the ice–solution interface.

Description of the freezing behaviour of real biofluids may require more information than can be provided on diagrams such as Figs. 7.3 and 7.4. In general, real behaviour is difficult to model because the multi-component, reactive system it is designed to represent is not often fully understood and cannot be accurately described in simple terms. A complete knowledge of the biochemistry is not absolutely essential for the understanding of nucleation and growth of the ice phase, but it is wise to assume that the presence of ice may be a disruptive, and perhaps limiting, factor in biochemical reactions. Changes in reaction kinetics, and in ultrastructural integrity, threaten the viability of the organism, at least in physiological terms. In Section 7.4 consideration will be given to the effect of ice in the intracellular and extracellular fluids of unicellular organisms. This section ends with a few brief comments on two especially-important characteristics of biofluids: unfrozen water and freeze-concentration.

Unfrozen water has been a controversial topic for many years, and in many disciplines, largely because the evidence is not often clear cut and conclusions have therefore been inferential. For example, statements have appeared both for and against the binding of water molecules to macro-molecules, and opinions appear to vary on the description of hydration as an ordering phenomenon. The truth of these events may or may not be helpful when faced with the general question: 'when does a water molecule, in stable equilibrium, behave differently from an ice molecule at the same temperature?' The answer must surely be: 'whenever the water molecule is prevented from forming part of an ice cluster by a stronger, more pervasive force field.' In a geoglaciological context, this may be seen in the effect of surface charge on a mineral substrate, as discussed in Section 6.3. In bioglaciology the cause is more likely to be found, if not at the surfaces of free macromolecules, quite possibly in the vicinity of macromolecular networks. In the pores of cell membranes, for example, translational, rotational and vibrational freedoms of molecules are limited not only by the dimensions of the pore but by the force fields of the molecules which line the pore. More generally, the work of Porter and coworkers (1983, 1987) suggests the existence of a cytomatrix, or microtrabecular lattice, the surface of which may influence the behaviour of water molecules in a film which would extend at least the thickness of the coherence length (10 Å) mentioned earlier, and may be 30–50 Å thick. This may account for 20–40% of the cytoplasm water freezing below the equilibrium freezing temperature of the bulk (Clegg, 1987).

Freezing point depression is perhaps the most obvious way in which water may remain in the liquid phase below 273 K. The falling liquidus of Fig. 7.3 is a reminder that such an effect continues under stable conditions to the point E, and under metastable conditions to the point G, beyond which all remaining water, in the metastable vitreous form, is unfreezable

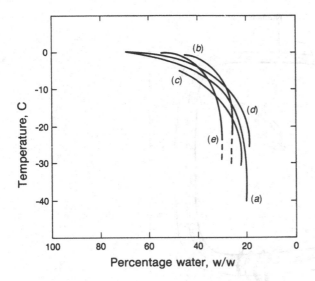

Fig. 7.5. Temperature–water content curves of concentrated aqueous suspensions: (*a*) yeast cells, (*b*) glutaraldehyde fixed yeast cells, (*c*) bovine psoas muscle, (*d*) whole human blood and (*e*) glutaraldehyde fixed human erythrocytes (following MacKenzie (1975)).

in the sense that it does not become crystalline; however, the viscosity may be large enough to render the material solid for all practical purposes. Apart from this, many aqueous suspensions which do not vitrify are capable of retaining a substantial amount of unfrozen water, as illustrated in Fig. 7.5. The reasons for this are not completely known but surface effects associated with small pores, intracellular and supracellular, provide the most likely explanation. Should the aqueous phase also be a concentrated solution it may give rise to pockets of even stronger solution included within the ice phase. Both of these prospects enable water to retain some of its mobility provided that the pores or pockets interconnect, thereby creating a diffusional network.

The biophysical and biochemical effects of freeze concentration may be dramatic. Ironically, they will be greater if the solution is dilute than if it is strong simply because the final concentration is fixed by the final temperature, at least for a series of stable states following the liquidus. Freezing alters tonicity, a fact which is important if the solution is to be used in the suspension of cells; the freeze-concentration factor of a 0.154 M sodium chloride solution, for example, exceeds 30 before the eutectic point is reached at $-23.13\,^\circ\text{C}$ (Franks, 1985). Equally important are the changes in reaction kinetics. In general, freezing produces two opposing

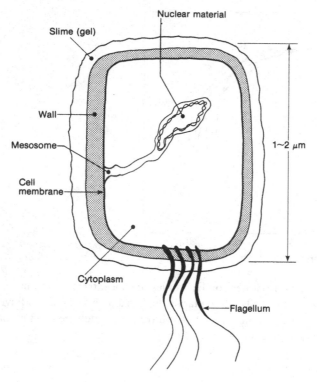

Fig. 7.6. Model of unicellular bacterium.

effects: an increase in concentration and a decrease in reaction rate, following the Arrhenius equation. For a second order reaction with two reactants A and B, the reaction rate for A is given by $dA/dt = -KA_0B_0$ where the subscript 0 refers to the unfrozen state. The reaction constant K is in fact proportional to both the Arrhenius effect (a decrease with decreasing temperature) and the freeze-concentration effect (an increase with decreasing temperature); K may therefore increase or decrease as the solution temperature is lowered. For dilute aqueous solutions, the freeze concentration effect tends to dominate initially as the temperature is lowered, but eventually has the lesser importance; the reaction constant may thus exhibit a maximum with respect to temperature.

7.3   The cell

The body of the human adult contains about 10^{14} cells. The average size of these cells therefore cannot be much more than $10\,\mu m$. In fact the size of human cells typically ranges from $3\,\mu m$ (the diameter of a leucocyte) to 1 m (the length of a nerve cell); most fall within the range $5-20\,\mu m$ (Swanson, 1964). In other organisms, one of the largest cells (volumetrically) is the ostrich egg which is about 5 cm diameter; the smallest cell is the mycoplasma which has a diameter of about $0.1\,\mu m$ (Swanson, 1964). (Smaller again are the viruses which are not classified strictly as cells.) It is thus evident that the size ratio of the largest to the smallest cell is of the order of 10^7, and from this fact alone we might anticipate some difficulty in developing a single physical model which would apply to every kind of cell. Physiological considerations can only add to the complexity.

It is also clear that even the smallest cells contain a great many molecules. The water molecule, for example, has an overall size of the order $10^{-4}\mu m$ which is only one-thousandth of the size of the mycoplasma. Always keeping in mind that some intracellular molecules may be much longer than a water molecule, it is evident that for many purposes cells are large enough to be treated as continua. Therefore, even though detailed knowledge of the architecture and behaviour of individual molecules may be vital to the understanding of biochemical behaviour, it may also be assumed that molecules are present in sufficiently large numbers to give statistical meaning to averaged phenomena: diffusion, osmosis, reaction rates, etc.

The simplest organism consists of a single cell, shown diagrammatically in Fig. 7.6. The surrounding wall, which provides both shape and strength, is coated with a slimy material which guards against chemical attack, dehydration and also facilitates motion driven by the flagella(um). The cell proper begins inside the wall with the thin, flexible membrane which contains the cytoplasm – the sum of all internal substances, excluding the nuclear material, which is housed in a membrane fold (mesosome). For such a simple cell[5] the cytoplasm contains no organelles. Bacteria and blue-green algae provide examples.

The nuclear material contains the genetic imprint, the DNA molecules, which identify the bacterium and are necessary for self-reproduction. Such molecules are usually coiled because of their length, and cannot pass through the cell membrane which is semi-permeable (or selectively permeable). On the other hand, small molecules, water in particular, have little difficulty migrating into or out of the cytoplasm, thus permitting the continuous supply of nutrients and the continuous discharge of wastes. The energy required for movement, growth and reproduction is usually chemical (either aerobic or anaerobic) but may be partly photosynthetic, depending upon the particular bacterium. Typical bacteria walls and membranes have thicknesses in the range 10^{-2}–$10^{-1}\mu m$ and 10^{-3}–$10^{-2}\mu m$, respectively, indicating that they are

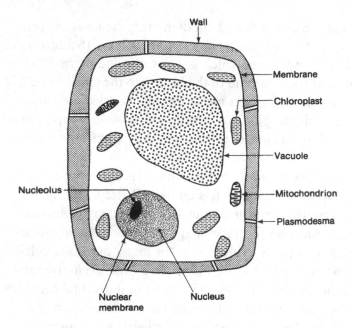

Fig. 7.7. Model of plant cell.

much more extensive than the size of a water molecule but only represent a few layers of the larger molecules which comprise them (Dyson, 1978).

Plants, of course, are multi-cellular organisms, but before considering anatomical arrangements it is important to be familiar with the single cell, shown diagrammatically in Fig. 7.7. It is immediately obvious that this is more complicated than the bacterium cell, even though it is not mobile, or even self-contained; along with similar and different cells comprising the same plant, it will be more specialized than a unicellular organism could afford to be. Plant cells are thus capable of much greater diversity.

Plant cells vary a good deal in size though not too much in shape. Typically, they are fairly rectangular and stack up much like bricks in a wall. In fully grown trees, and in some smaller plants, they can be fairly large (1 mm), but may also be found in sizes ranging down to 10 μm (Richardson, 1975). The cellulose making up the wall is usually fairly rigid, and may possess a number of plasmodesmata, which facilitate intercellular communication. Inside the wall is the semi-permeable membrane which in turn houses the cytoplasm.

The figure shows the cytoplasm of a plant cell as a composite of sub-units. The chloroplasts execute the well-known process of photosynthesis while the mitochondria oxidise various nutrients as the main source of intracellular energy. The vacuole on the other hand is an 'empty' space. In

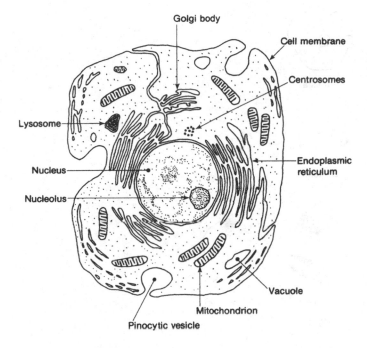

Fig. 7.8. Model of animal cell.

fact, it is a membrane-lined sac filled principally with aqueous, intracellular sap, the amount of which controls not only the store of water and other substances in the cell but the rigidity (turgor) of the structure, and hence of the plant. The nucleus, like the chloroplasts and mitochondria, is contained in its own membrane.

As might be expected, an animal cell is typically more complex and more specialized than a plant cell. Fig. 7.8 provides a schematic representation. The nucleus is again a complex region which contains, among other subnuclear constituents, the nucleolus. On close examination, the membrane of the nucleus is usually found to consist of two membranes each about $10^{-2}\mu$m thick and separated by a gap of the same order (Dyson, 1978); this double layer is punctuated by pores of about $10^{-1}\mu$m diameter. Various channels, or membranous folds, lead to and from the nucleus; among these are the Golgi body, a layering of flattened saccules about 2–$3 \times 10^{-2}\mu$m apart, and the endoplasmic reticulum which, as its name indicates, is a network of tubules or cisternae (many of these features may also be found in plant cells). Distributed throughout the cytoplasm are many membrane-bound subsystems each with a special function. Apart from the lysosomes, which contain digestive enzymes, and the vacuoles

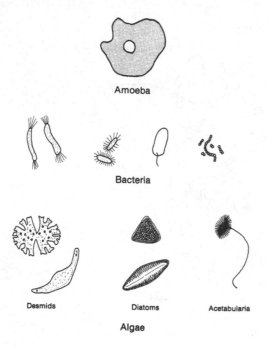

Fig. 7.9. Shapes of unicellular organisms.

(which are neither as large nor as common as in plant cells) the principal organelle is the self-producing mitochondrion[6].

Not shown in Fig. 7.8 are the many possible folds of the cell membrane. In a simple organism, the membrane can engulf and 'swallow' material in the aqueous environment through the processes of *pinocytosis* and *phagocytosis*: these 'drinking' and 'eating' mechanisms enable the membrane to capture large molecules and smaller cells for digestion in the cytoplasm. Apart from this active enfoldment, the plasma membrane may also be very convoluted, thus permitting an adjustable and flexible interconnection with adjacent cells, and increasing the scope of the cell's interaction with the aqueous environment; this interaction is facilitated by repeated, high density folds – the *microvilli* – whose purpose is evidently to increase the surface area available for molecular transport processes.

Figs. 7.6, 7.7 and 7.8 are merely schematic representations but they do serve to illustrate the main architectural features of cells, so far as these are important during freezing. It is evident that each cell is not a simple system but, in general, consists of a loosely-supported, geometrically-irregular, membrane (full of crevices and channels) surrounding an irregular assortment and arrangement of functional, interdependent subunits. The physical and chemical regulation of the cell, even forgetting growth and division, is very complex indeed. In particular, the flow of material into, out of, and throughout the cell is unlikely to be clearly understood or accurately described unless it is being treated within a narrowly confined region: near the plasma membrane, around mitochondria, within the endoplasmic reticulum, etc. It follows that a fully comprehensive study of cell freezing must eventually be viewed in the same way. In the meantime, simplified models must be used.

7.4   Freezing of unicellular organisms

Simple though they may be, anatomically and physiologically speaking, unicellular organisms do exist in a wide variety of sizes and shapes. In length, they range from about $0.2\,\mu m$ (bacteria) to about 0.1 m (algae), and in configuration they vary from the nondescript shapes of amoebae to the rods, spirals and spheres of certain bacteria; they also include the fascinating geometries of the desmids, diatoms and acetabularia among the algae (Swanson, 1964). Fig. 7.9 gives some indication of the variety.

Unicellular organisms are usually found in a liquid or gaseous environment. Although they display varying degrees of mobility, they are immersed in fluids which carry them about (or do not carry them about) in a way which is essentially beyond their control. A slimy gel coat may help with the lubrication or adhesion of their surfaces, and the flagella and cilia undoubtedly help with propulsion, but the organisms are ultimately dependent on their immediate environment for survival. They exist in a dynamic equilibrium. Sudden changes in the composition or phase of the environment upset that equilibrium.

Organisms living in an aqueous solution do not necessarily freeze as the solution is frozen. For example, changes in the concentration of solutes in the neighbourhood of an approaching ice surface may cause the organism to swim away; if it is not mobile, its small size may prevent it from being engulfed by the interface (see Section 4.1). In any event, the osmotic pressure difference resulting from increased solute concentration in the surrounding fluid may cause the loss of enough cell water to lower the cytoplasm freezing temperature close to the ambient temperature, thus reducing supercooling and inhibiting nucleation. Subzero temperatures are not necessarily injurious to cells, but it is to be expected that as the temperature is progressively lowered its effect will become increasingly important. In general terms, the initial lowering of temperature below $0\,°C$ produces no glaciological effect[7] at least until the freezing point depression has been exceeded, and perhaps not even then if the extracellular material is capable of withstanding supercooling. As noted in Section 7.2, lowering the temperature of water increases the rate of ice embryo formation. When the supercooling is slight, a few emergent nuclei may grow into large crystals, a process aided by the tendency of larger crystals to grow at the expense of smaller crystals; when the supercooling is great, the nuclei formed are more numerous but grow more slowly in a restricted space.

There is considerable experimental evidence documenting the existence of supercooling in biological systems. In addition to direct microscopic evidence, there are data obtained with DTA[8] and DSC methods. The use of seeding techniques has demonstrated that the onset of freezing invariably produces a sudden rise in temperature, and this fact may be used to measure the equilibrium freezing temperature[9] T^i, thus establishing the actual degree of supercooling. More generally, however, T^i is taken as the melting point, obtained by subsequent thawing. In cytoplasm, T^i is less than T^0, a fact which suggests caution in describing subzero intracellular temperatures automatically in terms of supercooling, which is a metastable state. On the other hand, the observations of Luyet (1960) on frozen gelatin, for example, reveal that a decrease in water concentration may eventually lead to a decrease in the crystal growth velocity, as noted earlier in the *TTT* curves. Once the effect of freezing point depression has been carefully subtracted, the reduction of diffusional freedom, and ultimately the onset of vitrification, may exert a considerable influence on ice growth at lower temperatures. The situation may also be complicated by the presence of unfreezable water in the cytomatrix.

After ice has appeared outside the cell, intracellular water may freeze *in situ* or may flow through the cell membrane and be frozen outside. In many circumstances, intracellular freezing is inhibited or delayed; unnucleated cytoplasm temperatures of −5°C are not uncommon, and much lower values have been observed (Mazur, 1966). The membrane tends to act as a temperature-sensitive barrier for reasons which, although not fully understood, include several interrelated possibilities: biochemical and biophysical processes within the membrane are dependent on temperature (Pringle and Chapman, 1981; McGrath, 1981); membrane shrinkage from dehydration creates mechanically-induced changes in morphology (Steponkus *et al*, 1981); and membrane pores may not transmit ice above a threshold temperature[10]. In any event, the appearance of intracellular ice is not dependent upon inoculation from outside of the membrane; homogeneous or heterogeneous nucleation may occur within the cytoplasm.

Because of their size, most microorganisms are at, or near, thermal equilibrium with their environment, but in most natural circumstances they are not in chemical equilibrium, except with respect to certain substances, notably water. In the event of an unexpected drop in temperature, thermal equilibrium is not much affected but the water equilibrium may be disturbed, in which case intracellular and extracellular events may then differ considerably. Outside of the cell, a falling temperature may or may not be accompanied by supercooling, but if and when ice does form, extracellular fluid returns close to stable, two-phase equilibrium. Should the temperature drop further, and the extracellular fluid follow the dictates of phase equilibrium, the intracellular fluid will then most likely become supercooled; in fact it may have been supercooled before ice appeared outside of the cell. This situation – metastable equilibrium within the cell and stable equilibrium on the outside – is of central importance in exploring the intrinsic freezing behaviour of a single cell. It is a framework within which glaciation may be usefully viewed: in determining the effect of extracellular freeze concentration on cell dehydration; in studying the effect of cooling rate; in examining the effect of cryoprotective additives; and in predicting the onset of intracellular nucleation.

In general, freeze concentration outside the cell will generate an osmotic pressure difference, and this causes diffusion of water outward through the membrane, which is usually much more permeable to water than to solutes. The membrane is thus assumed to be selectively permeable. To be precise, the chemical potential (or vapour pressure) of water deep within the cytoplasm differs from that in the bulk of the extracellular water. Resistance to the resultant water flow lies partly in the membrane and partly in the convective systems on either side of it. To determine the latter resistances would require a complete solution of the overall diffusion problem, but this will not be attempted here; nor is it really necessary at this stage. Instead, two simplifying assumptions will be made: firstly, that the chemical potential is nearly uniform throughout each of the intracellular and extracellular regions; secondly, that the resistance of the membrane is much greater than the convective resistances (the Biot number equivalent is very large). This immediately reduces the diffusion equation from a second order partial differential equation to a much simpler first order ordinary differential equation. Apart from the mathematical advantage, there is also a practical advantage because extremely difficult experiments on diffusive mass transfer may then be replaced by simpler experiments designed to measure the amount of water remaining in the cell. As noted in other cell studies (Lock, 1969b; Johnson, Smith and

Lock, 1969), a characteristic dimension, or the volume itself, often provides a convenient dependent variable.

This approach to the freezing behaviour of microorganisms was first introduced by Mazur (1963), whose work lays down the theoretical foundation. In essence, it consists of a statement of continuity: namely, that the rate at which cell water passes through the membrane, as a result of the chemical potential difference across it, must be equal to the rate at which the intracellular water changes. Thus

$$\frac{dV}{dt} = \frac{-kART}{v_w} \ln\left(\frac{P_c^V}{P_e^V}\right) \tag{7.2}$$

in which V is the volume of cell water, k is the membrane permeability and A is the membrane surface area: subscripts c and e refer to inside and outside the cell, respectively. Following Mazur, the difference in chemical potentials has been expressed in terms of vapour pressures but may also be expressed in terms of osmotic pressure, as used later in equation (7.10). In either representation, the difference across the membrane is established by freezing on the outside with supercooling on the inside; it must therefore accommodate both freeze-concentration (here represented by Raoult's law) and phase change (reflected in the Clausius–Clapeyron equation). If the external pressure applied to the system is assumed to be constant and the cell curvature is ignored, temperature-induced variations, in the vapour pressure ratio for a freezing, dilute aqueous solution are thus written

$$\frac{d}{dt}\left[\ln\left(\frac{P_c^V}{P_e^V}\right)\right] = \frac{d}{dt}(\ln x_w) - \frac{\lambda_{iw}}{RT^2} \tag{7.3}$$

Before combining this with equation (7.2) two more relations must be introduced: firstly, an expression for membrane permeability,

$$k = k_R \exp\left[b(T - T_R)\right] \tag{7.4}$$

which represents its temperature dependency (in relation to an arbitrary datum T_R); and secondly, the cell temperature–time relation specified for a constant cooling rate $B = dT/dt$: thus

$$\frac{d}{dt} = B\frac{d}{dT} \tag{7.5}$$

As noted earlier, the system is assumed to be spatially isothermal, starting at the initial freezing point of the extracellular fluid T_0.

When the cooling rate is vanishingly small, i.e., B→0, stable equilibrium conditions are maintained throughout: $P_c^V = P_e^V$. The volume of cell

water, and its dependency on temperature, may then be found directly
from equation (7.3), by substituting

$$x_w = \frac{V}{V + n_s v_w}$$

where n_s is the number of moles of solute in the cell and v_w is the molar
volume of water, and using the initial condition $V = V_0$ at $T = T_0$. More
generally, equations (7.2), (7.3), (7.4) and (7.5) may be combined to give

$$T \exp\left[b(T_R - T)\right] \frac{d^2V}{dT^2} - \left[(bT + 1) \exp\left[b(T_R - T)\right] - \right.$$
$$\left. \frac{ARk_R n_s T^2}{B(V + n_s v_w)V}\right] \frac{dV}{dT} = \frac{\lambda_{iw} A k_R}{B v_w} \tag{7.6}$$

which is to be solved subject to the conditions

$$V = V_0 \quad \text{and} \quad \frac{dV}{dT} = \frac{1}{B}\frac{dV}{dt} = 0 \quad \text{at } T = T_0$$

Despite its complexity, equation (7.6) does yield approximate analytic
solutions. By defining non-dimensional variables

$$\psi = \frac{V_0 - V}{V_0}, \quad \theta = \frac{T_0 - T}{T_0}$$

and introducing the non-dimensional parameters

$$\alpha = bT_0, \quad \gamma = \lambda_{iw}/RT, \quad \eta = n_s v_w/V_0$$
$$\epsilon = \frac{-ARk_R T_0^2}{B v_w V_0} \exp\left[b(T_0 - T_R)\right]$$

Ling and Tien (1969) were able to develop the following closed-form
solutions.

For stable equilibrium conditions ($B = 0$ or $\epsilon = \infty$) they found that

$$\psi = \left[1 - (1 - \theta)^\gamma\right]\Big/\left[1 - \frac{(1 - \theta)^\gamma}{1 + \eta}\right] \tag{7.7}$$

while by assuming that supercooling is never great ($\theta \ll 1$) and dehydration is also small ($\psi \ll 1$) they concluded that

$$\psi = \frac{\epsilon\gamma}{\alpha\beta(\beta - \alpha)}[(\beta - \alpha) + \alpha \exp(-\beta\theta) - \beta \exp(-\alpha\theta)] \tag{7.8}$$

where

$$\beta = (\alpha + 1) + \epsilon\eta/(1 + \eta)$$

For very high cooling rates ($\epsilon \ll 1$) they used a perturbation expansion in
ϵ to show that

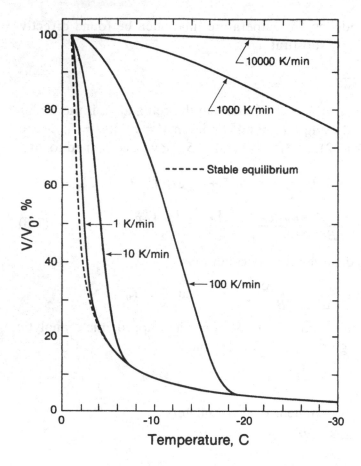

Fig. 7.10. Calculated percentages of supercooled intracellular water remaining at various temperatures in yeast cells cooled at indicated rates (following Mazur (1963)).

$$\psi = \psi_0 + \epsilon \psi_1 + \dots$$

or

$$\psi = \frac{\epsilon \gamma}{\alpha(\alpha + 1)}\{1 + \alpha \exp[-(\alpha + 1)\theta] - (\alpha + 1)\exp(-\alpha\theta)\} \quad (7.9)$$

since $\psi_0 = 0$.

The numerical results of Mazur (1963) are shown in Fig. 7.10 with V/V_0 plotted against temperature. These are for the yeast *Saccharomyces cerevisiae*, and reveal that cooling rates of less than $1\,\text{K}\,\text{min}^{-1}$ produce behaviour very close to the stable equilibrium curve shown dotted (natural biological cooling rates typically fall in the range $10^{-2} - 1\,\text{K}\,\text{min}^{-1}$): freeze-concentration may then be expected to prevent intracellular ice

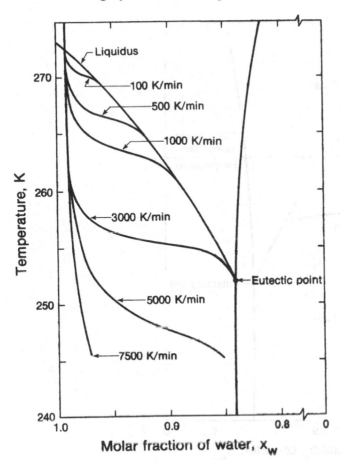

Fig. 7.11. Locus of states of intracellular solutions cooled at different rates with no external supercooling: locus of states in extracellular solution coincident with liquidus for all cooling rates (following Silvares *et al.* (1975)).

formation. At higher cooling rates, the permeability of the membrane prevents a close approach to stable equilibrium over a temperature range which increases with the cooling rate: for example, at $100\,\mathrm{K\,min^{-1}}$, stable equilibrium is not restored until the temperature has reached $-20\,°C$; above this temperature, intracellular water is supercooled by an amount equal to the horizontal distance between the actual and equilibrium curves. Within the metastable range, supercooled intracellular water may freeze before equilibration can be achieved. Experimental observations made on *S. cerevisiae* by Nei (1954, 1960, 1963) and Mazur (1961, 1963) confirm this possibility. Mazur noted that the cell survival rate first increases with cooling rate, suggesting that large increases in intracellular

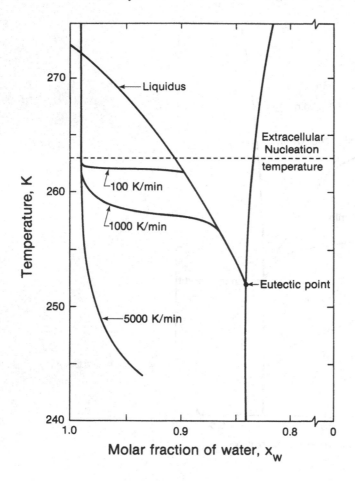

Fig. 7.12. Locus of states of intracellular solutions cooled at different rates with external supercooling of 10 K: for all cooling rates the locus of states of extracellular solution coincident with liquidus for temperatures less than nucleation temperature (following Silvares *et al.* (1975)).

solute concentration were the cause of death; at higher cooling rates, the observed decrease in survival rate suggests that higher water retention and supercooling led to the formation of lethal intracellular ice.

The original 'lumped parameter' model of Mazur is not capable of handling the intricacies of external and internal cell geometry, or of describing the details of diffusion. As Mazur noted, it is limited to ideal solutions and assumes that the cell membrane is not only permeable to water alone but has a constant surface area. However, like Newton's law of cooling in thermophysics, this theoretical model is an essential beginning and provides the basis on which to measure subsequent theoretical

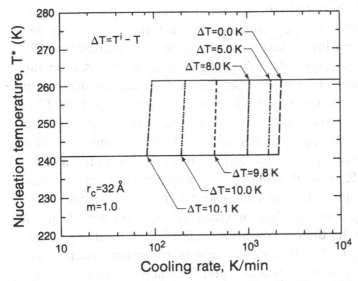

Fig. 7.13. Effect of supercooling $\Delta T = T^i - T$ on the heterogeneous freezing temperatures of red blood cells cooled at constant rates (following Toscano *et al.* (1975)).

contributions. Silvares, Cravalho, Toscano and Huggins (1975) have applied the model to human erythrocytes and have studied a more elaborate freezing protocol, including pre-cooling (ice-free), cooling with extracellular ice, cooling beneath the eutectic temperature, and prolonged holding of the final temperature. Apart from noting that the temperature dependence of the latent heat and non-ideality of the solution had little effect, they concluded that supercooling of the extracellular solution and a model of membrane permeability which included the effect of osmolality, both altered the results significantly in favour of intracellular ice formation. Incorporating the T–x phase diagram discussed in Sections 2.2 and 7.2, Fig. 7.11 shows the locus of states for erythrocytes suspended in an initially isotonic (0.154 mol l^{-1}) solution of sodium chloride cooled at different rates. For any given cooling rate, the unequilibrated cell water may be measured as the horizontal displacement between the actual cooling curve and the liquidus [11]; this water is seen to increase significantly as the cooling rate increases. Fig. 7.12 shows that the effect of supercooling the extracellular solution by 10 K is to increase the amount of unequilibrated water further and hence improve the probability of intracellular nucleation significantly; this is especially true at the lower cooling rates. These predictions are consistent with experimental observations (Diller, 1975; Schwartz and Diller, 1983).

The model may also be used in the prediction of intracellular nucleation. Toscano, Cravalho, Silvares and Huggins (1975), using the analysis discussed above, have developed curves for the nucleation temperatures T^* of erythrocytes as a function of cooling rate. These were based on the critical embryo models of homogeneous and heterogeneous nucleation discussed in Section 2.8. For either mechanism, it was found that T^* is virtually independent of cooling rate in each of two distinct regions: a low value for low cooling rates and a higher value for higher cooling rates. Not surprisingly, the heterogeneous nucleation temperatures were found to be higher (by about 30 K) than the homogeneous nucleation temperatures.

The effect of supercooling the cell suspension medium is shown in Fig. 7.13 for heterogeneous nucleation when the nucleant particle has a radius r_c of 32 Å and the degree of wetting m is 1.0 (zero contact angle). The two ranges of T^* are clearly evident, while the transitional cooling rate which separates them is seen to depend upon the amount of supercooling, in accord with the observations of Diller (1975). Increasing the supercooling progressively restricts the range of cooling rates for which the cells can tolerate temperatures less than 260 K without intracellular ice appearing. A larger catalyst radius, as would occur through cell agglomeration, would exacerbate this effect. It is interesting to note that seeding of the cytoplasm by the passage of extracellular ice through the membrane pores may be treated as a special case of heterogeneous intracellular freezing with the 'nucleant' radius simply being taken equal to the membrane pore radius. Given such inoculation, the two (heterogeneous) sources of nucleation are indistinguishable.

More recently, the Mazur model has been extended through the formulation of Kedem and Katchalsky (1958) to include the potential cryoprotectant effects of additives. In a series of papers, Diller and Lynch (1983, 1984a, 1984b) have considered the coupled flow of water and additive through the cell membrane, thus relaxing two restrictions: that the extracellular fluid be treated as a binary solution, and that the membrane be permeable to water alone. Using the formulation of irreversible thermodynamics (see Section 2.4) applied to dilute solutions, they were able to solve the equivalent of equation (7.2) and a companion equation for the amount of additive (glycerol). They employed a hypothetical, but representative, cell with a fixed surface area of $135 \times 10^{-8} \mathrm{cm^2}$ and a membrane permeability described by the Arrhenius form

$$k = k_R \exp\left(-\Delta E/RT\right)$$

in which ΔE is an activation energy; it was assumed that the permeability

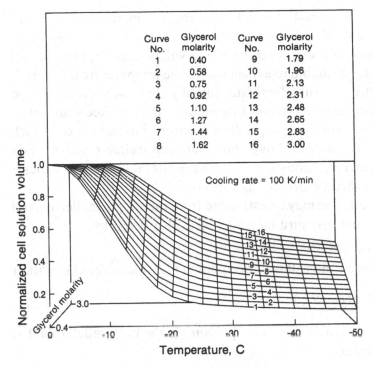

Curve No.	Glycerol molarity	Curve No.	Glycerol molarity
1	0.40	9	1.79
2	0.58	10	1.96
3	0.75	11	2.13
4	0.92	12	2.31
5	1.10	13	2.48
6	1.27	14	2.65
7	1.44	15	2.83
8	1.62	16	3.00

Fig. 7.14. Transient cell solution volumes for freezing in various glycerol concentrations at $100\,K\,min^{-1}$ (following Diller and Lynch (1983)).

ratio for water and the additive was fixed for all temperatures. Numerical solutions were used to show the effects of cooling rate and initial glycerol concentration on the flux and intracellular volume of water, electrolyte and additive. Fig. 7.14 illustrates how the addition of glycerol tends to maintain the volume of the cell solution when the cooling rate is $100\,K\,min^{-1}$. A similar effect occurs at other cooling rates.

In order to prevent injury, or reduce its severity, a great number of cryoprotectants have been studied. It has been found that their protective ability is not easily described, although it appears to be a function of their molecular structure, and their ability to form hydrogen bonds in particular; the simple effect of dilution is not enough to explain events. Ideally, the specific role of a cryoprotectant is best correlated against a specific form of injury, but this is easier said than done. In general, cryoprotectants should reduce the harmful effects of salt concentration through the obvious measure of dilution, but they should also delay or prevent nucleation and subsequent growth through freezing point depression, extended supercooling and ultimately through vitrification.

The chemical consequences of ice formation, which parallel the physical consequences, are essentially two-fold: the effect of increased concentration and the effect of decreased temperature. It is not possible to generalize on which of these opposing effects will be greater, but as noted in Section 7.2 it is reasonable to expect that as the temperature is lowered, concentration effects will dominate initially but will gradually be overtaken by temperature effects. In general, the two effects are interdependent, and may combine to produce reactions[12] within the cell which are irreversible. As noted above, however, intracellular activity lags behind the external conditions which cause the alterations; there is therefore a period of grace before equilibration.

The equilibration time may be estimated from the simple model of a cell in which the osmotic pressure (see Section 2.3) is given by

$$\Pi = RTx_s/v \approx \mathcal{R}Tn_s/V$$

The rate of change of the cell water (equation (7.2)) may thus be written

$$\frac{dV}{dt} = kA\,(\Pi - \Pi_e)$$

(7.10)

where Π_e is the equilibrated osmotic pressure in the extracellular fluid at temperature T. Hence

$$\frac{dv}{dt} = kA\mathcal{R}Tn_s \left(\frac{V_e - V}{VV_e}\right)$$

or, in normalized form,

$$\frac{dv}{d\tau} = a\left(\frac{v_e - v}{v}\right)$$

where $v = V/V_0$, $v_e = V_e/V_0$ and $\tau = t/t_c$: V_0 is taken as the characteristic (initial) cell water volume, t_c is the characteristic (equilibration) time and V_e is the final equilibration volume. The coefficient

$$a = kA\mathcal{R}Tn_s t_c/V_0 V_e$$

evidently has an order of magnitude of unity and hence the equilibration time may be estimated from

$$t_c = O\left(\frac{V_0 V_e}{kA\mathcal{R}Tn_s}\right)$$

(7.11)

in which V_e is determined from the stable equilibrium form of equation (7.3)

$$\frac{\lambda_{iw}}{RT^2} = \frac{d}{dT}\,(\ln x_{we}) = \frac{d}{dT}\ln\left(\frac{V_e}{V_e + n_s v_w}\right)$$

which integrates from $x_w = 1$ at T^0 to give

Table 7.3. *Effect of cell size on equilibration time (using yeast data from Mazur (1966))*

Cell diameter (μm)	Cell area (μm^2)	Equilibration volume (μm^3)	Equilibration time (min)
1	3.1	0.016	0.0061
6	110	3.4	0.036
25	200	270	0.16

$$V_e(T) = \frac{n_s v_w \exp\left[\lambda_{iw}(T - T^0)/RTT^0\right]}{1 - \exp\left[\lambda_{iw}(T - T^0)/RTT^0\right]}$$

where T^0 is the equilibrium freezing temperature of pure water. Using the yeast data of Mazur (1966), equilibration times are shown in Table 7.3 as a function of cell diameter. It is evident that even the largest cell will effectively equilibrate in less than one minute. In general, this implies that small cells respond faster to high cooling rates than large cells, and are therefore much less susceptible to intracellular ice formation but more vulnerable to concentration effects.

7.5 Freezing in plants

Unlike unicellular organisms, the cells of trees are bound together in a structure which, for many practical purposes, may be regarded as both fixed and rigid. From the standpoint of freezing, this introduces both advantages and disadvantages. The most obvious disadvantage is the fact that the organism may not flee from an approaching freezing front. Among the advantages are the shielding which neighbouring cells may provide for each other, and the fact that the death of some cells does not necessarily render the tissue, less still the entire plant, dysfunctional.

Like unicellular organisms, plants are totally dependent upon their largely fluid environment. Aquatic plants move at the whim of a wave, current and tide, and must accept both thermal and solute conditions prevalent within the water. They are, however, afforded some freeze protection by the latent heat capacity of the water which acts as a cold sink (typically $Ste \leqslant 1$ and the sensible heat capacity therefore contributes only marginally to this protection). In polar and sub-polar regions, many bodies of water do not freeze completely during winter and, despite the restrictions on respiration, life often continues under the ice at temperatures not too different from those of other seasons. In more temperate regions this is

Table 7.4. *Soil pore sizes and their functions (following Kohnke (1968))*

Pore diameter (mm)	Pore function classification	Biotic limits
0.00003	Hygroscopic surfaces	
0.0002		
0.0003	Storage of plant-available water	Bacteria
0.001		
0.003		
0.009		
0.02	Capillary conduction	Root hairs
0.03		Protozoa and algae
0.06		
0.1	Aeration porosity	Rootlets
0.3	—	
1.0	Fast drainage	

equally true; and in any region the freeze-resistance mechanisms employed by unicellular organisms may be operative. However, the sudden accretion of clouds of frazil particles is an additional hazard near an open stretch of water because it creates a suffocating blanket and, if sufficiently massive, may uproot the plant by virtue of its buoyancy. Fortunately, this is a rare event.

Atmospheric plants are much more vulnerable, largely because atmospheric temperatures may fall well below those underneath the ice cover which, apart from providing protection through its thermal capacity, is often insulated by a blanket of snow. Within the soil, the root system is afforded the best protection and indeed may be the only part of the plant capable of surviving unusually low air temperatures when the amount and availability of unfrozen water may be abnormally low. Table 7.4 shows the hydro-biotic limits of a moist soil in relation to pore size. As the soil freezes, the latent heat of the water offers some initial thermal protection

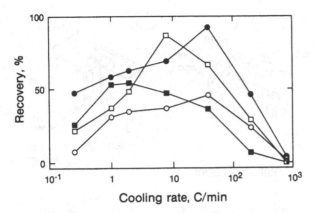

Fig. 7.15. Recovery (%) of *Chlorella emersonii* after cooling at different rates to −196 °C in the following concentrations of DMSO, 0.375 M (open circles), 0.75 M (filled circles), 1.5 M (open squares) and 2.5 M (filled squares) (following Morris (1980)).

to rootlets and root hairs but this temporary advantage is soon lost as the unfrozen water content drops and the soil is dehydrated. As noted in Sections 6.3 and 6.4, this dehydration is most pronounced in fine, frost-susceptible soils which are capable of generating substantial water sinks. In the absence of meltwater in the spring, such massive freeze-dehydration added to normal transpirational loss could be lethal. Moreover, the dehydration is often accompanied by ice segregation which may cause serious mechanical damage within the soil and root system.

As might be expected, the rapid freezing behaviour of plant cells in suspension closely follows that discussed in the previous section. Fig. 7.15 illustrates the effect of cooling rate and additive (dimethyl-sulphoxide DMSO) concentration on *Chlorella emersonii* cooled to −196 °C and rapidly thawed. For any given additive concentration, the survival rate exhibits the characteristic maximum with respect to cooling rate, but the curve as a whole is shifted vertically by varying the amount of additive. The survival rate evidently exhibits a maximum with respect to additive concentration. DMSO is capable of penetrating the cell membrane, and its effect is to reduce salt concentration both inside and outside the cell. In small amounts, the intracellular dilution effect reduces the extent of harmful salt concentration, and the extracellular effect reduces dehydration. In larger amounts, the additive evidently introduces its own toxic effects and, by reducing dehydration further, increases the probability of nucleation within the cell. The shift of the maximum survival rate to lower

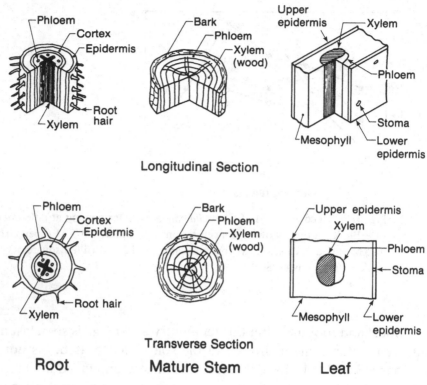

Fig. 7.16. Xylem/phloem system of a plant.

cooling rates by increasing additive concentration is clearly noticeable in the figure.

The cause of cell death is seldom the appearance of extracellular ice, particularly when the cell is at a phase with minimal metabolic needs. On the other hand, intracellular ice will cause death if it grows to a significant extent and compromises the viability of organelles through engulfment, or grows in such a way as to re-texture the cytoplasm, thus destroying fine structure. The principal site of injury is generally believed to be in the cellular membranes, but the cause of injury has yet to be established unequivocally. It has been suggested that the cause is a thermotropic phase transition in which the membrane changes from liquid–crystalline to solid–gel form (Morris 1980; Lyons, Raison and Steponkus, 1979). This suggestion stems from the interpretation of Arrhenius plots and appears to imply a chilling injury rather than a freezing injury. The conclusion, however, is at variance with the fact that supercooling *per se* is not an established cause of injury. More likely causes of membrane destabilization are: the freeze-induced increase in solute concentration, especially electrolytes; and freeze-induced electrical potentials (Steponkus, 1984; Steponkus, Stout, Wolfe and Lovelace, 1985). These effects introduce stresses in the membrane, the first chemically and the second electrically, and it is perhaps these stresses, together with the stress of membrane dehydration itself, that lead to destabilization and disruption.

Above the cellular level, the appearance and effect of ice is best seen in relation to the bulk movement of water in the plant. In root, stem and leaf, water movement takes place in two parallel systems: the *xylem*, which carries salt-rich water through, and from, the roots to, and through, the leaves (transpiration); and the *phloem*, which brings nutrient-rich water back down to the roots (translocation). These systems are not entirely independent of each other, but for the present purposes it will be assumed that they are. Fig. 7.16 is a schematic representation of the xylem/phloem system in root, stem and leaf. The xylem and phloem may be regarded as bundles of long plant cells running parallel to each other. The *cambium* is the region where new xylem and phloem cells are built each year. During normal growth in a mature stem, the cambium, phloem and bark move out a little further each year, thus leaving behind dead cells which may be divided into two categories: those more recently formed (and nearest to the phloem), and those formed years ago. Both of these categories of dead cells have undergone a transformation in which their ends have become perforated (even absent) and the protoplasm has been lost; this allows the longitudinal flow of water. The older cells in the xylem are strengthened by lignin and thus become wood. The more recent cells carry most of the water which can, however, be made to percolate through the woody core. The phloem is a similar arrangement of parallel, intact cells, but the end walls are not perforated thereby preventing the bulk flow of water that is so characteristic of the xylem.

Upward flow of water is driven by evaporation at the surface of the leaves: the partial vapour pressure outside the leaf is low enough to suck water from the leaf interior, by osmosis, thus producing large tensile stresses within the xylem which evidently behaves quite passively as a variable cross–section, porous bed capable of transporting the dissolved mineral salts to the leaves. Translocation in the phloem, on the other hand, is not so simple. Its principal purpose is to send the products of photosynthesis (carbohydrates such as sucrose, for example) down to regions of respiration, growth and storage, but the mechanism by which this is performed is not well established: pressure difference and protoplasmic streaming are two current theories (Richardson, 1975; Gilpin 1981c).

Should the ambient temperature drop below $0\,^\circ$C the plant has no exter-

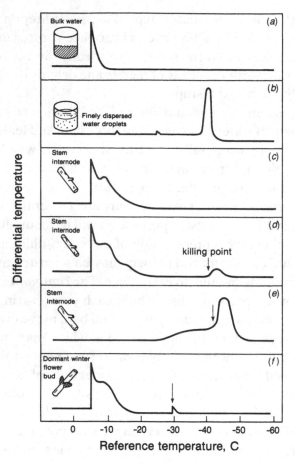

Fig. 7.17. Differential thermal analysis of water and plant tissues:
(*a*) heterogeneous nucleation of pure water; (*b*) homogeneous nucleation of
pure water; (*c*) acclimated dogwood stem; (*d*) acclimated apple stem;
(*e*) acclimated hickory stem; (*f*) acclimated peach flower bud (following
Burke *et al.* (1976)).

nal defence mechanisms. Metabolism and the sensible heat capacity of the
entire organism are minimal defences, but the emission of latent heat from
vascular and extracellular water may delay intracellular freezing for short
periods. In the event of an early frost the leaves can offer little resistance:
they are not thick enough to avoid freezing for very long, except through
supercooling. The frost itself, acting as a nucleator, tends to limit the
supercooling, and ice nucleation-active bacteria may eliminate it entirely
(Lindow, 1983). In any event, ice seeding will occur at the entry sites
provided by stomata, lenticels and wounds, and is therefore very depen-

dent on morphology [13] and recent events, especially those which control the moisture content and temperature in air and soil (Burke *et al.*, 1976).

Many plants freeze within a degree or two below 0 °C. The very hardy woody plants of the boreal forest of North America, for example, readily form extracellular ice, and are able to tolerate cell dehydration and corresponding levels of freeze-concentration to very low temperatures (in laboratory experiments down to −196 °C when fully acclimated (George, Burke, Pellett and Johnson, 1974)). In certain types of birch, dogwood, willow and aspen, such behaviour converts almost 70% of the total water content to extracellular ice, thus implying that some 30% is unfrozen, and perhaps unfreezable (Burke *et al.*, 1976). A comparison of an acclimated dogwood stem with pure water is given in Fig. 7.17 using DTA. The figure also shows the response of an acclimated apple stem which also undergoes extensive extracellular ice formation, as the first part of the curve indicates. However, it is apparent that another freezing event takes place in the apple stem at a lower temperature; this may mark the point where intracellular water is frozen but the proof is wanting (Quamme, Weiser and Stushnoff, 1973).

The formation of extracellular ice is not necessarily injurious to a plant, although the ice is capable of being disruptive by growing between adjacent cell walls or between the wall and the plasmalemma. In ice tolerant plants (often called frost tolerant), large ice masses may be formed in the cortex and mesophyll without apparent damage (Wiegand, 1906). Should the plant be intolerant of ice, it must adopt another strategy to survive: namely, supercooling, which circumvents dehydration and freeze-concentration. Supercooling down to −47°C has been recorded in the xylem of some cold-hardy, woody species of Asia and North America (George, Burke, Pellett and Johnson, 1982). In general, only part of the water content will supercool, but the acclimated hickory stem provides an example in which the entire water content freezes spontaneously. This is illustrated in Fig. 7.17 which reveals that the hickory stem, unlike the apple stem and peach flower bud, does not exhibit the earlier heterogeneous nucleation of extracellular water. It must be emphasized that the shape of the low temperature exotherm, and its relation to the temperature scale, are by no means fixed. The work of Sakai (1979, 1982), for example, suggests that dehydration may occur if the cooling rate is low enough; the water then freezes externally in a process which, overall, is not unlike the ice segregation process in soil discussed in Section 6.4.

The precise mechanism which permits large supercooling is not known, although it is clearly not a disguised form of freezing point depression (Franks, 1985). It is often attributed to solutes which inhibit ice nucleation, but may result, in part, from the ability of the plant to re-distribute water throughout its tissues in spaces small enough, or near surfaces electrically charged enough, to render the solution unfreezable: in the former instance by reducing the probability of homogeneous nucleation, and in the latter by preventing tetrahedral lattice formations. A particular study by Quamme and Gusta (1987) on dormant peach flower buds indicated that supercooling was accompanied by the migration of water to less strategic sites (in the flower bud scales and pith) where it froze. This process is reminiscent of the extraorgan freezing mentioned above, and leads quite naturally to the general suggestion that supercooling is part of an active re-distribution process in which biochemistry, biophysics and anatomy all play an integral part.

7.6   Freezing in animals

Shelter, generally regarded as a basic requirement of human life, is our principal means of avoiding freezing temperatures. The home, the workplace, and their extensions permitting us to commute, all serve to isolate human beings from unwanted ice. We rely heavily on the engineering design of shelters to prevent the ingress of ice and to mitigate the effects of frost and melted snow. Even outdoors, and especially when we are engaged in various winter sports, the same strategy of thermal isolation is pursued. A few fur-coated mammals duplicate this behaviour; others approximate it but must be content with a restricted range of physical activity during a period of hibernation. The alternative is to flee from the ice and associated low temperatures, but only birds have the ability to move economically over the distances covered by the 0 °C isotherm in the course of a year.

Despite their mobility, most animals – poikilotherm and homoiotherm alike – have a restricted ability to avoid the coming ice. They can, however, develop protective coats and cocoons, and seek out an appropriate habitat before resigning themselves to the final preparation of their bodies. Among the poikilotherms in particular, this preparation takes a variety of forms and has usually begun much earlier, long before the approach of the first frost. It is frequently called cold hardening or frost [14] hardening. In essence it consists of ridding the body of as much water as possible, thus reducing the probability of homogeneous or heterogeneous nucleation in bulk fluids; the remaining water is then re-organized so as to leave it mobile enough to permit a minimum level of metabolic activity

(Storey, 1983). Dehydration is thus a general strategy while self-imposed starvation ensures that no ice-nucleation particles remain in the gut (Salt, 1952). In addition, the water occupying extracellular and intracellular spaces is, in general, re-allocated and perhaps biochemically altered with results similar to those seen in plants: ice-nucleation catalysts in the extracellular water signal a strategy of intracellular ice avoidance through tolerable freeze-concentration; ice-nucleation anticatalysts signify a strategy of supercooling.

As a rule of thumb, it might be expected that the earlier the stage of development of an organism the better it would be able to cope with ice. (This is by no means an invariable rule, e.g., insects overwinter in only one specific stage: egg, larva, pupa or adult.) The organism is usually smaller then and has a lower water content, expressed as a fraction of total mass. Plant seeds, for example, not only tend to have low water contents but often supercool inversely with their water content (Juntilla and Stushnoff, 1977). Insects provide many examples of deep supercooling: the larvae of *Bracon cephi* supercool to −45 °C, the eggs of the lepidopteran *Seiraphera diniana* reach −50 °C, and the larvae of the gall fly *Eurosta solidaginis* may still be unfrozen at −55 °C (Ring, 1980; Salt, 1956, 1959).

To some extent, these low temperatures are achieved through a colligative form of freezing point depression (−15 °C in the case of *Bracon cephi* larvae) but there is evidence to suggest that most of the range is attributable to the presence of substances which inhibit nucleation and crystal growth (Salt, 1961; Duman, 1982; Duman, Horwath, Tomchaney and Patterson, 1982). Glycerol may be such an agent. It must also be borne in mind that supercooling is not a predictable event in insects, and carries with it the statistical requirement that the probability of spontaneous nucleation increases with the duration of the supercooling. Over and above this requirement, the insect faces the possibility of accidental nucleation through extraneous water, particularly that located in the vicinity of body openings, channels and cuticular pores (Ring, 1980). As with plants, morphology is one of the principal determinants of accidental inoculation, although this risk is reduced through mobility which plays a crucial role in the selection and protection of overwintering sites.

The supercooling ability of insects such as the bark beetle *Scolytus multistriatus* or the gall fly *E. solidaginis* (Ring, 1980) is usually achieved at the end of a period of dehydration; this is as an integral part of their physiology, and enables them to remain in a prolonged state of hibernation. Insects such as the Arctic carabid *Pterostichus brevicornis* (as adults) or the Antarctic chironomid midge *Belgica antarctica* (as larvae)

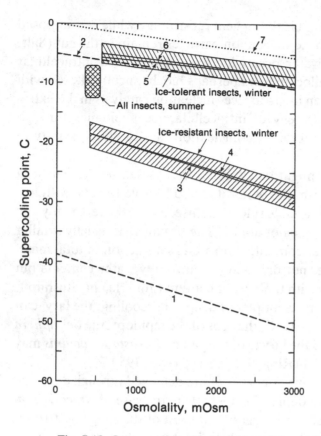

Fig. 7.18. Supercooling points of intact insects and aqueous solutions
as a function of osmolality. *Line 1*, highly purified solutions of glycerol,
representing homogeneous nucleation. *Line 2*, solutions of glycerol containing
potent ice nucleators from haemolymph of freeze-tolerant insect. *Line 3*,
regression line of data from freeze-avoiding bark beetle. *Line 4*, regression
line for nine species of freeze-avoiding beetles. *Line 5*, regression line from
freeze-tolerant beetle. *Line 6*, regression line for data from 4 species of
freeze-tolerant beetles. *Line 7*, corresponding melting points. *Shaded area*,
distribution of values of freeze-tolerant and freeze-avoiding beetles in winter
and summer.

(Block, 1982) cannot adopt such a strategy and must allow ice to form in
their extracellular fluid. Fig. 7.18 shows the typical range of conditions for
ice-resistant and ice-tolerant insects (Zachariassen, 1985). The latter,
which generally remain active, tend to avoid supercooling which, if interrupted by accidental nucleation, could lead to lethal intracellular ice or
high osmotic stress. Ice-nucleating catalysts have been found in the
haemolymph of several ice-tolerant insect orders, including Hymenop-

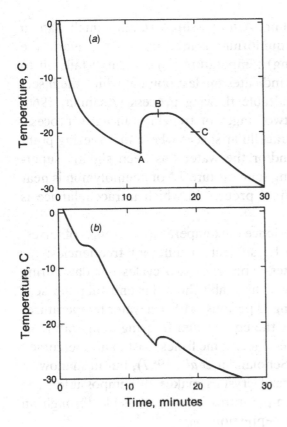

Fig. 7.19. Typical freezing curves in ice-resistant insects; (*a*) single superfreezing point e.g., *Scolytus rateburgi*; (*b*) double superfreezing points e.g., arctic carabid beetles (following Ring (1980)).

tera, Diptera and Coleoptera (Zachariassen, 1985). The catalysts appear to be proteinaceous (Duman *et al.*, 1985). However, it is important to note that ice-resistance and ice-tolerance are not mutually exclusive characteristics. In a thorough review of the cold hardiness of invertebrate poikilotherms, Block (1982) has listed nine orders which are ice-resistant and five which are ice-tolerant, while eight groups have both characteristics; morphology and ecology may well be important.

Under natural conditions, the cooling rates of insects are probably less than the rate of $1 \, \mathrm{K \, min^{-1}}$ typically used in laboratory experiments. Despite this, calorimetric techniques have provided useful data on the freezing characteristics of ice-tolerant insects (Ring, 1980). Fig. 7.19 shows two typical results from thermal analysis, revealing a close relation with the DTA data [15] for plants discussed in the previous section. In Fig. 7.19(a), freezing begins at the point A, the sudden rise in temperature confirming the prior existence of supercooling. After passing through a small plateau at B (which, under stable equilibrium conditions, would mark the equilibrium freezing (or melting) temperature T^i) the curve falls off to the uncertain location C which indicates the last point at which the insect could be revived through a suitable thawing process (Asahina, 1966, 1969). Fig. 7.19(b) indicates two stages of freezing: an early process, suggesting ice formation in extracellular spaces where the freezing point has been slightly depressed and/or the water has been slightly super-cooled (the spontaneous freezing temperature T* of haemolymph is near $-8\,°C$ (Baust, 1981)); and a final process in which intracellular ice is formed.

Aquatic animals seldom experience the temperature extremes of terrestrial animals. (They may also be subject to different frequencies: for example, intertidal invertebrates experience two cycles per day during winter.) None the less, fishes which inhabit the polar and subpolar seas must be able to cope, for prolonged periods, with sea water temperatures of $-1.85\,°C$, at least $1\,°C$ below the equilibrium freezing temperature of their body fluids. In the absence of ice, some fishes exist in a supercooled state during their entire lives (Scholander *et al.*, 1957), but in shallow or open water the presence of frazil crystals makes this impossible. The additional lowering of their temperatures is made possible through an unusual form of freezing point depression.

The work of DeVries and others (1970, 1980, 1982) has revealed that the freezing point depression of coldwater fishes may be attributed to the presence of antifreezes: glycopeptides are common in Antarctic fishes while peptides are found in northern fishes. However, these antifreezes do not operate in the usual colligative manner. Instead, they appear to inhibit the faster growth of ice crystals along their *a*-axes, but permit it along the *c*-axis (DeVries, 1982). The precise mechanism is unclear but has been modelled as adsorption-inhibition, in which the accommodation of water molecules at the ice–water interface is impeded by the incorporation of the antifreeze molecules. This is closely related to an antinucleation mechanism and may explain both the speed and the form of subsequent crystal growth which is reminiscent of growth in supercooled water except that the ice grows as long spicules or whiskers emanating from the basal plane; this may indicate a spiral dislocation (see Section 2.9).

An entirely different poikilothermic strategy is found in certain frogs which are capable of tolerating extracellular ice while they overwinter beneath leaves and snow where the temperature drops down to $-6\,°C$ (Layne and Lee, 1987). Typically, such frogs are able to withstand about

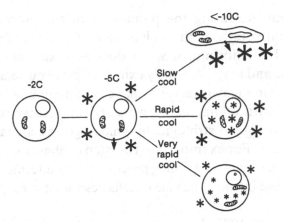

Fig. 7.20. Scheme of physical events in cells during freezing (following Mazur (1977)).

40% of their total water content being frozen while cryoprotectants such as glycerol and glucose appear in their blood (Storey and Storey, 1985b, 1986). It has also been observed that, unlike insects, the frogs do not undergo a period of frost hardening; the appearance of the ice itself evidently triggers the fast release of the cryoprotectants (Storey and Storey, 1985a).

Given the nature and function of homiothermy, it is not surprising that birds and mammals accommodate the presence of ice in their environment by pursuing a strategy of thermal isolation aided, in part, by control of metabolism. Consequently, one would not expect freeze adaptation to be an intrinsic aspect of their general physiology. Interest in the freezing behaviour of such animals therefore falls under a number of distinct subheadings: cryopreservation of human organs, accidental injury to human tissue e.g., frostbite, controlled necrosis in human tissue e.g., tumour destruction, and the freeze preservation of food. The second of these is beyond the scope of this book and the last two will be discussed later.

Cryopreservation is perhaps best approached by recalling the freezing behaviour of individual cells: Fig. 7.20 is a useful summary in schematic form. Most cells do not freeze under stable equilibrium conditions above $-2\,°C$, or thereabouts, and are capable of supercooling for a short period of time to around $-5\,°C$. Below this temperature, which generally heralds the onset of osmotic water loss, events depend strongly on the cooling rate. If stable equilibrium conditions are approached, as commonly occurs in natural circumstances, the cooling rate is very low ($\ll 1\,\mathrm{K\,min^{-1}}$) and the cell's lost water freezes externally, thus reducing the probability of intracellular ice formation but increasing the probability of concentration-induced damage. At unnaturally high cooling rates ($\gg 10^2\,\mathrm{K\,min^{-1}}$), transmembrane chemical equilibrium of water does not exist, and the cytoplasm first supercools and may eventually exhibit internal nucleation. These facts have given rise to the two-factor damage hypothesis of Mazur: freeze-concentration and intracellular ice. At worst, this hypothesis offers a choice of death; at best it suggest strategies by which a given cell may be treated to avoid both factors. For example, the two-step cooling process is a particular protocol which stages the cooling process so that intermediate equilibration can help avoid intracellular ice in cells destined for very low temperatures.

Mazur and others (Mazur, 1977; Leibo, Mazur and Jackowski, 1974) have successfully demonstrated that the two-factor hypothesis applies not only to suspensions of cells from different mammals but to simple, multi-cellular organisms, specifically two-cell and eight-cell embryos of mice. In general, however, the differences between cell suspensions and complete organs are significant (Jacobsen and Pegg, 1984). Although cryoprotectants have produced beneficial effects in intact organs, their addition and removal is complicated by the need to avoid a substantial rise in the vascular resistance. The theory of fluid flow through porous media impregnated with ice is reasonably well understood in many physical systems (see Section 6.3) and should be applicable to structured organ tissue, which has a low hydraulic conductivity. In addition to its effect on permeability, the bulk of the organ is also reflected in permissible cooling rates: the collective volume of the cells creates temperature gradients throughout the organ, thus implying cooling rates which vary spatially; similarly, the effective surface–volume ratio of the cells is closer to that of the organ as a whole, and therefore much less than that of a single cell. Organs thus demand lower cooling rates.

The fact that organs are not composed of identical cells complicates the difficulties introduced by vascular and interstitial architecture: the functions of individual cells cannot be neglected. To some extent, these difficulties are also found in plants, from which considerable experience has been gained, although their bioglaciological characteristics may be quite different. In general, the growth of interstitial and intravascular ice compromises cell intercommunication and limits the flow of nutrients and wastes; in particular, intravascular ice from cell-derived water may grow (another example of ice segregation in porous media) and cause gross vascular damage. The ice is also likely to cause disruption of vessel and membrane surfaces through dehydration and the purely mechanical effect of constriction; it may ultimately invade the cell itself (Mazur, Rigopoulos and Cole, 1982; Pegg, Diaper, Skaer and Hunt, 1984).

Despite the complexity of tissue and whole organ freezing, knowledge and experience have grown considerably during the past two decades (Ashwood-Smith, 1980; Jacobsen and Pegg, 1984). The cryopreservation of corneas was demonstrated in the early 1960s, and within a decade considerable advances had been made in the cryopreservation of heart valves, developing teeth and human skin. More recently, steady progress has been recorded with tissue from the pancreas, thyroid and parathyroid; the same is true for intact veins, kidneys and hearts. However, the appearance and effect of ice in the vascular and interstitial spaces continues to limit our ability to freeze whole organs without damage, and because of this it has been suggested (Jacobsen and Pegg, 1984) that strategies designed to tolerate ice may even have to be abandoned.

Among ice-free alternative possibilities is the proposal to induce vitrification (Fahy, MacFarlane, Angell and Meryman, 1984). In Fig. 7.20 it is suggested that very rapid cooling tends to reduce the severity of lethal intracellular ice, at least in terms of crystal size, but substantial amounts of extracellular ice are still created. However, if high concentrations of a non-toxic cryoprotectant are first introduced, very rapid cooling may not permit any nucleation at all. This prospect was noted earlier in the discussion of Fig. 7.3 from which it may be noted that for solute concentrations lower than at I, corresponding to the intersection of the vitrification curve and the homogeneous nucleation curve, nucleation may occur at any but the highest cooling rates, particularly if ice-nucleation catalysts are present. But for concentrations greater than this it appears that even heterogeneous nucleation may be suppressed if the cooling rate is high enough. This makes possible the attainment of ice-free low temperatures, subject to two important provisos: the concentration of the cryoprotectant should not be high enough to lead to toxic or adverse osmotic effects; and the devitrification process during thawing should not allow substantial growth of any ice nuclei formed during cooling. Thawing will be discussed in the next chapter.

Before leaving this section it is worth taking a brief look at an area of bioglaciology in which freezing is actually sought for the medical benefits it brings. Cryosurgery is a relatively recent development in medicine, although the first experiments date back over a century (Openchowski, 1883). In general, the technique may be used to induce a state of physiologic inhibition, to create a therapeutic lesion or destroy neoplastic tissue; it does so in a simple flexible and essentially haemostatic manner

(Cooper, 1967). Applications include the treatment of brain tumours, necrosis of cancer tissue and opthalmic surgery.

A typical cryosurgical probe consists of three long concentric tubes. The coolant passes down the inside of the centre tube and returns along its other side. Since the principal heat loss in the tissue is meant to be restricted to a small region at the probe tip, any secondary loss to the returning coolant must be strictly limited, often by means of a vacuum created between the outer and intermediate tubes. By varying the coolant temperature and flow rate, the surgeon controls the cooling rate and final temperature distribution in the tissue surrounding the tip. It is clear from earlier discussion that, apart from the bulk thermal properties of the tissue, its cytological and vascular make-up must be known if the effects of freezing are to be predicted with precision. Such a freezing process is non-isothermal whereas most cryobiological data have been obtained for spatially uniform (though transient) temperatures. Differences between fast and slow cooling rates have been observed in cryosurgery, and have been interpreted in terms of freeze concentration, intracellular ice formation and membrane destruction (Cooper, 1967). More recently, it has become apparent from studies in hepatic cryosurgery that the ice actually propagates through the vasculature, thus inducing dehydration and freeze-concentration in neighbouring cells in a manner which varies with the distance from the cold cryoprobe surface (Rubinski, Onik, Lee and Bastacky, 1987). These events may be accompanied by vascular expansion which is believed to be the principal cause of tissue destruction.

The extent of the freezing 'bulb' around the probe tip is usually well defined and may be estimated using the methods outlined in Chapter 3. Assuming that the latent heat is released at a single temperature, and that the sensible heat of the tissue is much less than the latent heat ($Ste \ll 1$), the progress of the lesion may be expressed in algebraic form for a wide range of probe temperatures (strictly the superheat ratio, see Chapter 3) (Cooper and Trezek, 1971). The results thus obtained are found to be in good agreement with numerical data, but the model neglects the effects of supercooling and moisture migration in the vasculature. Quite recently, the mapping of frozen lesions with ultrasound has been reported (Onik *et al.*, 1987).

7.7   Freezing of foods

In the light of discussion in the previous section, freezing of food might easily be thought of as an aspect of 'ice and death' rather than 'ice and life' were it not for several important facts: firstly, it is obvious that frozen food is ultimately used in support of human life; secondly, death of the organism as a whole by no means implies the death of its component parts, particularly at the cellular level (the enzyme systems of fruits, vegetables and meat are still active after freezing and thawing); and thirdly, the processes of nucleation and growth of ice after harvesting are essentially the same as, or extensions of, the processes which occur in the living organism. It is therefore appropriate to treat the freezing of foods in this chapter, albeit briefly. This will be done by first continuing the discussion of microorganisms presented in Section 7.4. The thermal behaviour of various foods – fluids, phytosystems (i.e., fruits and vegetables) and myosystems (i.e., edible muscle of mammals, fowl and fish) – will then be reviewed microscopically and macroscopically. The section, and the chapter, ends with an analysis and discussion of freeze drying.

The presence of microorganisms in food gives rise to two complementary problems: the determination of freezing conditions which will minimize the survival rate of pathogenic bacteria strains such as *Salmonella*, *Staphylococcus* and *Streptococcus*, while at the same time determining conditions which will maximize the survival rate of cells such as the yeast *Saccharomyces cerevisiae* in dough products. It is beyond the scope of this book to discuss the microbiology of foods, but it is worth noting that extensive laboratory work and wide-ranging industrial experience both suggest that the freezing behaviour of microorganisms is best understood in terms of the theoretical framework established in Section 7.4 and elaborated upon in Sections 7.5 and 7.6.

For any given microorganism, the cooling rate, and the surrounding medium, will largely determine the onset of nucleation and effect of extracellular ice. In general, this ice is not found to be harmful of itself, and the corresponding freeze-concentration effects may not be detrimental, at least not immediately. However, storage at low temperatures for long periods of time may permit a slow and gradual effect of freeze-concentration to become very significant (Clement, 1961). Similarly, the appearance of intracellular ice nuclei need not pose a threat unless, during thawing, the nuclei are allowed to grow disruptively. The number density of the cells may also be a significant factor. Injury of microorganisms in food is generally attributed to membrane damage, although metabolic changes have also been documented. A great many cryoprotectants, glycerol in particular, have been used to reduce this damage.

Ice itself, so widely used in the food and beverage industry, is by no means free of microorganisms (Marth, 1973), but most of the bacteria found are associated with the human and environmental sources encountered during handling and use. This is particularly true of commercial crushed ice (Foltz, 1953). Manufactured ice generally has a low microbial count, and even ice produced from polluted water may not present a health hazard if stored for several weeks (Jensen, 1943). To a large extent, the growth of microorganisms in food depends upon the water content of the food. Dried cereal grains, for example, often have insufficient moisture to support microorganisms whereas milk and many other fluid foods have it in abundance.

The freezing of fluid foods is complex both physically and chemically. Such foods range from dilute aqueous solutions, through suspensions, to thick, non-Newtonian pastes, and therefore embrace a wide range of freezing phenomena. An important special case is cow's milk, which contains about 87.2% water, 3.7% fat, 3.5% protein, 4.9% lactose and 0.7% ash (Watt and Merrill, 1963). Physically, the fluid consists of fat globules, with a diameter of the order of $4\,\mu$m ($1\,\mu$m if the milk is homogenized), and caseinate micelles having a diameter of about $0.1\,\mu$m and a mean distance of about $0.36\,\mu$m separating them: the micelles are believed to contain 'bound' water (Powrie, 1973). The equilibrium freezing temperature of milk is about $-0.55\,°$C, but supercooling is often observed, depending upon the precise conditions of cooling. At the container wall, the onset of (heterogeneous) nucleation begins the freeze-concentration of lactose and soluble salts, presumably under a condition of constitutional supercooling, but the growth of ice crystals in such a suspension of globules requires the modification of conclusions based on the simple equilibrium phase diagram. Entrapment of concentrated pockets within the advancing crystal fronts, and the prospect of lactose crystallization, may alter the details of the freezing process significantly (Powrie, 1973). For high cooling rates (i.e., freezing within a few minutes) the fat globules are essentially frozen in place (Lagoni and Peters, 1961) but low cooling rates permit many of them to be pushed ahead of the ice front into a region where their numbers may eventually become great enough to cause coalescence, commonly called *cryodemulsification*.

Freezing does not appear to have an immediate effect on the proteins in milk and milk products but storage at low temperature may lead to progressive destabilization. Over prolonged periods, and particularly over repeated freeze–thaw cycles, the destabilized protein will flocculate, and if the protein mass is great enough the flocs will then form a three-dimensional network characterized as a gel (Doan and Featherman, 1937). Freeze-induced alterations to physical properties have also been observed

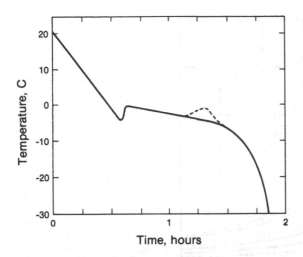

Fig. 7.21. Schematic freezing curve typical of foods and living matter
(following Fennema, Powrie and Marth (1973)).

in other fluid foods. Egg yolk, for example, which is widely used in the
manufacture of salad dressings, ice cream and bakery products, undergoes
a fundamental transition in the neighbourhood of $-6\,°C$. The equilibrium
freezing temperature of yolk is about $-0.6\,°C$, though supercooling down
to $-6\,°C$ (and lower) has been claimed for the contents of shell eggs. For
temperatures lower than $-6\,°C$, lipoproteins in the plasma interact with
dispersed granules to create a fluid condition described as *gelation*. Disruption of the granules is evidently caused by the freeze-induced concentration of salts (Chang, 1969). Gelation may be inhibited by using a cryoprotectant such as sucrose or glycerol.

The freezing behaviour of food myosystems is strongly influenced by
their bulk. Since they must be cooled from their surfaces, thermal conditions vary most rapidly near the surface itself, as may be seen from the
temperature distribution in a semi-infinite body whose temperature, initially uniform, is altered when the surface temperature is suddenly
lowered to, and maintained at, a value θ_0. It was noted in Chapter 3 that
the temperature distribution is then given by

$$\theta = \theta_0 \, \text{erf}\, [X/2(\varkappa t)^{\frac{1}{2}}]$$

where X is the distance beneath the surface. Differentiating this with
respect to X gives

$$\frac{\partial \theta}{\partial X} = \frac{\theta_0}{(\pi \varkappa t)^{\frac{1}{2}}} \exp\left(\frac{-X^2}{4\varkappa t}\right)$$

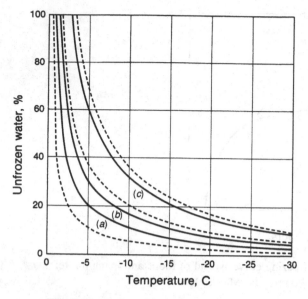

Fig. 7.22. Unfrozen water content of food phytosystems: (*a*) strawberries, 9% soluble solids; (*b*) peas, 13% soluble solids; (*c*) cherries, sweet, 22.5% soluble solids. Dashed lines – juices with 5, 15, 25% soluble solids (following Gutschmidt (1968)).

which is proportional to the heat flux, while differentiation with respect to *t* gives

$$\frac{\partial \theta}{\partial t} = \frac{-\theta_0 X}{2(\pi \varkappa t^3)^{\frac{1}{2}}} \exp\left(\frac{-X^2}{4\varkappa t}\right)$$

the cooling rate. It is thus evident that the heat flux (and, in general, a coupled mass flux) and the cooling rate are functions of time and location. The instantaneous cooling rate, in particular, exhibits a maximum with respect to X, the particular location of which gradually moves further beneath the surface as time progresses; at any fixed location, the cooling rate has a maximum with respect to time. Conclusions based upon data obtained at fixed temperatures or cooling rates, especially data at the cellular level, must therefore be applied with caution.

This comment is worth bearing in mind while interpreting freezing curve data from small, isothermal specimens, especially if the cooling rate is high and supercooling may occur. For cooling rates approaching stable equilibrium conditions, a curve such as that shown in Fig. 7.21 is usually obtained. As indicated, early withdrawal of heat at a steady rate produces a linear change in temperature down to the equilibrium freezing temperature (typically about −0.6 °C for milk, −2 °C for meats and −1 °C to −3 °C for fruits and vegetables (Hardenburg and Lutz, 1968)), and slightly beyond. Nucleation then occurs, and latent heat is withdrawn at a temperature which decreases slightly in response to gradual increases in solute concentration. Only when most of the water is frozen will the steep decline of temperature resume, quite possibly with the formation of eutectics[16]. Should any segment of the water behave differently from the bulk e.g., intracellular water, the curve may exhibit another small perturbation, as indicated by the broken line, and noted in the previous section. The cooling rates represented in Fig. 7.21 are considerably lower than those displayed in Fig. 7.19 but this is not surprising given the comparative mass of bulk food products and the need to avoid large temperature gradients.

At temperatures substantially below the freezing point, biochemistry and biophysics become especially complex, as reflected in the departure of the freezing curve from linearity. This curvature may be affected by concentration levels and reaction rates, but may also reflect the freeze inhibition attributable to pore surfaces, both intracellular and supracellular. Fig. 7.22 shows the unfrozen water content of several phytosystems plotted as a function of temperature. The form of the curves is remarkably similar to that exhibited by freezing soils (Section 6.4) and suggests that some water may still remain unfrozen in these systems at temperatures below $-30\,°C$. This behaviour is attributed to the concentration of soluble solids and suspended matter.

Freeze damage to food phytosystems frequently begins in the vascular and intercellular spaces. As would be expected, low cooling rates permit intracellular water to permeate the plasmalemma and freeze externally. Apart from shrinking the vacuole, this may produce large ice crystals capable of disrupting the tissue and creating textural damage. The situation changes at higher cooling rates. For the parenchyma tissue of tomato fruit, for example, when cooling rates exceed about $10\,°C\,min^{-1}$, intracellular ice crystals appear (Mohr and Stein, 1969), thus implying that the transmembrane water flux is unable to keep pace with an increasing degree of supercooling. Whether the destruction of organelles following cell wall disruption and invasion by extracellular ice is preferable to their destruction by intracellular crystals is a moot point, but perhaps both may be avoided through the choice of suitable freezing and thawing rates.

Useful insights into the freezing behaviour of bulk myosystems and phytosystems may be obtained from the discussion of aqueous solutions given in Sections 4.1 and 7.2. It will be recalled that cooling-induced temperature and concentration gradients may give rise to columnar or dendritic growth into the bulk. This will occur if the surface layer of the

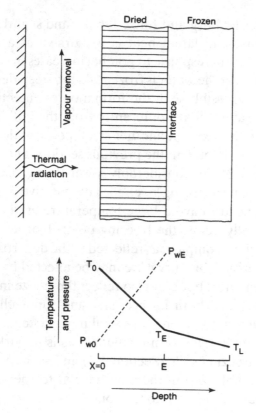

Fig. 7.23. The freeze-drying process.

system is first supercooled, nucleation then occurs, and the temperature of the fluid remains lower than the local equilibrium freezing temperature, at least in the region close to the advancing ice. The resulting cellular or dendritic growth provides a first order model of freezing beneath the surfaces of saturated porous media. An important example is the freezing of meat (Calvelo, 1981), especially when the direction of freezing lies parallel to the fibres.

For beef, it has been observed (Menegalli and Calvelo, 1979) that the particular form of equation (2.61) for a-axis growth in a constitutionally supercooled solution is given by

$$V_I = 1.57 \times 10^{-4} \, \theta^{2.15}$$

where $\theta = T^i - T$. Taking a representative value of $\theta = 4\,\mathrm{K}$, it is evident that the dendrites would take above $10\,\mathrm{s}$ to reach a typical depth of $2\,\mathrm{cm}$. Provided that the cooling rate is less than about $1\,\mathrm{K\,min^{-1}}$, little intracellular ice appears, and the extracellular dendrites grow alongside the fibres[17]. The freeze-concentration of the extracellular fluid leads to partial dehydration of the fibres and their consequent deformation. The lower the cooling rate the larger will be the extracellular dendrites and it might therefore be expected that tissue damage would then be greater. However, when the crystals form between the fibre bundles the damage is actually less (Taylor and Pegg, 1983). Another form of damage arises if the water vapour pressure of the ice near the surface is greater than that of the environment. This phenomenon – usually called freezer burn – occurs when the tissue is not protected by a moisture-proof barrier, and thus allows sublimation of the intercellular ice followed by further dehydration and shrinkage of the fibres.

As a technique, sublimation-dehydration – or freeze-drying – is widely used in both the food industry and the medical laboratory. To be effective, it requires the application of high vacuum and a source of heat to provide the latent heat of sublimation. The situation is described in Fig. 7.23 which shows a slice of material, thickness L, and originally frozen at temperature T_L. This is positioned next to a space across which radiant heat is supplied and from which water vapour is continually removed by vacuum equipment designed to maintain a water vapour pressure P_{w0} much less than the equilibrium value P_{wE} corresponding to interfacial sublimation at the temperature T_E. Ice thus sublimates at the interior plane $X = E$ which separates the frozen region from the dried region; the vapour generated diffuses towards the outer surface at $X = 0$. Although the process is transient, the time-dependence of the temperature field may be neglected if the Fourier number is large, in which case heat conduction in the frozen region is governed by

$$d^2 T_F / dX^2 = 0$$

while in the dried region it may be described by

$$d^2 T_D / dX^2 + A\, dT_D / dX = 0$$

where the advective coefficient $A = \rho c_p U / k$, in which ρ, c_p and U are the vapour density, specific heat and Darcian velocity, respectively, and k is the effective thermal conductivity of the porous matrix produced through drying.

Recalling the more general discussion in Section 6.3, the second of these equations implies that the tortuosity may be replaced by a scalar constant appropriate to these quasi-one-dimensional conditions. The equation also implies that the heat and vapour fluxes are uncoupled, despite the fact that a high vacuum may produce molecular mean free path lengths comparable with the width of the intercellular spaces left by the extracellular ice (Dyer and Sunderland, 1968).

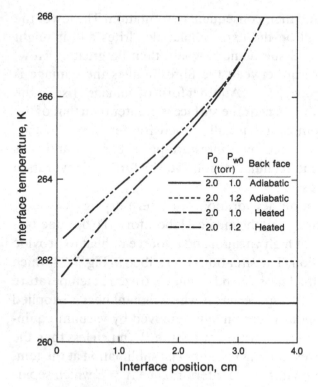

Fig. 7.24. Interface temperature versus interface position for bovine muscle (following Dyer and Sunderland (1968)).

In the dried region, the solution for the temperature has the general form

$$T_D = a \exp(-AX) + b$$

and since $T(0) = T_0$ and $T(E) = T_E$, the distribution is given by

$$\frac{T_D - T_0}{T_E - T_0} = \frac{1 - \exp(-AX)}{1 - \exp(-AE)}$$

In the frozen region, the linear profile must satisfy the conditions $T(E) = T_E$ and T_L, and is therefore given by

$$\frac{T_F - T_E}{T_L - T_E} = \frac{X - E}{L - E}$$

The sublimating interface condition may be satisfied by an overall energy balance applied to the entire layer $0 \leqslant X \leqslant L$. Thus

$$\rho U \lambda_{iv} + \rho c_p U (T_0 - T_E) = -k_D \left(\frac{\partial T_D}{\partial X} \right)_E + k_F \left(\frac{\partial T_F}{\partial X} \right)_E$$

in which the left hand side represents the rate at which thermal energy is

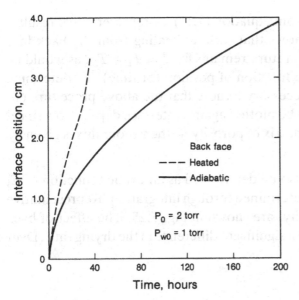

Fig. 7.25. Interface history for bovine muscle (following Dyer and Sunderland (1968)).

gained by the sublimating vapour (generated at T_E but leaving at T_0), while the right hand side represents the net rate at which the energy is supplied by conduction at both sides of the interface. Substituting the temperature derivatives, the energy balance may be written

$$\rho \lambda_{\text{iv}} U(1 + Ste_D) = \frac{k_F(T_L - T_E)}{L - E} - \frac{A k_D (T_E - T_0)}{\exp(AE) - 1} \qquad (7.12)$$

which $Ste_D = c_p(T_0 - T_E)/\lambda_{\text{iv}}$ is the Stefan number of the dried region. This transcendental equation gives the interface depth E as a function of the interface temperature T_E, and the vapour velocity U, neither of which are known *a priori* (compare this with the classical Stefan problem in which the interface temperature is known and fixed). Additional information is therefore required from a number of sources: notably the Clausius–Clapeyron equation, and an equation of state applied at T_E; and a mass transport equation for water vapour in air, incorporating not only the permeability of the dried region (which entails estimates of the shape and size of the 'pores' filled with extracellular ice) but, for the water vapour component, including a full description of diffusion extending into the non-continuum regime (Dyer and Sunderland, 1968).

For any particular interface temperature T_E, together with the corresponding equilibrium pressures and vapour concentrations, it is possible to determine the vapour flux, and hence both U and A. The interface depth E

then follows immediately from equation (7.12). A representative result is given in Fig. 7.24 which shows that without heating from the back face $(X = L)$ the interface temperature remains fixed at $T_E = T_L$, as would be expected; otherwise T_E is a function of position (or time). To determine the interface history it is necessary to note that the above procedure enables the vapour velocity to be plotted against interface depth. Now since, for sublimation creating a matrix of porosity $\mathcal{N}$, the vapour flux is given by

$$\rho_v U = \rho_i \mathcal{N} \, dE/dt$$

it follows that dE/dt may also be determined as an explicit function of E, thus permitting $E(t)$ to be determined through integration. Representative results, obtained numerically, are shown in Fig. 7.25. The effect of back face heating is seen to make a significant difference in the drying rate (Dyer and Sunderland, 1968).

8

Decay of ice

Growth and decay are twin subjects but they are not the same: in general, they are not equal and opposite; nor are they symmetric or anti-symmetric. It is common experience that the effect of freezing in a domestic water pipe is not always reversed by melting, and it is not difficult to appreciate that the growth processes leading to a snowflake are very different from the subsequent decay processes. Taken overall, these processes involve more than a simple phase change. Even interfacial behaviour generally reveals significant differences. Cinematographic images (Thomas and Westwater, 1963) of organic crystals in their melt, for example, are strangely reminiscent of medieval castles: orderly and systematic in their construction but irregular and fragmentary during decay.

It is often convenient to treat growth and decay as separate and distinct processes, each with a starting point and an end point. More generally, they must be combined as parts of an elaborate thermal history. The periodicities of diurnal and seasonal changes are but two natural examples from a list of many possibilities during which thermodynamic rate processes continue to operate and alternate in a variable manner. Growth is ultimately succeeded by decay, but it is apparent that the end point of one is not merely the beginning of the other; an end point may control the conditions under which subsequent events take place and thereby control the outcome.

This chapter is an attempt to reveal the principal similarities and differences found in the various aspects of growth and decay in ice. It begins with a review of mathematical and thermodynamic considerations from which a discussion of melting in crystals emerges. Sections 8.4–8.7 then give examples of ice decay processes in water, air, earth and organisms. Coverage is not meant to be exhaustive but is designed instead to provide a thematic introduction to decay, partly through contrast with growth but also through comparisons which may be made between the various contexts in which ice is found.

8.1   Reversibility and symmetry

When stable equilibrium conditions prevail, the thermodynamic description of phase change does exhibit a symmetry. In particular, the volume and entropy changes which occur during freezing are reversed during melting, there being no net effect on the environment. Time is not a relevant variable under these conditions which, in reality, may only be approximated if rates of change are very small. Theoretically, thermodynamic states are then limited to the stable equilibrium surfaces discussed in Chapter 2.

When metastable equilibrium conditions prevail, the state is defined by surfaces which extend beyond the saturation curves, as noted in Section 2.5. Macroscopically, there is no particular reason to believe that these additional surfaces should extend further from the saturation curve on one side than on the other. At first thought, symmetry might be expected, at least to the extent that the density difference between the phases is unimportant. In molecular terms, an excursion from saturated conditions is limited to the point where ice, water or water vapour clusters of a critical size have a net positive formation rate, according to the dictates of equilibrium thermodynamics. But the building of critical clusters is an order–disorder phenomenon which reflects the molecular structure of the phases present. This suggests that fluid phase embryos, which require less ordering, would form more readily than ice embryos; i.e., supercooling and supersaturation should range further than superheating. The subsequent growth of these critical embryos is a function of interface kinetics, the equations for which may, in principle, be stated in reversible form, again from the standpoint of equilibrium thermodynamics. However, the entropy generation associated with rate processes casts further doubt on the symmetry. At the limit of metastability, on the very threshold of instability, behaviour is clearly not symmetric; nor is it reversible.

The general requirements of mass and energy transfer during a phase change imply the existence of fluxes which are diffusive. This leads to the diffusion equation. The one-dimensional, transient form of the heat conduction equation, for example, is given by

$$\partial T/\partial t = \varkappa \, \partial^2 T/\partial X^2 \tag{8.1}$$

which may be compared with the corresponding representation of the wave equation

$$\partial^2 \phi/\partial t^2 = C^2 \, \partial^2 \phi/\partial X^2$$

It is apparent from these expressions that whereas the wave equation is symmetric in time (the equation is unchanged when t is replaced by $-t$) the conduction equation is not. The diffusion of heat, like all diffusive processes, is a dissipative phenomenon which, according to the entropy principle, implies an entropy generation rate. This rate is determined from the product of the flux and the gradient causing it. More precisely, the volumetric entropy generation rate is equal to the sum of the products of all of the simultaneous fluxes and their conjugate gradients (see Section 2.4). Thus, for the simultaneous diffusion of heat and a single component of mass,

$$\dot{s}_g = \mathbf{j}_{TH} \cdot \nabla\left(\frac{1}{T}\right) - \mathbf{j}_{M1} \cdot \frac{\nabla\mu_1}{T}$$

according to equation (2.26). Both terms on the right hand side are positive, as required by the entropy principle.

The continuous diffusion of material which is to undergo a phase transition at the interface implies a dissipation; the continuous supply or removal of latent heat requires a heat flux, thus carrying the same implication. Even though the magnitude of the dissipation rate may be the same for growth and decay, its effect is to destroy the symmetry of the phase transition by introducing hysteresis. That is, it is impossible to restore both the system undergoing transition and its environment back to their original condition. Clearly, the greater the rate of phase change the greater will be the hysteresis. Other hysteretic phenomena, whether microscopic or macroscopic, compound the effect.

The two fluxes most closely associated with the liberation or absorption of latent heat are not the only ones which may influence the process. Among others, the fluxes of thermal radiation and molecular momentum are perhaps worth mentioning briefly. Thermal radiation *at* a surface is not dissipative *per se*, but its transmission through, and absorption by, ice, water and water vapour induces internal heating which leads to augmentation of the conductive heat flux. In part, therefore, thermal radiation tends to be degenerative. Molecular momentum, on the other hand, diffuses throughout the fluid[1] phases as the result of a velocity field which both alters the other diffusive fields and is dissipative in its own right. While the effects of radiation and fluid motion are not intrinsic to phase transition, their presence is not uncommon and their role may be very significant, as will be demonstrated in later sections.

Finally, a word about temperature itself. In strictly thermodynamic terms, absolute temperature has no effect on phase transitions to or from ice Ih, over and above its influence on saturation pressures. Its principal role is played through differences or gradients. However, temperature levels do play a secondary role through their influence on material properties: latent heat, specific heat, thermal conductivity, etc. This is not a dissipative role, but by modifying the temperature and concentration (and velocity) distributions it does influence dissipation. In any event, differences in the thermal properties of the frozen and unfrozen phases will lend themselves to a departure from symmetry, at least mathematically; they may also alter the range of validity of analytic techniques, as noted in Section 3.4.

8.2   Melting of single ice crystals

The homogeneous nucleation of ice was discussed in Section 2.8 where attention was focussed on the growth of clusters having a critical radius r^* corresponding to a threshold energy ΔG^*. Implicit in the discussion was the notion of a fluid phase in which random molecular movements generate a certain probability that such clusters would exist, albeit in flickering forms. In applying this theoretical model to the nucleation of water from ice it is necessary to bear in mind that the molecular structure of a solid is more ordered and molecular motion is more restricted. The mobility attributed to self-diffusion is still present though the corresponding activation energy no longer connected with viscosity. If self-diffusion is taken to be the Fickian flow of lattice vacancies, the diffusion coefficient is given by

$$D = D_o \exp\left(-\Delta e/kT\right)$$

where D_o is a constant and Δe is the molecular activation energy. In this expression

$$\Delta e = \Delta e_v + \Delta e_m$$

in which the first term on the right hand side represents the energy required to form the vacancy and the second term represents the energy required for an adjacent molecule to fill it. Experiments indicate that Δe is about 0.65 eV as compared with 0.25 eV for water (Hobbs, 1974; McDonald, 1953).

Taken on its own, the lower molecular mobility in ice suggests that superheating would be more readily attainable than supercooling in water, but experimental evidence generally indicates otherwise (Kass and Magun, 1961; Kamb, 1970; Roedder, 1967, achieved a temperature of +6.5 °C in ice superheated with respect to water vapour). In the confines of a crystal interior the immediate effect of nucleation is the production of a much reduced pressure resulting from the density difference between the two phases. Extrapolating the ice–water saturation curve below the triple point, the maximum extent of such an effect is a negative (tensile) pressure of about 1.2×10^5 kPa corresponding to a temperature near +8 °C; this

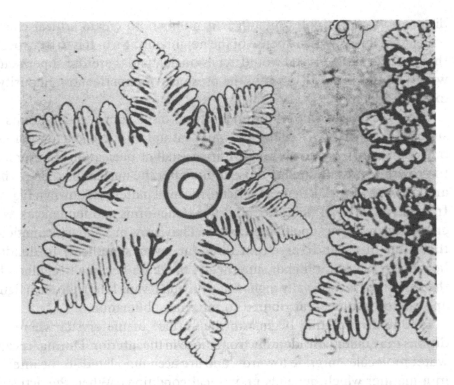

Fig. 8.1. An internal melt figure in ice: magnification ×7.6 (following Nakaya (1956)).

would occur at the extremity of the ruled (two-phase) surface, as noted in Fig. 2.10. In practice, a reduction in pressure would likely induce the appearance of the vapour phase (Mae, 1976) which then allows the pressure to rise, with the vapour and water in equilibrium with the ice: i.e., there would be a tendency to approach the triple point.

In Section 2.6 it was noted that the included angle of the free water molecule was not the same as that of an ice molecule incorporated in a tetrahedral matrix. This distortion introduces strain energy which is distributed throughout the crystal by dislocations, thereby creating its microstructure (Drost-Hansen, 1967). A theory of melting based on the idea that dislocations are the source of crystal disintegration has been devised by Kuhlmann-Wilsdorf (1965). At temperatures much less than the melting temperature, the free energy of the dislocation cores[2] is too great to permit many of them to exist, but as the melting temperature is approached this free energy vanishes, thus implying that the cores could be generated spontaneously throughout the crystal, thereby removing its shear strength. In this theory, nucleation consists of the formation of dislocation core dipoles in sufficient numbers to create critical clusters which then grow at the expense of the neighbouring ice. It is to be expected that defects in the crystal would act as nuclei, and therefore superheating would be less likely to occur in the presence of imperfection, impurity or external loading.

The presence of impurities and natural dislocations usually limits the superheat departure from the ice–water saturation curve to a few tenths of one degree. In this narrow range, intracrystalline nucleation of water leads to growth figures not unlike those found after the nucleation of ice. These are known as Tyndall figures. A dendritic melt pattern is shown in Fig. 8.1. In general, Tyndall figures exhibit the same forms and tendencies as ice crystals growing in supercooled water. They grow at a small inclination to the basal plane and vary from discs (at lower superheat) to dendritic hexagonal plates (at higher superheat); the latter become irregular when the superheat is especially high (Macklin and Ryan, 1966). No such figures appear to have been attributed to internal sublimation.

Frequently, melting begins on the surface of the crystal where conditions may differ considerably from those in the interior. During freezing, water molecules migrate towards, and are accommodated at, the interface in a manner which depends upon local conditions. When the activation energy is small in relation to diffusive resistance, and the interface is 'rough', accommodation is a distributed process in which water molecules are continuously added more or less uniformly over the interface. However, a higher activation energy is required for faceted growth in which molecules are deposited individually at specific sites located on terraces, steps or kinks. The general equilibrium situation may be described following Jackson (1958) who revealed the importance of the parameter $\alpha = \lambda \eta / k \nu T^i$, where η is the maximum number of nearest neighbour molecules which may be added during freezing, and ν is the number of nearest neighbours in the bulk of the bulk solid. In general, it is found that $\alpha > 2$ for faceting surfaces while $\alpha < 2$ for 'rough' surfaces. For ice, $\alpha > 2$ on the basal plane thus implying faceted, slower growth along the c-axis, as has been observed. However, $\alpha < 2$ for planes normal to the c-axis, and hence faster a-axis growth of ice should occur on 'rough' interfaces, whether they are incorporated in discs or dendrites; this is also consistent with observations.

Such an equilibrium theory should apply equally to melting, but it must, in general, be combined with the effect of interface geometry (Woodruff, 1968b, 1973). As predicted, non-faceting materials melt in the same way as they freeze with an interfacial zone which may be several molecules thick.

Such symmetry may be observed during the *a*-axis growth and decay of ice. For a faceting material, however, the sign of the interface curvature must also be incorporated because it controls the appearance of facet planes. If the growing phase is convex, facets will appear whether the phase change represents melting or freezing. On the other hand, if the growing phase of a faceting material is concave, facets will not appear for either melting or freezing. It is thus evident that equilibrium conditions, although complicated by curvature, do carry with them the classic idea of microscopic reversibility at the interface (Tammann, 1925). If so, the above conclusions should hold, at least in general terms, for sublimation and deposition, for which the evidence is very limited.

The sublimation process is similar to the melting process in that the molecular mobility again rises sharply as the free surface is approached[3] but the gradient is much steeper during sublimation as the difference between the latent heats of sublimation and fusion would suggest. On a crystal surface, the structural transition from solid to vapour may in fact occur in a two-zone film: the zone nearer the solid being liquid-like, the other approximating saturated vapour. The outer phase change would then be similar to evaporation, a behaviour which may be expected to persist until the temperature is well below $-20\,°C$ when the liquid-like film would probably have little effect. Estimates of the film thickness vary widely, the data of Beaglehole and Nason (1980) suggesting that at the melting point it is of the order of 100 Å on the basal face and 15 Å on the prism face. The thickness on either face appears to be less than a monolayer below $-20\,°C$.

In the absence of such a film, the sublimation process might be modelled as the reverse of the deposition process discussed in Section 5.1 were it not for the fact that the interface is typically concave from the vapour (growing) side and should not therefore be faceted. The Kuhlman–Wilsdorf model of disintegration suggests that sublimation originates in dislocations near the surface, although the overall process may be more complicated than melting because of the diffusive resistance imposed by other vapours, air in particular, and because of the coupling of heat and mass fluxes that may occur at very low pressures. It is to be expected that the greater the dislocation density, the greater will be the sublimation rate (Somorjai, 1968) and therefore the manner in which the crystal was formed and has been stressed may also be very important.

8.3 Melting of polycrystalline ice

It is reasonable to expect that the melting behaviour of an

extended body of ice formed from interlocking crystals or grains would be closely related to the behaviour of single crystals. In particular, the limited prospects or superheat, and the intragranular growth of Tyndall figures should also be observable in polycrystalline ice. It is only necessary to note that single crystals grown under specially-controlled laboratory conditions may have a different microstructure to naturally occurring or manufactured crystals: they may, for example, have different dislocation densities, depending on the stress field during, and subsequent to, their formation.

At the grain boundaries and in the intergranular spaces the situation is different. Firstly, if the original melt from which the ice was formed contained impurities these will normally be concentrated in the intergranular spaces where they create a localized freezing point depression. Secondly, at the junctions of three or more grains the ice has a convex interface of high curvature; the effect of this is also to lower the local melting temperature. It is therefore evident that as the bulk temperature is raised, the melting process in a polycrystalline body of ice will begin at the grain boundaries, most likely in the locations where more than two grains meet.

Intergranular melting is generally dependent upon the ice temperature distribution in the vicinity of the ice–water (or ice–vapour) interface, but the penetration and absorption of solar radiation may generate internal intergranular melting if the bulk of the ice is not much below its equilibrium freezing temperature. If this internal boundary melting becomes extensive, the intergranular spaces will become a labyrinth of interconnected passages, and the solid *in toto* will then exhibit a significant increase in porosity. Under such conditions, the intergranular melt material will flow in the presence of a pressure gradient (Nye and Frank, 1969). Should intergranular melting proceed sufficiently far, the grains will become disconnected and the structural integrity of the polycrystalline mass will then be lost. In many situations, intergranular melting simply augments and modifies the interfacial melting already taking place on those crystal faces which border the bulk fluid phase. As mentioned in the previous section, crystalline interfacial growth and decay rates are functions of the crystallography and prevailing thermal conditions; the details of melting or sublimation therefore reflect the prior details of formation.

In the discussion of formation of a natural ice cover it was noted that early growth on quiescent water took place in the plane of the water surface, proceeding along the (horizontal) *a*-axis when the supercooling was less than 0.3 °C. According to the Jackson model discussed in the previous section, this will create a faceted, horizontal interface attributable to a mosaic of crystals with vertical *c*-axes. Melting would remove these facets. In other circumstances, and particularly when the cooling rate is faster or a cut-off mechanism has restricted vertical *c*-axis growth, the ice–water interface consists of a mosaic in which the *a*-axes are vertical. (The same is true on the upper surface of the ice if melting of the top layers is followed by supercooling.) The author is unaware of any microscopic study of melting along the *a*-axis but the theoretical evidence presented earlier suggests that since growth was non-faceted so would be decay, regardless of interfacial geometry; in any event, the concave (growing) water phase would create rounding of any previous dendritic structures. Similar comments apply to sublimation: except perhaps in any local ice concavities, the interface will not be faceted and will decay as a 'rough' surface after a short-lived transitional stage when the corners of any previously-formed step edges are rounded.

The above observations and expectations are consistent with the effect of interface curvature noted in the previous section. Concavity of the fluid phase invariably implies a 'rough' interface during melting or sublimation. This is particularly evident at the grain boundaries where earlier melting naturally re-creates convex crystal surfaces. For a faceting material, the crystal surface velocity near the grain boundary during melting will be greater than during freezing. This implies that in the vicinity of the grain boundaries, freezing will be slower than melting for a polycrystalline ice sheet in which the *c*-axes lie parallel to the direction of growth.

As with single crystals, the conditions accompanying growth of polycrystalline ice differ from those during decay, and may therefore influence the outcome. Typical conditions for the formation of a natural ice cover lead to an interface velocity which decreases with depth. This situation will usually generate grain sizes and orientation which also change monotonically with depth. The crystallography, reflecting the interface motion, is thus a function of depth. The microscopic details of melting from beneath will therefore differ from those of freezing from above. Macroscopically, this difference may be unimportant: heat conduction within a crystal is anisotropic but the difference between the principal values is small and the granular structure often has an averaging effect. However, if the ice was formed from an aqueous solution the effect of the solutes may be significant. As noted in Section 4.1, the water Lewis number $Le = \varkappa/D \gg 1$, and hence the temperature field in the water extends much further from the interface than the concentration field. Since this generates constitutional supercooling in the water, the interface is potentially unstable and often develops the well-known column and cell pattern, at the common corners and borders of which the concentration levels are especially high. For ice in particular, the protruding columns are of the order of 1 cm long and 10^{-3} cm diameter (Tiller, 1958). Should the interface velocity be great enough, concentrated corner pockets are entrapped and left as inclusions. During melting, similar circumstances create 'constitutional superheating' in the solid phase if $Le \gg 1$; for ice, $\varkappa$ is much greater than in water while D is much less. In general, however, diffusion in the growing phase may not be neglected during melting and this feature, which is usually unimportant during freezing, has a stabilizing effect on the interface (Woodruff, 1968a). For ice, which is a poor solvent and difficult to superheat, this form of symmetry in the freeze–thaw behaviour of aqueous solutions is particularly noteworthy.

For a thickening horizontal sheet of natural polycrystalline ice, not only will the intergranular solute concentration likely change with depth, adding to the inhomogeneity of the ice, but solute rejection at the interface and its subsequent effect on buoyancy-induced convection may also be important. This behaviour, typified by the Rayleigh instability, is unlikely to be re-created to any significant extent during bottom melting which tends to maintain the less dense water above the more dense. Melting from the top, on the other hand, produces a layer of water which is potentially unstable (see Section 4.4). Should the limit of thermal stability be exceeded, thereby inducing natural convection, conditions and results are again very different from those found during the original freezing.

8.4   Decay of an ice cover

Among the many examples of melting in a hydroglaciological context, the decay of an ice cover is particularly instructive. The consequences of the process are observed every year from the shorelines of seas, lakes and rivers, and have been closely studied for decades. The decay process itself, in general terms, is found to be very complex and is still not fully understood despite the attention it has received. River, lake and sea ice each decay in a different manner and each reveals a significant range of variability attributed in part to the conditions of freezing as well as those of melting. This section is devoted to the description of the principal, and often common, elements of decay.

It is not unreasonable to expect that upper surface melting would begin when the net heat flux vector first points downwards. This is certainly true in an average sense but it is an expectation which ignores thermal capacity and tends to obscure fluctuations, especially diurnal variations, which often lead to freeze–thaw cycles. Such cycles may not be important in themselves but they do delay the thawing process and complicate its details. If the ice is covered in snow, the immediate effect of a downward net heat flux is to melt the upper layers of the snow. In the early stages of melting, the water produced is essentially bound by capillary forces within the crystalline snow matrix (ice, like many materials, is wet by its melt), and tends to collect at the junctions of dendrite arms. As the water film on the melting dendrite arms is thus removed, they are left exposed, thereby melting further until they vanish. The result is structural collapse (Knight, 1979) in the presence of drops of excess water which tend to percolate down through the snow pack and re-freeze in later stages of metamorphosis. In the absence of drainage, or sharply reduced temperatures, this melt water will ultimately produce white ice above the ice cover proper and often continuous with it; the process was discussed in Section 6.2.

Snow melt which does not re-freeze will either be removed through evaporation, thereby producing a cooling effect, or, in the absence of ice break-up, will remain in pools of decaying slush. Depending upon the local contribution of short-wave radiation and the speed and temperature of the wind, this excess water will fluctuate in temperature and amount. If the ice cover remains more or less intact, and is not overwashed by neighbouring water, the melt pools may persist throughout the summer period as an ablating heat 'source' supplied by Sun and warm winds. Should thaw holes or cracks appear, the water will drain (and in any event may percolate) through the ice to form under-ice melt ponds in regions of bottom depression; thus placed in direct contact with sea water at subzero temperatures, the fresh water re-freezes, as noted in Section 4.4. Wadhams (1988) has noted that multi-year sea ice has a bottom surface topography of blisters or bulges which distinguish it from the relatively smooth bottom surface of first year ice; this difference may reflect the prior existence of under-ice melt ponds, in which case it is attributable to melting on the upper surface.

Ablation at the undersurface also reduces the cover thickness. In certain circumstances this bottom melting may be caused partly by short-wave irradiation on the upper surface. Penetration through the ice and subsequent absorption in the water just below the interface creates a heat source which may or may not lead to convective instability, but always contributes to the interfacial heat balance. Bottom melting may thus be augmented by solar radiation provided that the upper surface cover will allow penetration[4]. This would not occur, for example, in the presence of a snow cover or if the white ice layer, which is an effective scatterer (Gilpin, Robertson and Singh, 1977), is very thick. More commonly, bottom melting is caused by a variety of convective phenomena which depend upon the bulk temperature and velocity of the water and, for sea ice, on the salinity. It is difficult to make a general statement about such phenomena except to note that their effect will depend upon the local details of the interface shape and roughness, and upon the adjacent velocity and temperature profiles. Clearly, the latter are functions of prevailing currents and the temperature of their source; they may also be dictated by attendant thermal or hydrodynamic instabilities. The reader is referred to Chapter 4 for a discussion of these phenomena.

Sensible heat addition to the ice cover may cause internal melting. Whether this energy is added by conduction or radiative absorption, its effect is to raise the bulk temperature of the ice to the point where preferential melting begins. As noted in the previous section, this melting occurs in the vicinity of impurities in general, and at the grain boundaries in particular. Its broad effect is to weaken the cover through the production of a labyrinth of interconnected regions filled with an aqueous solution having a density and osmotic pressure greater than those of the parent solution. The morphology of this 'rotten' ice is a reflection of its granular structure, in general, and the distribution of isolated inclusions in particular. If the ice cover was formed from supercooled, or constitutionally supercooled, water, the structure resembles an array of candles (grown with the a-axis vertical) and is easily broken in tension or flexure. Sea ice is an important example in which the more concentrated brine is flushed out, allowing the intergranular spaces to be filled subsequently with sea water, which has a lower bulk salinity, or with downward percolating snowmelt, which is essentially pure water (Weeks and Ackley, 1986). This purification process leads to distinctions between first year (or young) ice, second year ice, which has a greenish tint, and old ice, which has a bluish tint (Armstrong, Roberts and Swithinbank, 1973). (These distinctions and the associated colours are useful but not rigid classifications).

Intergranular melting serves to weaken the cover but is seldom the cause of its ultimate destruction which usually results from a multiplicity of mechanical actions arising out of wind and water behaviour. In the absence of nearby open water, the wind becomes the principal cause of destruction by generating enough shear stress over the surface of the flexible ice 'membrane' to create rafting, buckling and splitting forces in the downwind regions (turbulent gusts act as travelling loads which stress the cover dynamically). Augmented by the effects of thermal expansion and undercurrents, these forces are capable of opening fissures in the cover. Once the cover is thus fragmented the effect of the wind is magnified

Fig. 8.2. Break-up of fast ice in St. Anthony Bight, Newfoundland (following Squire (1979)).

through wave action, particularly in areas left open by wind shifts. Warming of the surface layer of water may also induce a convective instability[5], leading to the phenomenon of turnover, and this too helps to intensify the melting process. Warm winds are thus the principal cause of break-up on lakes, and may be a factor in river ice break-up.

In general, the break-up of ice on a river is strongly influenced by the river course and the water velocity profile. It is generally believed, for example, that an increased spring discharge accompanied by the uplifting effect of greater depths will create fractures in the ice cover and dislodge it from the shore. Turbulent convection on the bottom of the ice cover can produce a surface with transverse ripples, as noted in Section 4.5. This increases both the roughness and the heat transfer rate, and hence not only accelerates the melting process but creates a higher shear stress. Once break-up occurs in a river, the transverse velocity profile imposed on the fragmented surface cover generates a shearing, grinding and mixing action which further accelerates erosion of the fragments. The sound of this jostling, hissing mass can be quite awe-inspiring, and the downstream jam it can produce is often spectacular (Henderson and Gerard, 1982).

Open water may have a substantial effect on the break-up of a sea ice

Fig. 8.3. Weathered iceberg south-east of South Georgia (following Deacon (1984)).

cover (Wadhams, 1981). For example, an ocean swell will generate a flexural gravity wave motion in which the ice cover and the water beneath it are coupled together. The flexing action of the cover in response to the rise and fall of the ocean is manifest in a propagating wave with an amplitude which decreases with distance from the ice edge. Fracture usually occurs at points of weakness such as the tips of cracks previously formed through thermal contraction. In the shallows along a shoreline the effect of propagating waves is intensified by the decreased depth and may be further intensified by wave reflection from the shoreline or from nearby massive ice which is aground.

Other factors come into play near the edge of the ice cover. Thermal erosion, for example, may be attributed to the convective action of water which has absorbed solar radiation in its surface layers (Zubov, 1979). Mechanical breakage is equally important. On a comparatively small scale, 'ice edge waves' create additional bending which reaches a maximum at a distance of 15–30 m from the edge (Wadhams, 1973). Fig. 8.2 illustrates the regularity of the process (Squire, 1979); the fractures lying parallel to the edge are about 20 m apart. On a much larger scale lies the calving of glaciers, ice streams and ice shelves to create icebergs[6] (Robe, 1980). The size of these massive ice forms varies greatly but many occupy the range 1–100 km in length, and are formed from ice which may exceed 100 m in thickness. Calving may be simply described as the fracture of an ice slab, but the causes of fracture are various and usually complex: they include the effects of ice movement, weight, buoyancy, tides, waves and storms; and in addition to the stress field determinants, the effects of temperature and inhomogeneity are also important.

The life of an iceberg usually lasts 1–10 years, depending upon both its size and the route it follows. Deacon (1984) reports an iceberg first sighted in 1967 off the Antarctic coast. After spending the next five years aground, it was estimated to be 90 km long and 51 km wide. Afloat once more, it was sighted two years later, having calved and shrunk to a mere 80 × 43 km. Six years after that it had been reduced to 48 × 19 km. Fig. 8.3 shows an iceberg in an advanced stage of deterioration. In general, the deterioration mechanisms are complex but consist of two main processes; thermal and mechanical. Heat (and mass) transfer takes place as a result of solar absorption and convection. The former not only promotes progressive surface ablation but is capable of introducing the destructive effect of thermal shock in the early morning or after a period of cloud or fog cover. Convection is conveniently subdivided by the water line: sensible and evaporative heat exchange with the air; and sensible heat fluxes in the water attributable to both natural convection and the forced convection effects of waves and currents (Josberger, 1977). Wave action, in particular, is not only capable of providing the major-portion of water-line erosion, but generates an overhang which may lead to calving. These processes also apply if the iceberg is grounded, although their relative importances may then change (El-Tahan, Venkatesh and El-Tahan, 1984).

The mechanical effects at work during the deterioration of a large iceberg are similar to those responsible for the break-up of ice sheets. In the open ocean, the wave–structure interaction evidently creates bobbing and rolling motions, thus inducing flexural strains which may be large enough to trigger break-up (Wadhams, Kristensen and Orheim, 1983). It must be added, however, that the failure strain of an iceberg in a state of deterioration is not currently known precisely. This is hardly surprising, given the difficulties associated with instrumentation under such circumstances when metamorphosis is likely to be advanced. The above-water portion of a tabular iceberg, for example, consists entirely of the original layering of snow and firn, provided it has never capsized. Meltwater thus percolates easily through the freeboard volume, raising its temperature to the melting point when flexural stresses have their greatest effect (Weeks and Mellor, 1978).

8.5   Ablation of atmospheric and structural icing

Perhaps the simplest example of aeroglaciological melting is afforded by pure sublimation. As mentioned in Section 5.1, deposition may be initiated under metastable conditions when the vapour is supersaturated; growth is then controlled by vapour temperature and pressure. Under these conditions, the interface is often faceted. The corresponding metastable growth of vapour requires that the parent ice be superheated. This condition is difficult though not impossible to attain, as noted in Section 8.2. Wherever the c-axis is normal to the interface, the facets previously formed on ice convexities will round and sublimate away while facets should appear on ice concavities. The author is unaware of any test of this theoretical prediction or of any study of the interface which focuses on kinetics under such conditions.

Under quasi-steady conditions, when the vapour is superheated and the ice is either subcooled or at the equilibrium freezing temperature, sublimation is essentially deposition reversed (analogous to the reversal of freezing and melting with superheated water, though not, of course, with supercooled water). As diffusion-controlled phenomena, they both imply an interface which is rough on a molecular scale. Deposition may therefore be converted into sublimation, and vice versa, by simply altering the relative magnitudes of the heat fluxes in the interface equation: the process itself is unaltered. A similar comment applies to formation and decay above the triple point pressure. The entire process then entails melting/freezing at an interface nearer the ice and evaporation/condensation at an outer interface; it is thus more complicated, but it is still controlled by diffusion.

In Section 5.1 it was noted that the surface temperature of a growing frost layer often rises towards the triple point temperature where condensation will begin to occur. It is evident from earlier observations on melting snow crystals that the first water thus formed will be dominated by capillary forces which take it from the areas in which it is most likely to be generated, i.e., over the dendrite arms and tips, into the corners of branch junctions. Its growth there at the expense of the vacated dendrite arms leads to the formation of small drops which, as they grow larger, are subject to an increasing gravitational force. Should this force not ultimately remove them from the frost surface, or move them across it, they will percolate through the frost matrix until, upon reaching a zone of lower temperature, they will re-freeze. While the frost grows, a dendritic depositional process may thus be modified by a complex condensation/melting process in which the liquid may eventually become plentiful enough to form a continuous film with melting on one side and evaporation on the other. For circumstances in which the frost recedes, it is only necessary to add that the increased volume and mobility of this water in the presence of capillary, gravitational and aerodynamic forces adds further to the asymmetry between growth and decay.

There is some evidence to suggest that dry accretion may grow along the c-axis (Ackley and Itagaki, 1974), at least for temperatures near 0°C, but the process does not readily lend itself to reversal. This exemplifies the general rule that ablation of atmospheric and structural icing will be different, physically and mathematically, from accretion. A heavily-rimed snowflake perhaps comes nearest to reversibility[7] when, upon melting on a warm surface, it breaks up into droplets which are subsequently blown off by the wind. More typically, the ice accretion behaves more or less like a solid object the outer surface of which melts in a process which is appropriately described as thermal erosion. In broad terms, this process consists of warm air or radiant energy melting the surface layer which is then removed by aerodynamic forces.

For atmospheric ice in general, and hailstones in particular, the ablation process is especially complex because it is interactive: i.e., the shape, attitude and motion of the ablating body determine the local aerodynamic characteristics, but since these in turn control the rate of ablation at each point on the body, they determine the shape and thereby influence the motion of the body. This interactivity is also found during growth but the details, if not the entire pattern, will be different during ablation. The water present during hailstone ablation, for example, is limited not only by the rate and duration of the melting, but by the ability of the hailstone to avoid shedding. In the early stages, a film of water appears over the entire surface and re-circulates under the influence of gravity and wind shear stress but, eventually, the hailstone changes from its more or less spherical initial shape into an oblate spheroid which causes the water film to be re-organized as a toroidal ring around the stone equator. Should further melting take place, simple re-circulation in the enlarged water torus may be replaced by shedding attributable to surface instability or wind shear (Rasmussen, Levizzani and Pruppacher, 1984b).

For many examples of structural icing, total interactivity is replaced by a modifying activity in which the local ablation rates determined by the initial flow field cause only quasi-steady changes in the shape of the ice surface and thereby alter the local flow field gradually. Only when the flow regime reaches a point of instability, such as transition to turbulence or the onset of separation, does the local flow field alter abruptly. Such critical points may mark a substantial alteration in local ablation rates. In atmospheric icing, no less than in the conduit icing discussed in Section 4.6, the interface may become sensitive to flow alterations which cause abrupt changes in the local heat transfer coefficients. This is especially true when the ice growth rate is negative. In view of the stability characteristics of the ice–water interface discussed in Section 4.5, it is to be expected that ϵ and *Bi* would also play a role during ablation, when the heat flux at the water–air interface must also be taken into account. It appears that no studies of this situation have yet been published. Even so, it may be speculated that the closer the substrate temperature approaches the equilibrium freezing temperature, the more sensitive the interface will be to local variations in the flow conditions.

The determination of heat transfer rates on either side of the ice–water interface is usually not a simple matter. Within the ice, heat transfer is conductive but it takes place in a body whose geometry is not usually simple, nor constant, and must match with that of the substrate where the temperature is often unknown. Even when sensible heat effects may be ignored (*Ste* $\ll$ 1), conduction through the ice must account for in-homogeneities: pockets of air, unfrozen interstitial water and, for marine icing, pockets of brine. It must also account for internal heating resulting from the absorption of solar radiation. A knowledge of the temperature distribution within the ice is important not only in the determination of the heat flux at the interface but in the determination of adhesion at the substrate surface. Temperatures approaching 0 °C usually herald the loss of the ice's ability to withstand shear stress so that if the substrate has a smooth regular geometry, such as that of a circular cylinder, the accretion may become free to rotate and slide or, more hazardous still, to dislodge. Conduction in the substrate, frequently a metal structure, is thus a crucial aspect of decay.

Heat transfer on the water side of the interface is even more complex. The water film generally flows over a rough and irregular surface on which the static pressure distribution of the air is perhaps unknown and is seldom readily available. The film itself will be subject to a variety of instabilities including the onset of surface waves, internal turbulence and separation at the ice surface, the latter ranging from local re-circulation to complete detachment. The impact of drops blown from neighbouring accretions, or the distributed effect of warm rain, serve only to exacerbate the difficulties of formulation and calculation. The wind itself, shifting and gusting in absolute terms, or altering in relation to the structure[8] bearing the ice, may also exert a considerable influence. Finally, it is worth noting that the decay of accretion, as with any other natural ice form, seldom occurs in the presence of steady temperatures. Invariably, diurnal and other fluctuations cause a series of freeze–thaw cycles which generate a variety of temporary, though occasionally massive, ice forms.

Given the above difficulties, and the inherent threat of most forms of structural icing, it is not surprising that many attempts have been made to prevent the ice from forming or, failing that, to remove it expeditiously. Anti-icing strategies were discussed briefly in Section 5.5. The remainder of this section is devoted to a brief outline of deicing strategies. In broad terms, these fall into three categories: mechanical, chemical and thermal. The first of these springs quite naturally from traditional attempts to chip and fracture the ice with a hammer or an axe. Such efforts are limited to small scales, but may be enhanced with power-driven tools such as pneumatic drills and chain saws, wherever they are practicable. Obviously these measures are inadequate for an inaccessible surface such as an air-craft wing during flight; in that instance an inflatable surface cover, e.g., along the leading edge, may be more suitable. Such covers appear to be most appropriate and most convenient on large surfaces such as decks and platforms (Ackley, Itagaki and Frank, 1977). They also offer a practical form of ice prevention for small scale equipment which has been installed in a casing to which the inflatable cover may then be attached. Surface vibration is another useful technique which has been successfully applied to guy wires.

Chemical deicing methods have freezing point depression as their main objective but they are very limited in application. Clearly they are inap-plicable at temperatures lower than the obtainable depression, and work only on the exposed surface area of the accretion rather than on the bulk beneath. Toxic, corrosive and skid properties of such antifreeze solutions limit the method yet further.

Thermal methods generally offer the best prospects but they are by no means a perfect, or even widespread, solution (Makkonen, 1984a). The electromagnetic wave scattering characteristics of most types of accretion tend to restrict the scope of radiative heating (Gilpin *et al.*, 1977) but, where it is possible, this type of heating not only attacks the grain boundaries, thereby lessening the structural integrity of the ice, but also tends to warm the substrate surface and thus reduce adhesion. The supply of large amounts of heat to handrails, stanchions, masts, and strategic deck areas, would constitute an anti-icing strategy for marine structures, but an expensive one. Smaller amounts of heat, perhaps from waste heat sources, constitute a deicing strategy which would be more economical, especially if the heat was only directed towards the reduction of adhesion and was accompanied by simple mechanical removal. To be effective, however, such a combined method would require careful design of the shape and orientation of the surface elements likely to experience accretion so that ice could be removed not only economically, but simply and conveniently; perhaps even automatically and without danger.

8.6   Thawing of soil

Geoglaciological decay embraces a wide variety of situations and circumstances ranging from the cracking of a road to the slump of a hill side, and from the weathering of a rock face to the cataclysmic descent of an avalanche. In general, such situations contain the three most common effects of melting: a density change, alterations in the stress field and mobility of the water. Their common feature is found in the fact that the melting occurs within the confines of a porous medium and therefore reflects constraints and characteristics which are frequently absent in a hydroglaciological or aeroglaciological context (the melting of snow is an obvious exception).

As mentioned in the introductory remarks of this chapter, the thawing process may be important *per se* or may assume its importance as part of a cyclic process. Diurnal variations in the surface heat flux or the temperature of the environment may be substantial but their time scale is often too small to produce a lasting effect. Seasonal changes, on the other hand, have a greater amplitude as well as a longer time period, and therefore have a greater effect. Despite the rough periodicity of annual climatic effects, they are seldom sinusoidal; nor are they normally symmetric. For example, even if saturated soil were subject to a sinusoidal surface temperature, since the thermal properties during freezing are substantially different from those during melting, the depth of freezing will be much greater than the depth of thawing, thus tending to lower the depth of the permafrost table each year (until offset by the geothermal gradient) while maintaining a fixed and relatively shallow active (thaw) zone. Mathematical complications may also arise. For example, a small Stefan number solution applicable to freezing may no longer be valid during melting. This prospect arises not only because the specific heat of water is greater than that of ice but because the effective temperature difference is frequently greater during melting, when the snow cover is absent. The insulating effect of snow is usually felt more during the freezing period than during melting so that the asymmetry it produces tends to oppose that arising from thermal property differences. The prevailing climate often imposes its own asymmetry.

Freeze–thaw cycles in porous materials such as rock or concrete often have a destructive effect resulting from the rigidity of the material. Moisture, in either vapour or liquid form, may enter the pores under the influence of a pressure gradient, some of it being trapped in dead end pores or inside local ice envelopes. Unless air or another non-condensable gas is present, the subsequent freezing of water entrapments will lead to a substantial increase in pore pressure, the result of which may be fracturing of the surrounding mineral skeleton. The ability of the material and the ice–water mixture to relieve this pressure is compromised by the local rate of freezing, and this is closely related to the local cooling rate which, typically, is particularly high just beneath the free surface. It is therefore to be expected that fracturing would have its greatest effect close to the free surface. When the local fractures join together to form a single continuous fissure running roughly parallel to the free surface, the effect of thawing is to remove the ice adhesion forces and thus allow the outer layer of the material to flake off.

In an ice-saturated, granular soil, the immediate effect of thawing would be to induce a high degree of suction in the water thus formed, but only if the entire system were closed and unloaded. Typically, the system is not closed, and suction promotes entry of neighbouring air (or water) into the pores which exhibit no resultant change in volume: i.e., melting *per se* does not contribute to settlement unless the soil contains discrete ice inclusions or is subject to consolidation. Pore water released by thawing flows under the influence of the pore pressure field which, given a sufficiently low rate of thaw, will ensure that excess water is re-distributed, provided that no impermeable barriers stop it. However, if the rate of thaw is high enough, the hydraulic conductivity of the soil may not be great enough to permit the relief of this excess pore pressure. The soil effective stress σ_e at a depth Z, given by

$$\sigma_e = (P_0 + \varrho_s'gZ) - P$$

where P_0 is the applied surface stress and $\varrho_s'g$ is the submerged soil specific weight, will then be reduced as the excess pore pressure P rises.

In a compressible soil, the pore pressure distribution is described by consolidation theory, as developed in Section 6.3. For a one-dimensional situation, the excess pore pressure distribution satisfies the equation

$$C \, \partial^2 P / \partial Z^2 = \partial P / \partial t \tag{8.2}$$

in which $C = K/\varrho g \alpha$, and Z is measured vertically. Thawing at the Earth's surface may be expected to induce an excess pore-pressure distribution with the maximum occurring at the interface (Morgenstern and Nixon, 1971). Comparing equation (8.2) with that governing heat conduction, equation (8.1), it is evident that the relative rate at which heat and water 'diffuse' through the thawed soil is given by $\varkappa/C$. Since the heat and water transport occur simultaneously in the same region of space, the time and length scales are the same for each equation.[9] Hence, equation (8.2), written in normalized form, becomes

$$\left[\frac{Ct_c}{Z_c^2} \right] \frac{\partial^2 p}{\partial z^2} = \frac{\partial p}{\partial \tau}$$

or

$$\partial^2 p / \partial z^2 = \mathscr{R} \, \partial p / \partial \tau \tag{8.3}$$

where

$$\mathscr{R} = \frac{Z_c^2}{Ct_c} = \frac{k \, \theta_c}{m_i \lambda_{iw} C} = \frac{\varkappa}{C} \, Ste_i,$$

the time and length scales being obtained from the temperature scale in the interface equation, as discussed in Section 3.3.

It is thus clear that the solution to equation (8.3) may be expressed by

$$p = p(z, \tau, \mathscr{R})$$

in which the thaw–consolidation ratio $\mathscr{R}$[10], analogous to a capacity, is the sole parameter reflecting the intrinsic inability of the system to relieve excess interfacial water pressure. For $\mathscr{R} \ll 1$, transience is evidently unimportant, and the system is capable of relieving the excess pore pressure. For $\mathscr{R} > 1$, on the other hand, thawing produces a build-up of excess pore pressure and thus raises the prospect that the effective stress σ_e may become small enough to create unstable conditions. In this latter situation, the soil may be unable to withstand any but the smallest shear loads, and in the event that it lies on a slope, its own weight may be sufficient to cause it to slide *en masse* with potentially disastrous results (McRoberts, 1978). Should the ice be present in segregated form, the result of thawing may be equally spectacular. Ice wedges leave steep-sided slopes which are prone to collapse, and the mobility of the melt from ice lenses can lead to dramatic subsidence.

Equations (8.1) and (8.2), together with the interface equation representing phase change, are equally applicable during freezing or thawing, but the processes themselves are usually very different in manifestation. In a pore, phase change *per se* is essentially reversible, although temperature, pressure and concentration gradients do affect the symmetry, as noted earlier; and in certain circumstances, the instabilities created by surface tension produce a hysteretic effect which also denies reversibility. It was noted in Chapter 6 that freezing in a soil may induce lower water pressures but an attendant increase in the ice pressure, which is capable of causing ice segregation through the reduction of effective stress. In thaw consolidation, however, the reduction of effective stress is caused by the rise in water pressure attendant upon $\Re \geqslant 1$.

It thus appears that while an increase in the pressure of the forming phase is, in both instances, the principal cause of reductions in the effective stress, the precise mechanisms are not the same. In thawing, the water pressure field is coupled to the temperature field but not vice versa; interface movement essentially drives the pore pressure field. In freezing, on the other hand, the ice pressure field is intrinsically linked to the temperature field through the Clausius–Clapeyron equation applied to a thin freezing fringe immediately behind the interface.

The consequences of excess water pressure are very significant. In an extreme situation, it may lead to a catastrophic slide, as noted above. Less dramatic, but equally important, are circumstances where the lowering of the effective stress throughout the thawed layer alters its rheological behaviour, and transforms the soil to a material with properties somewhere between a plastic solid and a viscous liquid. Even in the absence of a phase change, the rheological behaviour of frozen soil is known to be sensitive to temperatures approaching the melting point. These thermally-induced alterations in stress and rheology are undoubtedly at the centre of many thawing phenomena observed in soils.

The mobility of groundwater is another important aspect of thaw phenomena, especially when they are viewed from a cyclic point of view. In general terms, the effect of freezing is to alter the distribution of groundwater and its flow pattern. Initially, this may led to diversion, upwelling and overflow revealed, for example, by the appearance of naleds (see Section 6.2), but there is a natural tendency towards re-stabilization in a new configuration compatible with winter conditions. The effect of thawing is to once again disturb the balance which then has a tendency to revert to the original summer 'equilibrium'; the transient aspect of these changes is seldom absent for long. Thawing is thus an agent which both creates a source of groundwater and alters the permissible patterns of flow. Naleds themselves can sometimes provide startling examples of the effects which man-made structures may introduce into seasonal melting behaviour. Should the naled grow to fill a culvert, for example, spring thaw in the surrounding terrain may produce more water than natural drainage can remove, thus causing extensive flooding. The very attempt to provide drainage will then have led to precisely the opposite. But the same result may occur naturally, ranking with other hazardous phenomena, such as slumps and avalanches, in terms of the overall damage which may be done. Strictly, the latter belong to a class of geological stability problems not necessarily associated with thawing conditions, although melt-induced structural decay can certainly act as a trigger, as noted earlier. The magnitude of the mass wasting which may occur in a short time is enormous, completely dwarfing the typical flow rates which occur in less spectacular circumstances. Velocities range from 1–$100\,\mathrm{m\,s^{-1}}$ (Longwell and Flint, 1969).

Nivation and slushflow provide examples of mass wasting which derive from melting. The first of these terms refers to erosion beneath and near snowdrifts. As noted in previous sections, sufficient melting in the upper snow layers will free water to percolate downwards. This melt water will eventually reach the base of the drift where it will accumulate and, if a slope exists, will begin to flow as a stream. Slushflow is essentially an extreme situation in which the entire snow cover is saturated with water, thus degenerating into a two-phase fluid-like layer flowing under the influence of gravity. These flows are capable of moving material ranging from fines to boulders, some of which may have been generated previously by frost action.

Many geoglaciological phenomena are created by neither freezing nor thawing, but by a repeated cycle which combines them. On a massive scale there is the example of slumping in high ice content cliff slopes (MacKay, 1966), while on a smaller scale we have frost creep and gelifluction. Frost creep on a slope is the process whereby a particle is subject to frost heave normal to the slope but falls almost vertically during the following thaw: the particle thus creeps further down the slope during each freeze–thaw cycle. Gelifluction, on the other hand, refers to the very slow, 'plastic' flow of water-saturated waste material downslope during the thawing period. It is common for these two events to occur together (Williams, 1966), and since both act in direct proportion to the sine of the slope angle it is often difficult to differentiate between them. As would be expected, downslope velocities are site specific, typically ranging up to $15\,\mathrm{cm\,yr^{-1}}$. Over many

Fig. 8.4. Gelifluction sheet encroaching on emergent strand, Mount Pelly, Victoria Island, NWT, Canada (following Washburn (1947)).

Fig. 8.5. Massive ice wedge (following Washburn (1973)).

years, this form of thaw-induced motion may create spectacular patterns on the landscape, as illustrated in Fig. 8.4.

Freeze–thaw patterns in which the bulk transport of material is less important are also widespread. Frost polygons and patterned ground are familiar sights in permafrost terrain. Ice wedges which develop from the freezing–cracking–thawing cycle, were discussed in Section 6.4. Although the seasonal cracks may be small, their cumulative effect over many years may produce a massive body of ice. An example from Yakutia is shown in Fig. 8.5. Thawing of such a wedge would clearly create a substantial change in the local landscape. Thawing of ground ice in general produces topographic features which are often embraced in the single term *thermokarst*, although the circumstances and processes involved, and the final effect, vary considerably. Thermokarst includes the thawing not only of wedges and polygons, but of such extensive and complex systems as a pingo or a lake basin which may involve a body of water, climate and vegetation, syngenetic[11] and epigenetic ice wedges, and the permafrost table; and all of these in a dynamic way (Washburn, 1973).

8.7   Thawing of organisms

As noted in Section 7.1, that fraction of planetary water which is actually contained in living systems is very small indeed. At any given moment in history it represents the most crucial contribution to the unfrozen water of this planet and, like the unfrozen water fraction of simpler physical systems, it decreases as the temperature decreases. On a geological time scale, the water available for biofluids undoubtedly decreased during each ice age and increased during the subsequent recession, but it is not known that these variations produced corresponding changes in the total amount of water contained within the organisms then alive. The effective geographical extent of the biosphere was no doubt influenced by these major shifts in climate, and it is reasonable to assume that adaptation mechanisms led to alterations in some life forms.

First thoughts on thawing in a bioglaciological context perhaps carry with them the expectation that it will be characterized by a complexity and variability unmatched by any other type of thawing. Living organisms, however minute, do not generally behave passively, and higher forms of life exhibit what might be described generally as an intelligent response to their environment, seeking to evade threats to their functional integrity. As noted in Sections 7.5 and 7.6, various strategies are adopted in order to inhibit or promote freezing, but once it has occurred it immediately raises

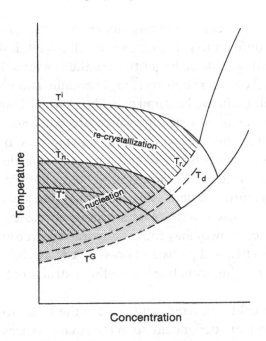

Fig. 8.6. Zones of ice nucleation and re-crystallization.

questions about impact: of freezing itself, of continued existence in the frozen condition, and of thawing.

In a laboratory context, this trio of questions is usually studied in relation to the *temperature protocol* which is followed during experimentation. Thus far, only the cooling process has been the subject of discussion during which it was noted that cooling rate and temperature are both important variables. When the cooling rate is low, the nucleation rate and intracellular supercooling are also usually low. On the other hand, intra-cellular solute concentration increases significantly. Under these conditions, ice forms only on the outside of the cell in temperatures which, although they may not be much below the equilibrium freezing temperature, can be maintained for a prolonged period. That is, a small number of extracellular ice embryos then has the opportunity to grow into relatively large crystals. A higher cooling rate is commonly accompanied by a higher nucleation rate but a smaller cooling period. Given a relative decrease in solute concentration and increase in supercooling inside the cell, it is not surprising that higher cooling rates produce ice embryos intracellularly as well as extracellularly, or that the embryos fail to grow as large as those formed with low cooling rates.

Once formed, ice nuclei grow in relation to the prevailing temperature and concentration fields. The former is typically uniform but the latter often varies significantly across the cell membrane under freezing and thawing conditions. While it is difficult to generalize from cell to cell, it is clear for both freezing and thawing that the larger the cell the slower will be its response to environmental changes; however, rapid equilibration is not necessarily an aid to survival. Diffusive heat and mass fluxes which lead to, or are accompanied by, biochemical reactions which are not wholly reversible, increase the possibility that recovery upon thawing may be remote. The cooling process determines the rate (and location) of nucleation, but the subsequent temperature and concentration fields, defining both gradients and absolute magnitudes, determine ice growth rates and biochemical reactivity. Nucleation itself may not be injurious but, as discussed in the previous chapter, ice growth may cause dehydration, freeze-concentration, disruption of biochemical paths and possibly the mechanical destruction of components ranging from fine intracellular structure to vascular tissue.

While the organism is being held at a fixed temperature the rate processes continue at velocities which are dependent upon the circumstances, as illustrated schematically in Fig. 8.6. Above the homogeneous nucleation curve T^*, ice will not appear spontaneously unless nucleants are present, in which case heterogeneous nucleation occurs at a higher temperature T_n. Many animals and plants lower T_n as part of a deliberate strategy to avoid freezing. Below the vitrification curve T^G, fluid viscosity drastically limits kinetic activity, thereby curtailing crystal growth and biochemical reaction rates. Between the heterogeneous nucleation curve and the vitrification curve is a region in which the organism is vulnerable to nucleation and ice growth, depending on the cooling rate. Fig. 8.6 assumes that heterogeneous nucleation is approached under metastable conditions for which cooling is infinitely slow. Non-zero cooling rates tend to depress the effective nucleation temperature, and may eliminate it entirely if the rate is high enough and the final temperature is below T^G.

The cooling rate strongly influences the effect of freezing on the organism by determining the moment of nucleation and the time available for crystal growth. It also helps determine the precise history of events by controlling the extent to which the locus of states approximates the liquidus and its metastable extension below the eutectic point. Under natural conditions, this approximation may be very close, implying massive ice growth, at least extracellularly. For ice-tolerant organisms, which deliberately manufacture nucleants and thereby avoid sudden cell dehydration and rapid extracellular ice growth, such quasi-steady events are usually reversible. The thawing process essentially re-traces the freezing process along the liquidus, apparently without damage (i.e., gross cell distortion of vascular expansion) being done; warming and cooling rates are both very low. Ice-avoiding organisms, by definition, actively suppress nucleation; clearly, they do not require thawing if successful. The penalty for failure at low temperatures is rapid ice growth, at the very least extra-cellularly; this is likely to be irreversible and lethal, especially if it appears within the cell. Thawing is of little consequence, unless the organism is being harvested as food.

The effect of storage at low temperature varies greatly. If the intra-cellular water has not nucleated, a single cell may be expected to survive, provided concentration effects are tolerable and general functional requirements continue to be met. The cell may have arrived at the storage temperature in a supercooled condition with the intracellular water con-centration being higher than the extracellular value. It will then tend to equilibrate at a low rate, given the temperature dependency of a typical membrane permeability, but will benefit from the diluting effect of any cryoprotectant present. If the storage temperature is low enough, nucle-ation may have occurred during cooling but ice growth will be limited: this is particularly true for storage temperatures below T^G. There are thus several possible starting points in the warming process: firstly, above T_n, where there is a complete absence of ice, in any form, and therefore the absence of a thawing process; secondly, below T_n, when the presence of stable extracellular ice is accompanied by intracellular water which is also at its freezing point or, if not, either supercooled or nucleated at a tem-perature low enough to restrict crystal growth; or thirdly, below T^G, where there may be a complete absence of crystalline ice. In the second of these situations, a single cell may only be affected by internal ice if nuclei are present in which case the adverse effects of crystal growth can be limited by reducing both the storage temperature and the storage time[12]. Since warm-ing at temperatures beneath T^i will exacerbate the difficulties caused by crystal growth during storage, it follows that rapid warming is to be pre-ferred over slow warming (Mazur *et al.*, 1970; Farrant, 1980; Bajaj, 1976), although care must be taken to avoid rapid changes in osmotic pressure and the mechanical damage which may occur in intercellular spaces (Leibo, Mazur and Jackowski, 1974; Pegg *et al*, 1984).

The third situation mentioned above may produce no glaciological effects in the conventional sense if the temperature remains below T^G: nuclei may not form, crystals may not grow. There may be biochemical implications, however. Warming and cooling effects are not generally symmetric about the vitrification temperature, as indicated in Fig. 8.6

which shows two additional curves above T^G. The curve of de-vitrification T_d marks the onset of an increase in kinetic effects as the temperature rises, but the time scales for molecular interaction are evidently still too large to permit a significant degree of nucleation or, if nuclei are already present, to allow significant alterations in the diffusion and accommodation rates at the ice–solution interface. Noticeable crystal growth does not occur until the curve of re-crystallization T_r has been reached. Above this temperature, the organism is vulnerable to the effects of ice growth until the equilibrium melting temperature T^i is reached. Between the re-crystallization temperature and the melting temperature crystals will grow at a rate dependent on both the temperature and the solution concentration. Slow warming thus permits relatively extensive crystal growth, and may lead to the formation of an ice matrix which, as the melting temperature is finally approached, will collapse.

In tissue, organs and organisms as a whole, the thawing process is much more complex than that described above. At the vascular level, as at the cellular level, dimensions and thermal resistances are both small enough to justify the assumption of uniformity in the local temperature. However, membranes and walls generally have a higher resistance to the diffusion of water and solutes than to the flow of heat, and thus create concentrations and concentration gradients which, at the very least, are dependent on the local temperature. As noted previously, the local cooling and freezing rates in bulk tissue are functions of both time and location, and hence no prescribed external cooling rate can generate internal uniformity unless that rate is infinitely slow. In general, therefore, the starting state in the thaw process may be different at different locations in the tissue. In an extreme situation, cells at the surface could be vitrified while those deep in the interior were surrounded by massive amounts of extracellular ice.

Even though spatial variations in cooling rate generate corresponding variations in the concentration of solutes and extracellular water, storage at low (uniform) temperature would permit a gradual redistribution. At any time after this has been achieved, thawing may begin and should proceed rapidly. The higher the warming rate the higher the de-vitrification temperature (Fahy *et al.*, 1984) and hence the smaller will be the effect of re-crystallization, should it occur. In fact, re-crystallization will not occur at all if the warming rate is sufficiently high because the ice–solution interface velocity exhibits a maximum[13] with respect to temperature: for temperatures above the optimum, interface velocity increases as the temperature is lowered, whereas below the optimum the increased tendency towards vitrification as the temperature is lowered produces the opposite effect. Thus to avoid the effects of re-crystallization the warming rate B must satisfy the relation

$$B \gg (V_I)_{opt} \, \theta/d_c$$

where $\theta = T^i - T_0$ (with T_0 as the storage temperature). $(V_I)_{opt}$ is the maximum crystal interface velocity for the biofluid in the range $T_0 < T < T^i$ and d_c is the characteristic dimension of the space occupied by the biofluid: the cell diameter, for example, would be appropriate for intracellular fluid. For many organs, these warming rates cannot be obtained by surface heat transfer which would induce unacceptably large temperature gradients. A more promising technique is the use of electromagnetic heating (Marsland, 1987).

Notes

Chapter 1

1 Ice I is most common on Earth. The other forms exist at higher pressures, as noted in Chapter 2.

2 It is assumed that the body rotates rapidly enough to create a balance between the radiation incident on its projected area and that leaving its entire surface area.

3 21 June and 21 December. On a plane normal to the solstices lie the equinoxes: 20 March and 22 September.

Chapter 2

1 Up to this point, the time derivative d/dt has referred to extensive material properties integrated over the body. Strictly, this *material derivative* should be distinguished, e.g. by writing it D/Dt, but widespread application of the fundamental equations (2.1), (2.2), (2.3), (2.5) and (2.6) seldom employs such notation.

2 The symbol đ has been used to denote changes in path functions, which are not thermodynamic properties and do not possess exact differentials.

3 Strictly, this is only true at plane interfaces in the absence of electric, magnetic and gravitational fields.

4 Defined by the discontinuities in the first derivatives of the Gibbs function. Higher order changes are beyond the scope of this book.

5 This stems from the Euler relation (Callen, 1960)

$$U = TS + \sum_i F_i X_i$$

which combines with equation (2.14) to give

$$S dT + \sum_i X_i d F_i = 0$$

which is equation (2.17).

6 $1/r = (1/r_1 + 1/r_2)/2$ where r_1, r_2 are the principal radii of curvature. Curvature is taken as positive when the centre of curvature is on the liquid side.

7 Analogous treatments using the Helmholtz function $F = U - TS$, or the potential $\Lambda = PV + \sigma A$, are more rigorous but yield results in the same form (Hirth *et al* ., 1970; Dufour and Defay, 1963.

8 In addition to 'classical' treatments, the use of molecular models or fluctuation theory provides alternative formulations (Pruppacher and Klett, 1980).

9 It will become apparent in later chapters that a third important component is the rate at which latent heat can be conducted away from the interface.

10 Normalization is the process in which dependent and independent variables are each divided by a reference quantity which indicates the range over which the variable is expected to vary. Normalized variables are thus dimensionless and have an order of magnitude of 1.

11 When $Fo \gg 1$, the equation simplifies to the quasi-steady form appropriate to slow growth.

Chapter 3

1 Radiation and convection effects may be lumped together in an overall heat transfer coefficient which is physically inconsistent but empirically convenient.

2 dm_i is positive during freezing and negative during melting. The reverse is true for the water mass fraction since $m_i + m_w = 1$. Hence the alternative form $dh = c_p(T)dT + \lambda dm_w$.

3 The characteristic quantities indicated by the subscript C are the intrinsic scales.

4 Clearly $X_{Ci} = O(\hat{X})$, but X_{Cf} cannot be defined generally; it may be the diameter of a pipe, the depth of a river, infinity, etc.

5 A quadratic is the simplest form incorporating the curvature associated with sensible heat under planar, transient conditions.

6 More generally, thermal properties are not constant and the grid is non-uniform, in which case the normalized equation $c(\partial\phi/\partial\tau) = (\partial/\partial x)(k\partial\phi/\partial x)$ may be integrated between $x_{n-\frac{1}{2}}$ and $x_{n+\frac{1}{2}}$ to yield

$$\sigma(\phi_n^{p+1} - \phi_n^p) = 2\Delta\tau \frac{[k_n(\phi_{n+1}^p - \phi_n^p)/\Delta x_n - k_{n-1}(\phi_n^p - \phi_{n-1}^p)/\Delta x_{n-1}]}{c_n\Delta x_n + c_{n-1}\Delta x_{n-1}}$$

which reduces to equation (3.51) when k, $c = \rho c_p$ and Δx_n are constants. Note that the element Δx_n is on the right hand side of the node x_n, and properties are evaluated within the element.

7 This stability criterion ($\Delta\tau/\sigma\Delta^2 x \leqslant 1/2N$ for N dimensions) is approximate and applies strictly to simple heat conduction.

8 Various other explicit/implicit schemes have been tried: see, for example, the discussion in Crank and Nicolson (1947), Lockwood (1966) and Goodrich (1982b).

9 That is, by assuming $\phi = a + bx$, with $\phi(x_1) = \phi_1$ and $\phi(x_2) = \phi_2$. Higher accuracy may be obtained by assuming $\phi = a + bx + cx^2$, with $(\partial\phi/\partial x)_1$, $\phi(x_2)$ and $\phi(x_3)$ given.

10 Alternatively, $\partial\phi/\partial x$ may be expanded in a Taylor expansion about ξ. Thus

$$\left(\frac{\partial\phi}{\partial x}\right)_\xi \approx \left(\frac{\partial\phi}{\partial x}\right)_{n-1} + (\xi - x_{n-1})\left(\frac{\partial^2\phi}{\partial x^2}\right)_{n-1} \approx \frac{\phi_n - \phi_{n-2}}{2\Delta x} +$$

$$(\xi - x_{n-1})\frac{(\phi_n + \phi_{n-2} - 2\phi_{n-1})}{\Delta^2 x}$$

11 In general, several nodal temperatures are required on either side, depending on the technique and the accuracy demanded.

12 $$\int_{\theta_1}^{\theta_2} f(\psi)\delta(\psi - \psi_0)d\psi = f(\psi_0) \text{ if } \theta_1 < \psi_0 < \theta_2$$
$$= f/2 (\psi_0) \text{ if } \psi_0 = \theta_1, \theta_2$$
$$= 0 \text{ if } \psi_0 < \theta_1 \text{ or } > \theta_2$$

(Korn and Korn, 1961).

13 Temperature is the only dependent variable of interest here. More generally, other scalars or vectors would also be specified.

14 Note that equation (3.57) has been evaluated at $\tau = p\Delta\tau$, thus implying an explicit formulation.

15 This residual definition is the negative of that given earlier. Since the residuals are ultimately set to zero the sign is a matter of convention.

Chapter 4

1 Fletcher (1973) used an exponential form.

2 Here it is occasionally described as *black* ice, consistent with the optical axis being vertical.

3 Ashton (1986) notes a value of $\mathcal{J} = 10^4 \, \mathrm{m}^{-3} \, \mathrm{s}^{-1}$.

4 Although less common, needle frazil occurs where turbulence levels are lower and supercooling is greater i.e. when dendritic growth occurs (Hanley and Rao, 1982).

5 For discoid frazil, the rise velocity is given by $\mathcal{V} \approx 10^4 d^2$, where d is the disc diameter (Ashton, 1983). With $d = 1 \, \mathrm{mm}$, $\mathcal{V} \approx 1 \, \mathrm{cm} \, \mathrm{s}^{-1}$, but a spectrum of diameters implies a range of differential velocities.

6 This appears to be the least unsatisfactory English word to distinguish this form of growth from accretion and the more general term freezing. Some authors prefer congelation.

7 In supercooled water it would lose heat, but such a prospect is limited to covers underswept by the open water of rapids or leads.

8 Sometimes referred to as the latent and sensible heat contributions, respectively.

9 An estimate of effective thermal conductivity may be found from

$$k_s = 0.021 + 4.2 \times 10^{-6} \rho_s + 2.2 \times 10^{-9} \rho_s^3$$

where ρ_s is the density (Ashton, 1986). Strictly, the effects of advection and evaporation/condensation should be incorporated.

10 Typically,

$$\frac{\partial k_i}{\partial X} \frac{\partial T}{\partial X} \ll k_i \frac{\partial^2 T}{\partial X^2}$$

11 It will also be transparent if gases expelled from solution are not trapped as bubbles within the ice.

12 The fact that $0.730 \neq 0.332 \, Pr^{1/3}$ exactly stems from the boundary layer approximations.

13 Experimental error of 10–20% in the data is attributed to free stream turbulence.

14 Having longitudinal periodicity as compared with Taylor–Goertler waves which have a transverse periodicity and take the form of longitudinal vortices.

15 Typically between $-4\,^{\circ}\mathrm{C}$ and $-7\,^{\circ}\mathrm{C}$, the lower temperatures being exhibited by water previously heated to help dissolve nucleating particles (Dorsey, 1948).

16 Graetz is credited with the first study of forced convection in the entrance of a pipe (1885).

17 In fact the discovery led to the transitional studies of ice sheets reported in the previous section.

18 $K = \dfrac{\nu}{U} \dfrac{\mathrm{d}U}{\mathrm{d}X} > 3.3 \times 10^{-6}$

where U is the mean water velocity, leads to re-laminarization (Moretti and Kays, 1965).

Chapter 5

1 Supersaturation, referred to a specific temperature, is the more usual description.

2 If f is the vapour deposition flux density and X_s is the collection distance (on each side of the step) over which surface diffusion operates to bring deposited material to the step, then a steady mass balance requires that $Uh\rho_i = 2fX_s$, when $X_s \gg h$.

3 Normal to the face. Sometimes called the linear velocity.

4 Ice saturation is a stable equilibrium curve whereas water saturation represents a metastable extrapolation below the triple point, as indicated on Fig. 2.8.

5 So is the temperature gradient but this plays only a small role in the interfacial heat balance, which determines the frost growth rate (see Section 5.6).

6 See, for example, Langmuir and Blodgett (1946). An alternative approach is given by Pearcey and Hill (1956).

7 This does not require that $d v_d / d\tau = 0$.

8 The droplet size spectrum upon impact thus varies around the cylinder, the mean radius increasing with distance from the forward stagnation point, where impact must always occur.

9 This term is ambiguous, being used for both a pure vapour environment and one consisting of very small droplets.

10 Strictly, this might also be described as an air advective flux but has been called the kinetic flux to avoid confusion.

11 λ_{wv} is replaced by λ_{iv} if water is absent on the accretion and sublimation takes place.

12 Bearing in mind variations in crystal habit, it must also involve the intertwining of elaborate, branched structures.

13 Assuming the core to be at T^i implies that the ice conductive flux is zero.

14 This approximates Whitaker's data which has been extended by Achenbach (1978). The Prandtl number has been taken as 0.71 and the properties of air are assumed to be constant.

15 Influencing both the cooling rate and the approach to homogeneous nucleation.

16 It is worth noting that the temporary shielding or enclosure of high collection efficiency objects by large nets or bags may be a useful alternative (Jaakkola, Laiho and Vuorenvirta, 1983; Gerger, 1974; Lock, 1972).

17 A small manufacturing plant discharging harmless water vapour into a capricious wind, for example, may produce poor visibility and a road icing hazard.

18 In fact, air temperatures may be slightly in excess of $0\,°C$ if evaporative or radiative cooling is effective.

19 The coefficient of proportionality was 0.1 when δ was measured in inches and U_∞ in miles per hour.

20 With the usual proviso that $Ste \ll 1$, and ignoring variations of k and ρ with temperature. The solution also implies that the interfacial temperature T_I is fixed or slowly varying.

Chapter 6

1 It is important to note that global isotherms do not generally follow the lines of latitude, mainly because of continentality.

2 The indices in the illustration are given in $°F$-days. Application of the factor 5/9 converts them to $°C$-days.

3 As discussed in Chapter 3, the air temperature may be used instead, but the resistance of any cover and associated convective system must then be known.

4 Incident on the earth's envelope i.e., the solar constant ($1373 \, W \, m^{-2}$) modified by latitude (ASHRAE, 1985). About one-third of this is reflected back into space.

5 Sometimes known as the direct beam energy which, when subtracted from the insolation, gives the diffusive contribution.

6 From a body temperature around $5800 \, K$.

7 A measure of surface roughness: Z_0 in the law of the wall (Prandtl, 1957)

$$U = \frac{U_*}{k} \ln \left(\frac{Z}{Z_0} \right)$$

8 The ratio of sensible heat to latent heat when water freezes, the Stefan number, was discussed in Chapter 3. An analogous definition applies to evaporation and condensation.

9 This assumes that the net global heat transfer rate $\dot{Q} = \dot{Q}_R + \dot{Q}_C = 0$, implying that the Earth's mean temperature has reached a steady value.

10 Glacier ice contains some water (0.1–1% by volume) in a network of intergranular veins several tens of microns in diameter; the permeability is therefore very low (Raymond and Harrison, 1975).

11 As noted earlier, the presence of meltwater has the effect of increasing the temperature of the layer (Muller, 1976); in general, freezing and melting tend to dampen temperature fluctuations.

12 $1.15 \times 10^{-6} \, m^2 \, s^{-1}$ is a representative value for glacier ice (Schwerdtfeger, 1963; Weller, 1967).

13 Other than quasi-steady variations in V^s, T_s or G.

14 This Russian term appears to be more general than the German *aufeis* or the non-descript English term *icing*.

15 Experience with ice itself indicates that molecular mobility within a perfect crystal, or as a result of various crystal faults, is negligible in relation to the fluxes recorded in most porous media.

16 As noted in the previous section, this process is also important throughout the cryosphere where it is reflected in metamorphosis.

17 Alternatively, the void ratio $e = \mathcal{N}/(1 - \mathcal{N})$ defines the ratio of the pore volume to the matrix volume.

18 More generally, the equation of motion of the fluid in an isotropic medium (Bear, 1969)

$$m \frac{\partial V}{\partial t} + \rho(V \cdot \nabla)V = -\rho g \nabla H - \frac{\mu}{k} \mathbf{V},$$

in which the mass of fluid per unit volume $m \neq \rho$, the density, may be reduced to equation (6.17) if inertia is negligible. Otherwise, the shear term, which generally contains both a viscous (Darcian) effect and an inertial (Reynolds stress) effect, must be written in the Forchheimer form $\mu V/k + aV^2/k^{\frac{1}{2}}$. The intrinsic permeability k is discussed in the text: a is an empirical constant.

19 It has been suggested (Colbeck, 1978, Shimizu 1970) that for snow,

$$k = 0.077 \, d_g^2 \exp \left(-7.8 \rho_s/\rho_w \right)$$

where ρ_s is the snow density and d_g is the grain diameter.

20 It is worth noting that this is strictly a tensor relation given by $\mathcal{D}_{sr}^m = D_m \tau_{sr}$ where D_m is a scalar.

21 The use of this word simply implies any diffusion process; e.g., heat conduction.

22 More precisely,

$$\mathscr{D}_{sr}^d = \alpha_{sr} V_s f_s.$$

In an isotropic medium, α has three principal axes, one lying parallel to the velocity V_s. Under such conditions, Bear (1969) suggests that

$$f_s = V_s/(V_s + A)$$

where $A = (2 + 4S^2)D_m/l$, l is the average pore length and $S = l/d$ is the pore slenderness ratio.

23 Assuming a forced flow field. In certain circumstances, natural convection may be important.

24 This approach implies that the medium is treated as a substance rather than a system. While this is often adequate for field purposes, it does neglect systemic characteristics. A mass weighted average is used here. That is, if the j-th component volume V_j is filled with material of density ρ_j, the amount of any extensive property contained in this volume will be $\zeta_j \rho_j V_j$ where ζ_j is the amount of ζ per unit mass. The total amount of ζ is therefore $\sum_j \zeta_j \rho_j V_j$. When ζ is mass itself, the total mass is given by $\sum_j \rho_j V_j$. Hence the mass average of ζ per unit mass is $\sum_j \zeta_j \rho_j V_j / \sum_j \rho_j V_j$ or $\sum_j \zeta_j \rho_j v_j / \sum_j \rho_j v_j$. For a porous medium consisting of a solid matrix s and a single fluid f, there are only two components: f and s.

25 For a sphere, $S_s = 3/\rho r$, where r is the radius and ρ is the material density, typically $2.7 \, g, cm^{-3}$ for mineral soils. For a circular plate, $S_s = 2/\rho t$, where t is the plate thickness.

26 More precisely, for all the pore water to be thus frozen, $\mathcal{N}\lambda_{iw} = (1 - \mathcal{N}) \, c_{pi}\theta$; hence $\theta \approx 160\mathcal{N}/(1 - \mathcal{N})$ is the amount of subcooling. For frost, the porosity $\mathcal{N}$ is typically greater than 0.5.

27 Alternative weighting schemes are discussed by Farouki (1981). Systemic characteristics are thus ignored.

28 When K is constant, $C_m = K/C$, where C is the coefficient of consolidation defined in equation (6.19).

29 Eliminating pressure between these two equations yields $\bar{h} = \bar{h}(T, v_w)$ and hence

$$d\bar{h} = \left(\frac{\partial \bar{h}}{\partial T}\right)_{v_w} dT + \left(\frac{\partial \bar{h}}{\partial v_w}\right)_T dv_w = \bar{c}dT + \frac{\lambda_{iw}}{\bar{\rho}} dm_w,$$

the approximation used previously (see equations (6.36) and (6.45)).

30 $Ste \ll 1$, and therefore supercooling cannot remove all the latent heat. On a larger scale, this situation was discussed in Sections 4.4 and 4.6, the only difference being that pore ice nucleates in the water interior, not on the substrate.

31 When $T_1 \approx T_{cr}$ the fringe Stefan number is small and $h_{cr} - h_1 \approx \lambda_{iw}$.

32 This concept was first suggested by Konrad (1980) who used the ratio $V_H/(\partial\theta/\partial Z)$. Gilpin (1982) later used the non-dimensional form of equation (6.55).

33 Clearly, other functions $K(\theta)$ would lead to other forms of S_p (Gilpin, 1980).

Chapter 7

1 The approximate distance over which the constraining influence of the tetrahedral cluster geometry is felt.

2 Crystallization of pure solvent water from solution; not including crystallization of solutes.

3 Since $T^*_0 > T^b_0$ it would appear to be difficult to form the vitreous solid directly from pure water. Early attempts used vapour deposition as an intermediate step but this is not necessary at very high cooling rates. See, e.g. Stewart and Vigers (1986).

4 Up to the peak crystallization rate, corresponding to 45% of the freezable water (Angell and Senapati, 1987).

5 Such cells are usually classified as prokaryotic or primitive. Eukaryotic cells, reflecting the process of evolution, are usually larger and more complex.

6 Mitochondria and chloroplasts each have their own DNA.

7 It may, however, produce biochemical effects where reaction rates are sensitive to temperature, as they usually are.

8 Differential thermal analysis (DTA) and differential scanning calorimetry (DSC) are calorimetric techniques for studying latent heat release (or absorption) as the specimen is cooled.

9 If the rate of heat withdrawal from the system does not exceed the latent heat liberation rate.

10 This feature was also noted in soil pores, as discussed in Section 6.4. Mazur (1966) estimates that a pore radius of 7–30 Å corresponds to a threshold temperature of about $-10\,°C$, but this must be regarded as a crude estimate, given the radius of the water molecule and the uncertain values of the surface tension and contact angle.

11 The liquidus is, of course, the stable equilibrium freezing curve of Fig. 7.10 which is an alternative form of presentation.

12 It is worth noting that Arrhenius plots are often non-linear in biochemical reactions.

13 For example, a thick, continuous cuticle may constitute an effective barrier to external seeding (Single and Marcellos, 1974).

14 Cold hardening simply implies an ability to survive lower temperatures, whatever their level; frost hardening might be more accurately called ice hardening or ice tolerance.

15 Bearing in mind that the abscissa may be time or temperature for a given cooling rate. Note also the relationship with freezing curves for complex physical systems containing unfrozen water (see, for example, Sections 2.5 and 6.4).

16 In systems with three or more components, there is no single eutectic point. In a ternary system, for example, any pair of components exhibits a continuous series of eutectic 'points' whose location depends upon the amount of the third component present. Only where the three curves thus generated actually meet is there a true point.

17 The fibres are multi-nucleated cylindrical cells containing sarcoplasma and bounded by a sarcolemma.

Chapter 8

1 The plastic deformation of ice also produces hysteresis.

2 The long-range elastic distortion energy is ignored, thus requiring opposite dislocation pairs or loops, a condition likely to be satisfied at the melting temperature.

3 Near the surface of a crystal, the number of lattice vacancies increases substantially, thus increasing mobility and reducing the activation energy barrier to a negligible level (Ubbelohde, 1965).

4 The extinction coefficient for both ice and water varies over at least one order of

magnitude and is a function of wavelength and impurities, including air bubbles (Devik 1932; Lyons and Stoiber, 1959).

5 As noted in Sections 4.3 and 4.4, warm water above cold water is potentially unstable when the temperature lies between 0°C and 4°C.

6 The evolving microstructure of the ice was discussed in Section 6.2. Equally important is the macrostructure resulting from its flow characteristics.

7 With respect to the original water molecules. The environment suffers an increase in entropy.

8 As a result of the structure moving: witness a galloping cable or the change of heading of a ship, aeroplane, etc.

9 The equations are analogous but the initial and boundary conditions are not, in general; this is especially true at the moving interface.

10 This ratio was first described by Morgenstern and Nixon (1971) in the form

$$R = \left(\frac{\mathscr{R}}{2}\right)^{\frac{1}{2}}$$

11 Syngenetic and epigenetic refer to formation in the presence or absence of sedimentation, respectively.

12 These measures may or may not be practical in cryopreservational terms.

13 This corresponds to the minimum freezing time on the *TTT* curves discussed in Section 7.2.

Selected Bibliography

Achenbach, E. (1977) The effect of surface roughness on the heat transfer from a circular cylinder to the cross flow of air, *Int. J. Heat and Mass Trans.* **20:** 359–69.

Achenbach, E. (1978) Heat transfer from spheres up to $Re = 6 \times 10^6$, *Proc 6th Int. Heat Transfer Conf., Toronto,* **5:** 341–6.

Ackley, S. F. (1982) Ice scavenging and nucleation: two mechanisms for incorporation of algae into newly-formed sea ice, *EOS* **63:** 54–5.

Ackley, S. F. and Itagaki, K. (1973) An evaluation of passive de-icing, mechanical de-icing, and ice detection, *US CRREL, Internal Report* 351.

Ackley, S. F. and Itagaki, K. (1974) The crystal structure of a natural freezing rain accretion, *Weather,* **29:** 189–92.

Ackley, S. F., Itagaki, K. and Frank, M. D. (1977) De-icing of radomes, and lock walls using pneumatic devices, *J. Glaciol.,* **19:** 467–78.

Ackley, S. F. and Templeton, M. K. (1979) Computer modelling of atmospheric ice accretion, U.S. Army Cold Regions Research and Engineering Laboratory Rep. 79–4.

Adams, W. P. (1981) Snow and ice on lakes. *In Handbook of Snow,* (eds. Gray, D. M. and Male, D. H.), Pergamon Press, Toronto: 437–74.

Ahti, K. (1978) On factors affecting rime formation in Finland, M.Sc. thesis, Dept. Meteorology, University of Helsinki.

Akai, A. (1979) Freezing avoidance mechanism of primordial shoots of conifer buds, *Plant & Cell Physiol,* **20**(7): 1381–90.

Akerman, J. (1982) Studies on naledi (icings) in the West Spitzbergen, *Proceedings 4th Canadian Permafrost Conf.,* Nat. Res. Co. Canada, Ottawa: 189–202.

Akitaya, E. (1975) Studies on depth hoar, Snow Mechanics Symp. International Association of Scientific Hydrology, Publ. **114:** 42–8.

Allaire, P. E. (1985) *Basics of the finite element method,* W. C. Brown, Dubuque, Iowa.

Allen, C. W. (1973) *Astrophysical quantities* 3rd ed. University of London, Athlone Press, London.

Altberg, V. I. (1936) Twenty years of work in the domain of underwater ice formation (1915–1935), *Int. Assoc. for Sci. Hydrol. Bull.,* **23:** 373–407.

Andersland, O. B. and Anderson, D. M. (1978) *Geotechnical engineering for cold regions,* McGraw-Hill.

Anderson, D. M. (1963) Latent heat of freezing soil water, *Proc. First Int. Permafrost Conf ,* Nat. Acad. Sci., Washington Pub. 1287.

Anderson, D.M. and Morgenstern, N.R. (1973) Physics, chemistry and mechanics of frozen ground: a review, 2nd Int. Conf. on Permafrost, Nat. Acad. Sci; Washington: 257–88.

Anderson, D.M. and Tice, A.R. (1972) Predicting unfrozen water contents in frozen soils from surface area measurements, *Highway Res. Record* **393**: 12–18.

Anderson, D.M. and Williams, P.J. (1985) *Freezing and thawing of soil-water systems*, American Society of Civil Engineers, New York.

Angell, C.A. (1982) Supercooled water. In *Water – A Comprehensive Treatise* (ed. Franks, F.), Plenum Press, London **7**: 1–82.

Angell, C.A and Senapati, H. (1987) Crystallization and vitrification in cryoprotected aqueous solutions. In *The biophysics of organ preservation*, (eds. Pegg, D.E and Karow, A.M.) Plenum Press, London: 147–62.

Arakawa, K. (1955) Studies on the freezing of water: III; crystallography of disc crystal and dendrites developed from disc crystals, *J. Fac. Science, Hokkaido Univ., Series II (Physics)* **4**(6): 355–8.

Aris, R. (1962) *Vectors, tensors and the basic equations of fluid mechanics*, Prentice-Hall, Englewood Cliffs, New Jersey.

Aristotle (1952) *Meteorologica* (trans. H.D.P. Lee) Harvard University Press, Cambridge, Mass.

Armstrong, T., Roberts, B and Swithinbank, C. (1973) *Illustrated glossary of snow and ice*, Scott Polar Research Institute, Cambridge.

Artem'yev, A.N. (1973) Annual and daily variations of the components of the heat balance of the underlying surface on the Antarctic plateau, *Sov. Antarc. Exped. Inf. Bull.* **87**: 497–9.

Asahina, E. (1966) Freezing and frost resistance in insects. In *Cryobiology* (ed. Meryman, H.T.). Academic Press, New York: 451–86.

Asahina, E. (1969) Frost resistance in insects, *Advances in Insect Physiology* **6**: 1–49.

ASHRAE Handbook of fundamentals (1985) ASHRAE, Atlanta, Ch. 2–7.

Ashton, G.D. (1972) Turbulent heat transfer to wavy boundaries, *Proc. Heat Transfer & Fluid Mech Inst.*: 200–13.

Ashton, G.D. (1980) Freshwater ice growth, motion and decay, In *Dynamics of snow and ice masses* (ed. Colbeck, S.C.) Academic Press, New York: 261–304.

Ashton, G.D. (1983) Frazil ice, In *Theory of dispersed multiphase flow*, Academic Press, New York: 271–289.

Ashton, G.D. (ed) (1986) *River and lake ice engineering*, Water Resources Publications, Littleton, Colorado.

Ashwood-Smith M.J. (1980) Low temperature preservation of cells, tissues and organs, In *Low temperature preservation in medicine and biology* (eds. Ashwood-Smith, M.J. and Farrant, J.) Pitman Medical Ltd, Tunbridge Wells: 19–44.

Bajaj, Y.P.S. (1976) Regeneration of plants from cell suspensions frozen at −20, −70, and −196°C, *Physiol. Plant* **37**: 263–8.

Bardarson, H.R. (1969) Icing of ships, *Jokull*, **19**: 107–20.

Bardarson, H.R. (1974) *Ice and fire*, published by the author, Reykjavik: 94.

Barnes, H.T. (1928) *Ice Engineering*, Renouf Publishing, Montreal.

Bartlett, J.T. Van Den Heuval, A.P. and Mason, B.J. (1963) The growth of ice crystals in an electric field, *Z. angew. Math. Phys.* **14**: 599–610.

Bauer, D. (1973) Snow accretion on power lines, *Atmosphere* **11** (3): 88–96.

Baust, J.G. (1981) Biochemical correlates of cold-hardening in insects, *Cryobiology*, **18**: 186–98.

Beaglehole, D. and Nason, D. (1980) Transition layer on the surface of ice, *Surf. Sci.*, **96**: 357–363.

Bear, J. (1969) Hydrodynamic dispersion. In *Flow through porous media* (ed. De Wiest, R.J.M.) Academic Press, New York: 109–99.

Bear, J. and Bachmat, Y. (1965) A unified approach to transport phenomena in porous media, underground storage and mixing of water, *Prog. Rep.* 3, Technion – Israel Institute of Technology, Haifa.

Bear, J. and Bachmat, Y. (1966) Hydrodynamic dispersion in non-uniform flow through porous media taking into account density and viscosity differences, *PN* 4/66, Technion – Israel Institute of Technology, Haifa.

Beard, K.V. and Pruppacher, H.R. (1971) A wind tunnel investigation of the rate of evaporation of small water drops falling at terminal velocity in air, *J. Atmos. Sci.* **28**: 1455–64.

Beatty, J.K., O'Leary, B. and Chaikin, A. (eds.) (1982) *The New solar system*, 2nd edn, Cambridge University Press, Cambridge.

Bell, G.H., Davidson, J.N. and Scarborough, H. (1961) *Textbook of physiology and biochemistry*, 5th edn, E & S Livingstone, Edinburgh.

Beltaos, S. (1984) A conceptual model of river ice breakup, *Can. J. Civil Eng.* **11** (3): 516–29.

Bentley, W.A. and Humphreys, W.J. (1962) *Snow crystals*, Dover Publications, New York.

Berger, A. (1977) Support for the astronomical theory of climatic change, *Nature* **269**: 44–5.

Berry, M.O. (1981) Snow and climate. In *Handbook of snow*, (eds. Gray, D.M. and Male, D.H.), Pergamon Press, Toronto.

Blanchard, D.C. and Woodcock, A.H. (1957) Bubble formation and modification in the sea and its meteorological significance, *Tellus* **9**: 145–58.

Block, W. (1982) Cold hardiness in invertebrate poikilotherms, *Comp. Biochem. Physiol.* **73A** (4): 581–93.

Bonacina, C., Comini, G., Fasano, A. and Primicerio, M. (1973) Numerical solution of phase-change problems, *Int. J. Heat Mass Trans.*, **16**: 1825–32.

Borisenkov, E.P. and Pchelko, I.G. (eds) (1972) *Indicators for Forecasting Ship Icing*, Leningrad (US Army Cold Region Research and Engineering Laboratory, Hanover, translation 481, AD A030 113).

Brandt, J.C. (1982) Comets. In *The new solar system* (eds. Beatty, J.K., O'Leary, B. and Chaikin, A.,) 2nd edn, Cambridge University Press: 177–86.

Brower, R.H. and Miner, E. (1961) *Japanese court poetry*, Stanford University Press, Stanford.

Brown, J. (1966) Massive underground ice in northern regions, *Army Science Conf. Proc.*, Dept., US Army, Washington; **1**: 89–102.

Brown, R.J.E. (1970) *Permafrost in Canada*, University of Toronto Press, Toronto: 126.

Budyko, M.I. (1958) Atlas teplovogo balansa, Leningrad (H. Flohn, *Erdk*, **12**: 233–7, 1958).

Burke, M.J., Gusta, L.V., Quamme, H.A., Weiser, C.J and Li, P.H. (1976) Freezing and injury in plants, *Ann. Rev. Plant Physiol*, **27**: 507–28.

Burns, J.A. (1982) Planetary rings, In *The new solar system* (eds. Beatty, J.K., O'Leary, B. and Chaikin, A.,) 2nd edn, Cambridge University Press, Cambridge: 129–42.

Burton, W. K., Cabrera, N. and Frank, F. C. (1951) The growth of crystals and the equilibrium structure of their surfaces, *Phil. Trans. R. Soc.* A **243**: 199–358.

Callen, H. B. (1960) *Thermodynamics*, J. Wiley & Sons, New York.

Calvelo, A. (1981) Recent studies on meat freezing, In *Development in meat science*, (ed. Lawrie, R.) Applied Science Publishers, London, **2**: 125–58.

Cannon, J. R. (1984) *The One-Dimensional Heat Equation*, Addison-Wesley Pub. Co, Reading, Mass.

Cansdale, J. R. (1984) Helicopter rotor ice secretion and protection, Sixth European Rotorcraft and Powered Lift Forum, Bristol, England.

Carey, K. L. (1970) Icing, occurrence, control and prevention: an annotated bibliography. *US Army Cold Regions Research and Engineering Laboratory, Hanover, Spec. Rep.* 151.

Carey, K. L. (1973) Icings developed from surface water and groundwater. US Army Cold Regions Research and Engineering Laboratory, Mono III -D3.

Carslaw, H. S. and Jaeger, J. C. (1959) *Conduction of heat in solids*, 2nd edn. Oxford University Press, Oxford.

Chaîné, P. M. and Skeates, P. (1974) Wind and ice loading criteria selection, Ind. Meteor. Study III, Environment Canada, Ottawa.

Chang, C. M. (1969) Studies on egg yolk, Ph.D thesis, University of Wisconsin, Madison.

Chappell, M. S. (1972) Stationary gas turbine icing problems: the icing environment, *Rep. DME/NAE (4)* Nat. Res. Canada, Ottawa.

Cheng, K. C., Inaba, H. and Gilpin, R. R. (1981) An experimental investigation of ice formation around an isothermally cooled cylinder in crossflow, *J. Heat Trans.* **103**: 733–8.

Cheng, K. C., Inaba, H, and Gilpin, R. R. (1987) Effect of natural convection on ice formation around an isothermally cooled horizontal cylinder, *Proc. Int. Symposium on Cold Regions Heat Transfer Conference, Edmonton*, ASME, New York.

Cheng, K. C. and Sabhapathy, P. (1985) An experimental investigation of ice formation over an isothermally cooled vertical circular cylinder in natural convection, ASME Paper 85-HT-1.

Cheng, K. C., Takeuchi, M. and Gilpin, R. R. (1978) Transient natural convection in horizontal water pipes with maximum density effect and supercooling, *Numerical Heat Trans* **1**: 101–15.

Cheng, K. C. and Wong, S. L., (1977) Liquid solidification in a convectively cooled parallel plate channel, *Can. J. Chem. Eng.* **55**: 149–55.

Chung, P. M. and Bywater, R. (1984) Role of the liquid layer in ice accumulation on flat surfaces, *J. Heat Trans.* **106**: 5–11.

Clancy, J. P. (1965) *Medieval Welsh Lyrics*, MacMillan Co, New York: 150.

Clegg, J. S. (1987) Cytoplasmic organization and the properties of cell water: speculations on animal cell cryopreservation, In *The biophysics of organ cryopreservation* (eds. Pegg, D. E. and Karow, A. M.) Plenum Press, New York: 79–88.

Clement, M. T. (1961) Effects of freezing, freeze-drying, and storage in the freeze-dried and frozen state on viability of *escherichia coli* cells, *Can. J. Microbiol*, **7**: 99–106.

Colbeck, S. C. (1973) Theory of metamorphism of wet snow. US Army Cold Regions Research and Engineering Laboratory, Hanover, Report 313.

Colbeck, S. C. (1974) Grain and bond growth in wet snow, *Snow Mechanics*

Symp., Publ. 114. International Association of Scientific Hydrology. The Hague: 51–61.

Colbeck, S. C. (1978) Physical aspects of water flow through snow *Adv. Hydrosci.* **11:** 165–206.

Colbeck, S. C. and Davidson, G., (1973) The Role of Snow and Ice in Hydrology, *Proc. Banff Symp.*, UNESCO, Geneva 1: 242–57.

Coles, W. D. (1950) Rollin, V. G. and Mulholland, D. R., Icing protection requirements for reciprocating-engine induction system, NACA Rep. 982.

Comini, G., Del Guidice, S., Lewis, R. W. and Zienkiewicz, O. C. (1974) Finite element solution of non-linear heat conduction problems with special reference to phase change, *Int. J. Num. Methods in Eng.* **8:** 613–24.

Cooper, I. S. (1967) Cryogenic surgery. In *Engineering in the practice of medicine*, (eds. Segal, B. L. and Kilpatrick, D. G.) Williams and Wilkins Company, Baltimore: 122–41.

Cooper, T. E. and Trezek, G. J., (1971) Rate of lesion growth around spherical and cylindrical cryoprobes, *Cryobiology* **7** (4–6): 183–90.

Cox, G. F. N. and Weeks, W. F. (1975) Brine drainage and initial salt entrapment in sodium chloride ice, US Army Cold Regions Research and Engineering Laboratory, Rep. 354.

Crank, J. and Nicolson, P., (1947) A practical method for numerical evaluation of solutions of partial differential equations of the heat conduction type, *Proc. Camb. Phil. Soc.* **43:** 50–67.

Cremers, C. J. and Mehra, V. K. (1982) Frost formation on vertical cylinders in free convection, *J. Heat Transf.* **104:** 3–7.

Crocker, G. B. (1984) A physical model for predicting the thermal conductivity of brine-wetted snow, *Cold Regions Sci and Tec.*, **10:** 69–74.

Croll, J. (1875) *Climate and time*, Appleton & Co., New York.

Davenport, A. G. (1986) Interaction of ice and wind loading on guyed towers, *Proc. 3rd Int. Workshop on Atmospheric Icing of Structures, Vancouver*, Can. Electrical Association, Ottawa.

Davis, S N. (1969) Porosity and permeability of natural materials, In *Flow through porous media* (ed. De Wiest, R. J. M.) Academic Press, New York: 53–90.

de Groot, S. R. and Mazur, P. (1962) *Non-equilibrium thermodynamics*, North Holland, Amsterdam.

De Wiest, R. J. M. (1965) *Geohydrology*, J. Wiley & Sons, New York.

Deacon, G. (1984) *The antarctic circumpolar ocean*, Cambridge University Press, Cambridge.

Debye, P. (1914) Zustansgleichung und Quantenhypothese mit einem Anhang uber Warmeleitung, In *Vortrage uber kinetische Theorie der Materie und der Elektrizitat* (ed. M. Plank), Teulner, Leipzig: 19–60.

Defant, F. (1951) Local winds, *Compendium of Meteorology*, Am. Meteor. Soc., Boston: 655–72.

Defay, R., Prigogine, I., Bellemans, A. and Everett, D. H. (1966) *Surface Tension and Adsorption*, Wiley and Sons, New York.

Devik, O. (1932) Thermal and dynamic requirements for ice formation in streams under Norwegian conditions (in German), *Geografysiske Publakasjoner* **9** (1): 5–100.

DeVries, A. L. (1980) Biological antifreezes and survival in freezing environments, In *Animals and environmental fitness*, (ed. Gilles, R.) Pergamon Press, Oxford.

DeVries, A. L. (1982) Biological antifreeze agents in coldwater fishes, *Comp. Biochem. Physiol*, **73A** (4): 627–40.

DeVries, A. L., Komatsu, S. K and Feeney, R. E. (1970) Chemical and physical properties of freezing point-depressing glycoproteins from Antarctic fishes, *J. Biol. Chem.*, **245** (11); 2001–2008.

DeVries, D. A. (1963) Thermal properties of soils. In *Physics of the Plant Environment* (ed. van Wijk, W. R.) North Holland, Amsterdam.

Diem, M. (1948) Zur Struktur der Niederschlage III, *Arch. Met. Geophys. Bioklim.*, **B16**, 347–55.

Dillard, D. S. and Timmerhaus, K. D. (1966) Low temperature thermal conductivity of solidified H_2O and D_2O, *Pure Appl. Cryogen.* **4**: 35–44.

Diller, K. R. (1975) Intra-cellular freezing: effect of extra-cellular supercooling, *Cryobiology* **12**: 480–5.

Diller, K. R. and Lynch, M. E. (1983) An irreversible thermodynamic analysis of cell freezing in the presence of membrane permeable additives. I. Numerical model and transient cell volume data, *Cryo-Letters*, **4**: 295–308.

Diller, K. R. and Lynch, M. E. (1984a) An irreversible thermodynamic analysis of cell freezing in the presence of membrane permeable additives. II. Transient electrolyte and additive concentrations, *Cryo-Letters*, **5**: 117–30.

Diller, K. R. and Lynch, M. E. (1984b) An irreversible thermodynamic analysis of cell freezing in the presence of membrane permeable additives. III. Transient water and additive fluxes, *Cryo-Letters*, **5**: 131–44.

Dillon, H. B. and Andersland, O. B. (1966) Predicting unfrozen water contents in frozen soils, *Can. Geotech. J.* **3** (2): 53–60.

Dirksen, C. and Miller, R. D. (1966) Closed system freezing of unsaturated soil, *Soil Sci. Soc. Amer. Proc.* **30** (2): 168–73.

Doan, F. J., and Featherman, C. E. (1937) Observations on concentrated frozen milk, *The Milk Dealer*, **27** (3): 33–5.

Dolgushin, L. D., Yevteyev, S. A. and Kotlyakov, V. M. (1962) Current changes in the Antarctic ice sheet, *IAHS-AISH* **58**: 286–94.

Dorsey, N. E. (1948) The freezing of supercooled water, *Trans. Am. Phil. Soc.* **38** (3): 245–325.

Drost-Hansen, W. (1967) The water–ice interface as seen from the liquid side, *J. Colloid Interface Sci.*, **25**: 131–60.

Dubochet, J. and McDowall, A. D. (1981) Vitrification of pure water for electron microscopy, *J. Microsc*, **124**, RP3–4.

Dufour, L. and Defay, R. (1963) *Thermodynamics of clouds*, Academic Press, New York.

Duman, J. G. (1982) Insect anti-freezes and ice-nucleating agents, *Cryobiology* **19**: 613–27.

Duman, J. G. Horwath, K. L., Tomchaney, A. and Patterson, J. L. (1982) Anti-freeze agents of terrestrial arthropods, *Comp. Biochem. Physiol.*, **73A**: 545–55.

Duman, J. G., Neven, L. G., Beals, J. M., Olson, K. R. and Castellino, F. J. (1985) Freeze-tolerance adaptations, including haemolymph protein and lipoprotein nucleators, in the larvae of the cranefly *Tipula Trivitatta*, *J. Insect. Physiol.*, **31** (1): 1–8.

Dye, J. E. and Hobbs, P. V. (1968) The influence of environmental parameters on the freezing and fragmentation of suspended water drops, *J. Atmos. Sci.* **25**: 82–96.

Dyer. D. F., and Sunderland, J. E. (1968) Heat and mass transfer mechanisms in sublimation dehydration, *J. Heat Transfer*, **90C**: 379–84.

Dyson, R. D. (1978) *Cell biology: a molecular approach.* 2nd edn. Allyn & Bacon, Boston, Mass.

El-Tahan, M., Venkatesh, S. and El-Tahan, H. (1984) Validation and quantitative assessment of the deterioration mechanisms of Arctic icebergs, *Proc. 3rd Int. Offshore Mech. and Arctic Engin.*, ASME, New York III: 18–25.

Evans, G. W. (II), Isaacson, E. and MacDonald, J. K. L. (1950) Stefan-like problems, *Q. J. Appl. Math.* **8:** 312–19.

Fahy, G. M., MacFarlane, D. R., Angell, C. A. and Meryman, H. T. (1984) Vitrification as an approach to cryopreservation, *Cryobiology*, **21:** 407–26.

Fang, L. J., Cheung, F. B., Linehan, J. H. and Pederson, D. R. (1984) Selective freezing of a dilute salt solution on a cold ice surface, *J. Heat Transfer*, **106:** 385–93.

Farhadieh, R. and Tankin, R. S. (1975) A study of the freezing of sea water, *J. Fluid Mech.* **71** (2): 293–304.

Farouki, O. T. (1981) Thermal properties of soils, US Army Cold Regions Research and Engineering Laboratory, Mono 81–1.

Farouki, O. T. (1982) Evaluation of methods for calculating soil thermal conductivity, US Army Cold Regions Research and Engineering Laboratory, Rep. 82–8.

Farrant, J. (1980) General observations on cell preservation. In *Low temperature preservation in medicine and biology*, (eds. Ashwood-Smith, M. J. and Farrant, J.) Pitman Medical, Tunbridge Wells.

Feeney, R. E. (1975) A biological antifreeze, *American Scientist* **62:** (6) 712–19.

Fennema, O. R. Powrie, W. D. and Marth, E. H. (1973) *Low temperature preservation of foods and living matter*, Marcel Dekker, New York.

Fernandez, R., and Barduhn, A. J. (1967) The growth rate of ice crystals, *Desalination*, **3:** 330–42.

Fisher, R. A. (1926) On the capillary forces in an ideal soil; correction of formulae given by W. B. Haines, *J. Agric. Sci.*, **16:** 492–505.

Flato, G. M. (1987) Calculation of ice jam profiles, M.Sc. Thesis, University of Alberta, Edmonton.

Fleischer, R. (1953–4) Der Jahresgang der Strahlungsbilanz U. ihrer Komponenten, *Ann. d. Met.* **6:** 357–64.

Fletcher, N. H. (1970) *The Chemical Physics of Ice* Cambridge University Press, Cambridge.

Fletcher, N. H. (1973) The surface of ice, In *Physics and chemistry of ice*, (eds. E. Whalley, S. L. Jones and L. W. Gold) Roy. Soc. Canada, Ottawa.

Fletcher, N. K. (1978) Water: a unique substance, In *Water: Planets, Plants and People*, Austral. Acad. Sci., Canberra: 4–17.

Flohn, H. (1974) Background of a geophysical model of the initiation of the glaciation. *Q. Res.* **4:** 386–404.

Foltz, V. D. (1953) Sanitary quality of crushed and cubed ice as dispensed to the consumer, *Public Health Reports*, **68**(10): 949–54.

Forbes, R. E. and Cooper, J. W. (1975) Natural convection in a horizontal layer of water cooled from above to near freezing, *J. Heat Transf.* **97:** 47–53.

Franks, F. (1985) *Biophysics and biochemistry at low temperatures*, Cambridge University Press, Cambridge.

Franks, F., Mathias, S. F., Parsonage, P. and Tong, T. B. (1983) Differential scanning calorimetric study on ice nucleation in water and in aqueous solutions of hydroxyethyl starch, *Therm. Act.*, **61** (1–2): 195–202.

Franks, F., Mathias, S. F. and Trafford, K. (1984) The nucleation of ice in undercooled water and aqueous polymer solutions, *Colloids and Surfaces* II (3–4): 275–85.

Freeman, M. M. R. (1984) Contemporary Iniut exploitation of the sea-ice environment. In *Sikumiut: People who use the sea-ice* (eds. Cooke, A. C. and Van Elstine, E.) Can. Arctic Res. Comm., Ottawa: 74–96.

Froehlich, W. and Slupik, J. (1982) River icings and fluvial activity in extreme continental climate: Khangai Mountains, Mongolia, *Proc. 4th Can. Perm. Conf. Nat. Res. Co. Canada*, Ottawa. 203–11.

Frost, R. (1969) *The Poetry of Robert Frost*, lst edn (ed. E. C. Lathem), Holt, Rinehardt and Winston, New York.

Fukuda, H. (1936) Uber Eisfilamonte im Boden, *J. College of Agric., Tokyo*, 13: 453–81.

Fukusako, S. and Seki, N. (1987) Freezing and melting characteristics in internal flow, *Proc. Int. Symp. on Cold Regions Heat Transfer* ASME, New York: 25–38.

Fukuta, N. (1969) Experimental studies on the growth of small ice crystals, *J. Atmos. Sci.* 26: 522–31.

Gardner, L. and Moon, G. (1970) Aircraft carburettor icing studies, Mech. Eng. Rep. LP536, Nat. Res. Co. Canada, Ottawa.

Gardner, W. R. (1960) Soil water relations in arid and semi-arid conditions, UNESCO, Geneva, 15: 37–61.

Gardner, W. R., Hillel, D. and Benyamini, Y. (1970) Post irrigation movement of soil water: I. Redistribution, *Water Resources Res*, 6: (3) 851–861; II. Simultaneous redistribution and evaporation, *Water Resources Res*, 6 (4): 1148–53.

Gates, E. M. (1985) Simulated marine icing in an icing wind tunnel, Dept. Mech. Eng. Rep. 49, University of Alberta, Edmonton.

Gates, E. M., Liu, A. and Lozowski, E. P. (1988) A stochastic model of atmospheric rime icing *J. Glaciol.* 34 (116): 26–30.

Gates, E. M., Narten, R., Lozowski, E. P. and Makkonen, L. (1986) Marine icing and spongy ice, *Symposium on ice*, International Association for Hydraulic Research Iowa, II: 133–63.

Gavrilova, M. K. (1972) Radiation and heat balances, thermal regime of an icing, In *The Role of Snow and Ice in Hydrology: Proc. Banff Symp.* UNESCO, Geneva, 1: 496–504.

Geiger, R. (1965) *The climate near the ground*, Harvard University Press, Cambridge, Mass.

Gent, R. W. and Cansdale, J. T. (1985) The development of mathematical modelling techniques for helicopter rotor icing, *23rd Aerospace Sci. Meeting*, AIAA, Reno.

George, M. F., Burke, M. J., Pellett, H. M. and Johnson, A. G. (1974) Low temperature exotherms and woody plant distribution *Hort. Science* 9: 519–22.

George, M. F., Burke, M. J., Pellett, H. M. and Johnson, A. G. (1982) Freezing avoidance by deep undercooling of tissue water in winter-hardy plants, *Cryobiology*, 19: 628–39.

Gerger, H. (1974) Methods used to minimize, prevent and remove ice accretion on meteorological surface instruments, World Meteor. Org. Tech. note 135: 1–6.

Giauque, W. F. and Stout, J. W. (1936) The entropy of water and the third law of thermodynamics. The heat capacity of ice from 15 K to 273 K, *J. Amer. Chem. Soc.*, **58**: 1144–50.

Gibbs, J.W. (1948) *Collected Works*, **1**, Yale University Press, New Haven, Conn.

Gilpin, R.R. (1976) On the influence of natural convection on dendritic ice growth, *J. Cryst. Growth* **36**: 101–8.

Gilpin, R.R. (1977) The effect of cooling rate on the formation of dendritic ice in a pipe with no main flow, *J. Heat Trans*, **99**(3): 419–24.

Gilpin, R.R. (1978a) A study of factors affecting the ice nucleation temperature in a domestic water supply, *Can. J. Chem. Eng.* **56**: 466–71.

Gilpin, R.R. (1978b) The effects of dendritic ice formation in water pipes, *Int. J. Heat Mass Trans*, **20**: 693–9.

Gilpin, R.R. (1979a) A model of the 'liquid-like' layer between ice and a substrate with applications to wire regelation and particle migration, *J. Colloid & Interface Sci.* **68**: (2): 235–51.

Gilpin, R.R. (1979b) The morphology of ice structure in a pipe at or near transition Reynolds number, *Heat Transfer – San Diego, Symp. Ser 189*, **75**: 89–94.

Gilpin, R.R. (1980a) Theoretical studies of particle engulfment, *J. Colloid & Interface Sci*, **74** (1): 44–63.

Gilpin, R.R. (1980b) A model for the prediction of ice lensing and frost heave in soils, *Water Resources Res.* **5** (16): 918–30.

Gilpin, R.R. (1981a) Ice formation in a pipe containing flows in transition and turbulent regimes, *J. Heat Trans*. **3**: 363–8.

Gilpin, R.R. (1981b) Modes of ice formation and flow blockage that occur while filling a cold pipe, *Cold Regions Sci. & Tech*, **5**: 163–71.

Gilpin, R.R. (1981c) Surface spreading as a biotransport mechanism, *J. Biol. Phys*, **9**: 109–32.

Gilpin, R.R. (1982) A frost heave interface condition for use in numerical modeling, *Proc. 4th Can. Perm. Conf*. Nat. Res. Co. Canada, Ottawa.

Gilpin, R.R. Hirata, T. and Cheng, K.C. (1978a) Longitudinal vortices in a horizontal boundary layer in water, including the effects of the density maximum at 4C, ASME paper 78-HT-25.

Gilpin, R.R. Hirata, T. and Cheng, K.C. (1980) Wave formation and heat transfer at an ice-water interface in the presence of a turbulent flow, *J. Fluid Mech*. **99** (3): 619–40.

Gilpin, R.R., Imura, H. and Cheng, K.C. (1978b) Experiments on the onset of longitudinal vortices in horizontal Blasius flow heated from below, *J. Heat Trans* **100**: 71–7.

Gilpin, R.R., Robertson, R.B. and Singh, B. (1977) Radiative heating in ice, *J. Heat Trans*. **99** (2): 227–32.

Gilpin, R.R. and Wong, B.K. (1973) A study of some factors that influence ground temperature, *Proc. 4th Can. Conf. Appl. Mech.*: 771–2.

Gilpin, R.R. and Wong, B.K. (1976) 'Heat valve effects' in the ground thermal regime, *J. Heat Trans*, **98**(4): 537–42.

Gonda, T. and Komabayashi, M. (1971) Skeletal and dendrite structures of ice crystals as a function of thermal conductivity and vapour diffusivity, *J. Met. Soc. Japan* **49**: 32–41.

Goodman, T.R. (1958) The heat-balance integral and its application to problems involving a change of phase, *Trans. ASME*, **80**: 335–42.

Goodman, T.R. (1964) Application of integral methods to transient nonlinear heat transfer. In *Advances in Heat Transfer*, Academic Press, New York, **1**: 51–122.

Goodrich, L. E. (1978) Efficient numerical technique for one-dimensional thermal problems with phase change, *Int. J. Heat Mass Trans.*, **21**: 615–21.

Goodrich, L. E. (1982a) The influence of snow cover on the ground thermal regime, *Can. Geotech. J.* **19**: 421–32.

Goodrich, L. E. (1982b) An introductory review of numerical methods for ground thermal regime calculations, DBR Paper 1061, Nat. Res. Co. Canada, Ottawa.

Graham, J. *et al* (1964) Nuclear magnetic resonance study of interlayer water in hydrated layer silicates, *J. Chem. Phys*, **40**: 540–50.

Grey, B. J. and MacKay, D. K. (1979) Aufeis (overflow ice) in rivers. *Canadian Hydrology Symp.* 79: *Cold Climate Hydrology*, Nat. Res. Co. Canada, Ottawa: 139–63.

Gunderson, J. R. (1966) A study of heat conduction with change of phase, M.Sc. thesis, Dept. of Mechanical Engineering, University of Alberta.

Gutschmidt, J. (1968) Principles of freezing and low temperature storage with particular reference to fruit and vegetables. In *Low Temperature Biology of Foodstuffs* (eds. Hawthorn, J. and Rolfe, E. J.) Pergammon Press, Oxford: 299–318.

Guymon, G. L. and Berg, R. L. (1976) Galerkin finite element analog of frost heave, *Proc. 2nd Conf. on soil–water problems in cold regions*, Hydrol. Div. American Geophys. Union, Washington: 111–13.

Haines, W. B. (1925) Studies in the physical properties of soils, II, A note on the cohesion developed by capillary forces in an ideal soil, *J. Agric. Sci.* **15**: 529–35.

Haines, W. B. (1930) Studies in the physical properties of soils V. The hysteresis effect of capillary properties and the modes of moisture distribution associated therewith, *J. Agr. Sci*, **20**: 97–116.

Hallett, J. (1961) The growth of ice crystals on freshly-cleaved covellite surfaces, *Phil. Mag.* **6**: 1073–87.

Hallett, J. and Mason, B. J. (1958) The influence of temperature and supersaturation on the habit of ice crystals grown from the vapour, *Proc. Roy. Soc.* **A 247**: 440–53.

Hallett, J. and Mossop, S. C. (1974) Production of secondary ice crystals during the riming process, *Nature*, **249**: 26–8.

Hamill, T. D. and Bankoff, S. G. (1964) Similarity solutions of the plane melting problem with temperature-dependent thermal properties, *Ind. & Eng. Chem. Fundamentals* **3** (2): 177–9.

Hanley, T. O. and Rao, S. R. (1982) Acoustic detector for frazil, *IAHR Symp.* **1**: 101–10.

Hardenburg, R. E. and Lutz, J. M. (1968) In *ASHRAE guide and data book, Applications*, ASHRAE, Atlanta: 453–468.

Hariyama T. (1975) The riming properties of snow crystals, *J. Met. Soc. Japan* **53**: 384–92.

Hariyama, Y. (1977) Study of frost formation based on a theoretical model of the frost layer, *Heat Transfer – Japanese Res.*, **6** (3): 79–94.

Harlan, R. L. (1973) An analysis of coupled heat-fluid transport in partially frozen soil, *Water Resources Res* **9**: 1314–23.

Hatsopoulos, G. N. and Keenan, J. H. (1965) *Principles of general thermodynamics*, Wiley & Sons, New York.

Hawking, S. W. (1988) *A brief history of time*, Bantam Books, London.

Hayashi, Y., Aoki, K. and Yuhara, H. (1977) Study of frost formation based on a theoretical model of the frost layer, *Heat Transfer – Japanese Res.* **6** (3): 79–94.

Hays, J.D. Imbrie, J. and Schackleton, N.J. (1976) Variations in the earth's orbit: pacemaker of the ice ages, *Science* **194:** 1121–32.

Henderson, F.M. and Gerard, R. (1982) Flood waves caused by ice jam formation and failure, *IAHR*, **1:** 277–97.

Hess, J.L. and Smith, A.M.O. (1967) Calculation of potential flow about arbitrary bodies In *Progress in Aeronautical Sciences*, Pergamon Press, Oxford, **8:** 1–138.

Hillel, D. (1980a) *Fundamentals of soil physics*, Academic Press, New York.

Hillel, D. (1980b) *Applications of soil physics*, Academic Press, New York.

Hillig, W.B. (1958) The kinetics of freezing of ice in the direction perpendicular to the basal plane, In *Growth and Perfection of Crystals*, (ed. Doremus, A.), J. Wiley, New York: 350–60.

Hirata, T. (1987) Recent advances in the study of formation of ice-band structure in water-flow pipe, *Proc. Int. Symp. on Cold Regions Heat Transfer*, ASME, New York: 39–45.

Hirata T., Gilpin, R.R. and Cheng, K.C. (1979a) The steady state ice layer profile on a constant temperature plate in a forced convection flow I. Laminar regime, *Int. J. Heat & Mass Trans*, **22:** 1425–33.

Hirata, T., Gilpin, R.R. and Cheng, K.C. (1979b) The steady state ice layer profile on a constant temperature plate in a forced convection flow II. The transition and turbulent regimes. *Int. J. Heat and Mass Trans*, **22:** 1435–43.

Hirth, J.P., Pound, G.M. and St. Pierre, G.R. (1970) Bubble nucleation, *Metallurgical Trans* **1:** 939–45.

Hobbs, P.V. (1973) Ice in the atmosphere: a review of the present position. In *Physics and chemistry of ice*, Roy. Soc. Can., Ottawa: 308–19.

Hobbs, P.V. (1974) *Ice Physics*, Oxford University Press, Oxford.

Hobbs, P.V. and Alkezweeny, J. (1968) The fragmentation of freezing water droplets in free fall, *J. Atmos. Sci.*, **25:** 881–8.

Hobbs, P.V. and Mason, B.J. (1964) The sintering and adhesion of ice, *Phil. Mag.*, **9:** 181–97.

Hobbs, P.V. and Scott, W.D. (1965) A theoretical study of the variation of ice crystal habits with temperature, *J. Geophys. Res.* **70:** 5025–34.

Hoinkes, H. (1954) Beitrage zur Kenntris des Gletscherwindes, *Arch. f. Met. (B)* **6:** 36–53.

Horiguchi, K. and Miller, R.D. (1983) Hydraulic conductivity functions of frozen material, *Proc. 4th Int. Perm. Conf.* Nat. Acad. Sci., Washington: 504–8.

Horjen, J. (1981) Ice accretion on ships and marine structures, Marine Structures and Ships in Ice, Norwegian Hydrodynamic Laboratories, Report No. 81–02.

Horjen, J. and Vefsnmo, S. (1983) Offshore icing-phase II, extended theory of sea spray icing, Offshore Tech. Testing and Res. Group Rep. STF88–83050.

Horner, R.A. (1985) (ed) *Sea Ice Biota*, CRC Press, Boca Raton, Florida.

Hwang, G.J. and Yih, J. (1973) Correction on the length of ice-free zone in a convectively-cooled pipe, *Int. J. Heat Mass Trans.*, **16:** 681–3.

Idle, D.B. (1968) Ice in plants, *Science*, **4:** 59–63.

Illingworth, V. (ed) (1985) *The MacMillan dictionary of astronomy*, 2nd edn, MacMillan Press, London.

Imbrie, J. and Imbrie, K.P. (1979) *Ice ages: solving the mystery*, Harvard University Press, Cambridge, Mass.

Isono, K. and Iwai, K. (1969) Growth mode of ice crystals in air at low pressure, *Nature*, 1149–50.

Itagaki, K. (1984) Icing rate on stationary structures under marine conditions, US Army Cold Regions Research and Engineering Laboratory, Rep: 84–12.

Jaakkola, Y., Laiho, J. and Vuorenvirta, M. (1983) Ice accumulation on tall radio and TV towers in Finland, Proc. of First International Workshop on Atmospheric Icing of Structures (Ed: Minsk, L.D.) US Army Cold Regions Research and Engineering Laboratory, Hanover, Special Rep. 83–17: 249–60.

Jackson, F. (1964) The solution of problems involving the melting and freezing of finite slabs by a method due to Portnov, *Proc. Edin. Math. Soc.* **14** II (2): 109–28.

Jackson, K.A. (1958) *Liquid Metals and Solidification*, American Soc. for Metals, Cleveland: 174.

Jackson, K.A., Uhlmann, J. and Hunt, J.D. (1967) The nature of crystal growth from the melt, *J. Crystal Growth* **1** (1): 1–36.

Jacobsen, I.A. and Pegg, D.E. (1984) Cryopreservation of organs: a review, *Cryobiology*, **21**: 377–84.

Jayaweera, K.O.L.F. and Mason, B.J. (1965) The behaviour of freely falling cylinders and cones in a viscous fluid, *J. Fluid Mech.* **22**: 709–20.

Jayaweera, K.O.L.F. and Mason, B.J. (1966) The falling motions of loaded cylinders and discs simulating snow crystals, *Q.J. Roy. Met. Soc.* **92**: 151–6.

Jellinek, H.H.G. (1959) Adhesive properties of ice, *J. Colloid Sci.* **14**: 268–79.

Jennings, A. (1977) *Matrix computation for engineers and scientists*, J. Wiley & Sons, London.

Jensen, L.B. (1943) Bacteriology of ice, *Food Research* **8**: 265–77.

Johnson, G.R., Smith, R.S. and Lock, G.S.H. (1969) Accumulation of material at the severed ends of myelinated nerve fibres, *Amer. J. Physiol.*, **217** (1): 188–91.

Johnson, T.V. (1982) The Galilean satellites, In *The new solar system* (eds. Beatty, J.K., O'Leary, B. and Chaikin, A.,) 2nd edn Cambridge University Press, Cambridge: 143–160.

Joliffe, I. and Gerard, R. (1982) Surges released by ice jams, *Proc. Workshop on the Hydrology of Ice-covered Rivers*, Edmonton Nat. Res. Co. Canada, Ottawa: 253–9.

Josberger, E.G. (1977) A laboratory and field study of iceberg deterioration. In *Iceberg utilization*, Proc. 1st Int. Conf. (ed. Husseiny, A.A.) Pergamon Press, New York: 245–64.

Jumikis, A.R. (1962) *Soil mechanics*, Van Nostrand, Princeton, New Jersey.

Juntilla, O. and Stushnoff, C. (1977) Freeze avoidance by deep supercooling in hydrated lettuce seeds, *Nature*, **269**: 325–7.

Kamb, B. (1970) Superheated ice, *Science*, **169**: 1343–4.

Kane, D.L. and Slaughter, C.W. (1972) Seasonal regime and hydrological significance of stream icings in central Alaska. In *The role of snow and ice in hydrology: Proc. Banff Symp.* UNESCO, Geneva **1**: 528–40.

Kass, M. and Magun, S. (1961) Zur uberhitzung am, phasenubergangfest – flussig, *Z. Kristall.*, **116**: 354–70.

Kay, B.D., Hons, D.B. and Coit, J.B. (1975) Thermophysical characterization of the surface tier of an organic soil, *Proc. Conf. on Soil-Water Problems in Cold Regions*, Div. Hydrol., American Geophysical Union, Washington.

Kedem, O. and Katchalsky, A (1958) Thermodynamic analysis of the permeability of biological membranes to non-electrolytes, *Biochim Biophys Act*, **27**: 229–46.

Kepler, J. (1611) A new year's gift, or on the six-cornered snowflake, G. Tampach, Frankfurt, *The six-cornered snowflake*, trans. C. Hardie, Oxford University Press, Oxford (1966).

Ketcham, W.M. and Hobbs, P.V. (1968) Step growth on ice during the freezing of pure water, *Phil. Mag.* **18**: 659–61.

Kinosita, S. and Fukuda, M., (eds) (1985) *Ground Freezing*, Proc. 4th Int. Symp. on Ground Freezing (Sapporo), A.A. Balkema, Rotterdam.

Knight, C.A. (1962) Studies of Arctic lake ice, *J. Glaciol.*, **4**: 319–35.

Knight, C.A. (1968) On the mechanism of spongy hailstone growth, *J. Atmos. Sci.*, **25**: 440–4.

Knight, C.A. (1979) Observations of the morphology of melting snow, *J. Atmos. Sci.*, **36**: 1123–30.

Knight, C.A. and Knight, N.C. (1968) Spongy hailstone growth criteria I orientation fabrics, *J. Atmos. Sci.*, **25**: 445–52.

Knight, C.A. and Knight, N.C. (1970) Lobe structure of hailstones, *J. Atmos. Sci.*, **27**: 667–71.

Kobayashi, T. (1957) Experimental researches on the snow crystal habit and growth by means of a diffusion cloud chamber, *J. Met. Soc. Japan* **75**: 38–44.

Kobayashi, T. (1961) The growth of snow crystals at low supersaturation, *Phil. Mag.* **6**: 1363–70.

Kobayashi, T. (1967) On the variation of ice crystal habit with temperature. In *Physics of Snow and Ice*, Institute of Low Temperature Science, Hokkaido University, Sapporo: 95–104.

Kohnke, H. (1968) *Soil Physics*, McGraw-Hill, New York.

Konrad, J-M. (1980) Frost heave mechanics, Ph.D Thesis, Dept. Civil Eng., University of Alberta.

Konrad, J-M. and Morgenstern, N.R. (1980) A mechanistic theory of ice lense formation in fine-grained soils, *Can. Geotech. J.*, **17**: 473–86.

Koopmans, R.W.R. and Miller, R.D. (1966) Soil water and soil freezing characteristic curves, *Soil Sci. Soc. Amer. Proc.*, **30**: 680–5.

Korber, C. and Ray, G. (1987) Ice crystal growth in aqueous solutions. In *The biophysics of organ preservation* , (eds. Pegg, D.E. and Karow, A.M.,) Plenum Press, London: 173–200.

Korn, G.A. and Korn, T.M. (1961) *Mathematical handbook for engineers and scientists*, McGraw-Hill, New York: 742.

Krass, M.S. (1984) Ice on planets of the solar system, *J. Glaciol.*, **30** (106): 259–74.

Krog, J.O., Zachariassen, K.E., Larsen, B. and Smidsrod, O. (1979) Thermal buffering in Afro-Alpine plants due to nucleating agent-induced water freezing, *Nature*, **282**: 300–1.

Kuhlmann-Wilsdorf, D. (1965) Theory of melting, *Phys. Rev.*, **140**: 5A 1599–1610.

Kuroiwa, D. (1965) Icing and snow accretion of electric wires, US Army Cold Regions Research and Engineering Laboratory, Res. Rep. 123.

Kushner, A.S. and Walston, W.H Jr. (1977) Conduction and natural convection heat transfer in a phase change region, *Symp. on Comput. Technol. for Interface Prob.* App. Mech Div. ASME, New York, **30**: 1–18.

Lacey J.J.Jr. (1973) Turbine engine icing and ice detection, ASME paper 72–GT–6.

Lagoni, H. and Peters, K.H. (1961) Die getrierdestabilisierung von milch und rahm, ein problem der warmeleitfahigkeit, *Milchwissenschaft* **16**: 197–200.

Lamb, D. (1970) Growth rates and habits of ice crystals grown from the vapour phase, Ph.D thesis, Univ. Washington, Seattle.

Lamb, D. and Hobbs, P.V. (1971) Growth rates and habits of ice crystals grown from the vapour phase, *J. Atmos. Sci.* **28**: 1506–9.

Lamb, H.H. (1969) Climatic fluctuations. In *World survey of climatology* , 2, *General climatology* (ed. Flohn, H.), Elsevier, Amsterdam: 173–249.

Lamé, G. and Clapeyron, B.P. (1831) Sur la solidification par refroidissement d'un globe liquide, *Annal. der Phys. und Chem.*, **47**: 250–6.

Landau, H.G. (1950) Heat conduction in a melting solid, *Q.J. Appl. Math.* **8** (11): 81–94.

Langham, E.J. (1981) Physical properties of snowcover. *In Handbook of snow* (eds. Gray, D.M. and Male, D.H.), Pergamon Press, Toronto: 275–337.

Langhorne, P.J. (1982) Crystal alignment in sea ice, Ph.D. thesis, University of Cambridge.

Langleben, M.P. (1972) The decay of an annual cover of sea ice, *J. Glaciol.*, **11**: 337–44.

Langmuir, I. and Blodgett, K.B. (1946) A mathematical investigation of water droplet trajectories, US Army Air Forces Tech. Rep. 5418.

Larkin, B.S. and Dubuc, S. (1976) Self-deicing navigation buoys using heat pipes, European Space Agency, ESA SP112 I: 529–35.

Laws, R.M. (ed) (1984) *Antarctic Ecology* 1 and 2, Academic Press, New York.

Layne, J.R. and Lee, R.E. Jr., (1987) Biophysical and environmental parameters influencing ice formation in the body fluids of anurans, *Proc. 24th Meeting*, Society for Cryobiology, Edmonton.

Leibo, S.P., Mazur, P. and Jackowski, S.C. (1974) Factors affecting survival of mouse embryos during freezing and thawing, *Exp. Cell. Res.* **89**: 79–88.

Levi, L., Achaval, E. and Aufdermauer, A.N. (1970) Crystal orientation in a wet growth hailstone, *J. Atmos. Sci* . **27**: 512–13.

Levi, L. and Aufdermauer, A.N. (1970) Crystallographic orientation and crystal size in cylindrical accretions of ice, *J. Atmos. Sci.* **27**: 443–52.

Levi, L. and Prodi, F. (1978) Crystal size in ice grown by droplet accretion, *J. Atmos. Sci.*, **35**: 2181–9.

Levin, L.M. (1954) Distribution function of clouds and raindrops by sizes, *Dokl. Akad. Nauk. SSSR* **94**: 1045.

Lewis, J.S. (1982) Putting it all together. In *The new solar system* (eds. Beatty, J.K., O'Leary, B. and Chaikin, A.) 2nd edn. Cambridge University Press, Cambridge, 205–12.

Lindow, S.E. (1983) The role of bacterial ice nucleation in frost injury to plants, *Ann. Rev. Phytopathol.*, **21**: 363–84.

Ling, G.R. and Tien, C.L. (1969) Analysis of cell freezing and dehydration, ASME paper 69–WA/HT–31.

List, R. (1960) *Growth and Structure of Graupels and Hailstones* Geophys. Mono. 5, American Geophysical Union Washington: 317.

List, R. (1965) Components contributing to the heat exchange of growing hailstones *Proc. Cloud Phys. Conf.*, Met. Soc. of Japan, Tokyo: 286–90.

List, R., MacNeil, C.F. and McTaggart-Cowan, J. (1970) Laboratory investigations of temporary collisions of raindrops, *J. Geoph. Res.* **75**: 7573–80.

List, R. and Schemenauer, R.S. (1971) Free-fall behaviour of planar snow crystals, conical graupel and small hail, *J. Atmos. Sci.* **28**: 110–15.

Locatelli, J.D. and Hobbs, P.V. (1974) Fallspeeds and masses of solid precipitation particles, *J. Geophys. Res* **79**: 2185–97.

Loch, J.P.G. (1980) Frost action in soils: state-of-the-art, *2nd Int. Symp. Ground Freezing, Eng. Geol*, **18**: 213–24.

Lock, G.S.H. (1969a) On the use of asymptotic solutions to plane ice-water problems, *J. Glaciol.* **8** (53): 285–300.

Lock, G.S.H. (1969b) Elastic expulsion from a long tube, *Bull. Math. Biophys*, **31**: 295–306.

Lock, G.S.H. (1971) On the perturbation solution of the ice-water layer problem, *Int. J. Heat Mass Trans.*, **14**: 642–4.

Lock, G.S.H. (1972) Some aspects of ice formation with special reference to the marine environment, *Transactions, North East Coast Institution of Engineers & Shipbuilders*, **88**: 175–84 and D57–D62.

Lock, G.S.H. (1974) The growth and decay of ice, *Proc. 5th Int. Heat Transfer Conf.*, Tokyo, Japan Soc. of Mech. Engineers, Tokyo.

Lock, G.S.H. (1986) The features of ice, Letter to the editor, *J. Glaciol*, **32** (110): 135.

Lock, G.S.H. and Foster, I.B. (1989) Observations on the formation of spongy ice from fresh water, *Proc. 2nd Int. Symp. on Cold Regions Heat Transfer*, Hokkaido University, Hokkaido.

Lock, G.S.H., Freeborn, R.D.J. and Nyren, R.H. (1970) Analysis of ice formation in a convectively – cooled pipe, *Proc. 4th Int. Heat Trans. Conf. Versailles*, Elsevier, Amsterdam: Cu 2.9.

Lock, G.S.H., Gunderson, J.R., Quon, D. and Donnelly, J.K. (1969) A study of one-dimensional ice formation with particular reference to periodic growth and decay, *Int. J. Heat Mass Trans.* **12**: 1343–52.

Lock, G.S.H. and Kaiser, T.M.V. (1985) Icing on submerged tubes: a study of occlusion, *Int. J. Heat Mass Trans.* **28** (9): 1689–98.

Lock, G.S.H. and O'Callaghan, W.J. (1989) Flow through a row of closely-spaced circular cylinders normal to a cold water stream, *Trans. Can. Soc. Mech. Eng.* **12** (3): 129–47.

Lockwood, F.C. (1966) Simple numerical procedure for the digital computer solution of non-linear transient heat conduction with changes of phase, *J. Mech. Eng. Sci.* **8** (3): 259–63.

Longwell, C.R., and Flint, R.F. (1969) *Physical Geology*, J. Wiley, New York.

Lorius, C. (1973) Les calottes glaciaires remoins de l'environnement, *La Recherche* **4**: 457–72.

Low, P.F., Hoekstra, P. and Anderson, D.M. (1968) Some thermodynamic relationships for soil at or below the freezing point: 2. Effects of temperature and pressure on unfrozen soil water, *Water Resources Res* **4** (3): 541–4.

Lowe, R. (1983) *Kangiryuarmiut Uqauhingita Numiktiltitdjutingit*, Committee for Original Peoples Entitlement, Inuvik.

Lozina-Lozinskii L.K. (1974) *Studies in cryobiology*, (trans. P. Harry), J. Wiley & Sons, New York.

Lozowski, E.P. and Gates, E.M. (1984) Ice accretion on structures in a marine environment, ASME, New York, 84–Pet–2.

Lozowski, E.P. and Oleskiw, M.M. (1983) Computer modelling of time-dependent rime icing in the atmosphere, US Army Cold Regions Research and Engineering Laboratory Rep. 83–2.

Lozowski, E. P., Stallabrass, J. R. and Hearty, P. F. (1983) The icing of an unheated, non-rotating cylinder, Parts I and II, *J. Clim. and App. Met.* **22** (12): 2053–62, 2063–74.

Ludlam, F. H. (1958) The hail problem *Nubila* **1:** 12–96.

Lunardini, V. J. (1987) Some analytical methods for conduction heat transfer with freezing/thawing, *Proc. Int. Symp. on Cold Regions Heat Transfer*, ASME, New York: 55–64.

Luyet, B. J. (1960) On the mechanism of growth of ice crystals in aqueous solutions and on the effect of rapid cooling in hindering crystallisation, In *Recent Research in Freezing and Drying*, (eds. Parkes, A. S. and Smith, A. U.) Blackwell, Oxford: 3–22.

Luyet, B. J. (1966) Anatomy of the freezing process in physical systems, In *Cryobiology*, (ed. Meryman, H. T.) Academic Press, New York: 115–38.

Luyet, B. J. and Rapatz, G. (1958) Patterns of ice formation in some aqueous solutions, *Biodynamica* **8:** 1–68.

Lyons, J. M., Raison, J. H and Steponkus, P. L. (1979) The plant membrane in response to low temperature: an overview, In *Low Temperature Stress in Crop Plants*, Academic Press, New York: 1–24.

Lyons, J. B. and Stoiber, R. E. (1959a) Orientation fabrics in lake ice, *J. Glaciol.*, **4** (32): 367–370.

Lyons, J. B. and Stoiber, R. E. (1959b) The absorptivity of ice: a critical review, Scient. Rep. #3, Air Force Cambridge Research Centre Tech. Note 59–656., Dartmouth College, Hanover.

MacArthur, C. D. (1983) Numerical simulation of airfoil ice accretion, AIAA paper 83–0112.

MacFarland, J. D. and Dranchuk, P. M. (1976) Visualization of the transition to turbulent flow in porous media, *Technology*, April–June: 71–8.

MacFarlane, D. R. Kadiyala, R. K. and Angell, C. A. (1983) Homogeneous nucleation and growth of ice from solutions. *TTT* curves, the nucleation rate and the stable glass criterion, *J. Chem. Phys.* **79:** 3921–7.

MacKay, J. R. (1965) Gas-domed mounds in permafrost, Kendall Island, NWT, *Geog. Bull.*, **7:** 105–15.

MacKay, J. R. (1966) Segregated epigenetic ice and slumps in permafrost, MacKenzie Delta area, NWT, *Geog. Bull.*, **8:** 59–80.

MacKay, J. R. (1971) The origin of massive icy beds in permafrost, Western Arctic Coast, Canada, *Can. J. Earth Sci.*, **8** (4): 397–422.

MacKay, J. R. (1972) The world of underground ice, *Ann. Assoc. Amer. Geog.*, **62** (1): 1–22.

MacKay, J. R. (1973) The growth of pingos, Western Arctic Coast, Canada, *Can. J. Earth Sci.*, **10:** 979–1004.

MacKay, J. R. (1987) Some mechanical aspects of pingo growth and failure, Western Arctic Coast, Canada, *Can. J. Earth Sci.*, **24** (6): 1108–19.

MacKenzie, A. P. (1975) The physico-chemical environment during the freezing and thawing of biological materials, In *Water Relations of Foods*, (ed. Duckworth, R. B.) Academic Press, New York: 477–504.

Macklin, W. C., and Ludlam, F. E. (1961) The fall speeds of hailstones, *Q.J. Roy. Met. Soc.* **87:** 72–81.

Macklin, W. C. and Ryan, B. F. (1966) Habits of ice grown in supercooled water and aqueous solutions, *Phil. Mag.* **14:** 847–60.

Mae, S. (1976) Perturbations on disc-shaped internal melting figures in ice, *J. Crystal Growth* **32:** 137–8.

Magono, C. (1953) On the growth of snowflake and graupel, *Sci. Rep. Yokohama Nat. Univ.* **1** (2): 18–40.

Magono, C. and Lee, C. (1966) Meteorological classification of natural snow crystals, *J. Fac. Sci. Hokk. U. Ser. VIII*, **2:** 321–35.

Makkonen, L. (1984a) Atmospheric icing on sea structures, US Army Cold Regions Research and Engineering Laboratory, Monog. 84–82.

Makkonen, L. (1984b) Heat transfer and icing on a rough cylinder, *Cold Reg. Sci and Tech.*, **10:** 105–16.

Makkonen, L. (1984c) Modeling of ice accretion on wires, *J. Clim. and App. Meteor.* **23:** 929–39.

Makkonen, L. (1986a) Salt entrapment in spray ice, *Symposium on Ice*, Int. Ass. for Hydraulic Research, Iowa, II: 165–78.

Makkonen, L. (1986b) The effect of conductor diameter on ice load as determined by a numerical icing model, *3rd Int. Workshop on Atmospheric Icing of Structures, Vancouver*, Can. Electrical Ass., Ottawa.

Marsland, T. P. (1987) The design of an electro-magnetic rewarming system for cryopreserved tissue. In *The Biophysics of Organ Cryopreservation* (eds. Pegg, D. E. and Karow, A. M.) Plenum Press, London: 367–86.

Marth, E. H. (1973) Behaviour of food micro-organisms during freeze-preservation. In *Low Temperature Preservation of Foods and Living Matter*, Marcel Dekker, New York: 386–436.

Martin, S. (1981) Frazil ice in rivers and oceans, *Ann Rev. Fluid Mech.* **13:** 379–97.

Martin, S. and Kauffman, P. (1974) The evolution of under-ice melt ponds, or double diffusion at the freezing point, *J. Fluid Mech.*, **64** (3): 507–27.

Mason, B. J. (1953) The growth of ice crystals in a super-cooled water cloud, *Q.J. Roy. Met. Soc.*, **79:** 104–11.

Mason, B. J. (1960) Ice-nucleating properties of clay minerals and stony meteorites, *Q.J. Roy. Met. Soc.* **86:** 552–6.

Mason, B. J. (1971) *Physics of Clouds*, Oxford University Press, Oxford.

Mason, B. J., Bryant, G. W. and Van Den Heuval, A. P. (1963) The growth habits and surface structure of ice crystals, *Phil. Mag.* **8:** 505–26.

Masursky, H. (1982) Mars, In *The New Solar System* (eds. Beatty, J. K., O'Leary, B. and Chaikin, A.,) 2nd edn. Cambridge University Press, Cambridge: 83–92.

Mathias, S. F., Franks, F. and Trafford, K. (1984) Nucleation and growth of ice in deeply undercooled erythrocytes, *Cryobiology* **21** (2): 123–32.

Maykut, G. A. (1986) The surface heat and mass balance, In *The Geophysics of Sea Ice* (ed. Untersteiner, N.) Plenum Press, New York.

Maykut, G. A. and Untersteiner, N. (1971) Some results from a time–dependent thermodynamic model of sea ice, *J. Geophys. Res.* **76** (6): 1550–75.

Mazur, P. (1961) Manifestations of injury in yeast cells exposed to subzero temperatures, I. Morphological changes in freeze-substituted and 'frozen-thawed' cells, *J. Bact.* **82:** 662–72.

Mazur, P. (1963) Kinetics of water loss from cells at sub-zero temperatures and the likelihood of intra-cellular freezing, *J. Gen. Physiol.*, **47:** 347–69.

Mazur, P. (1966) Physical and chemical basis of injury in single-celled micro-organisms subjected to freezing and thawing, In *Cryobiology*, (ed. Meryman, H. T.) Academic Press, New York: 213–316.

Mazur, P. (1977) Slow freezing injury in mammalian cells. In *The Freezing of Mammalian Embryos* (eds. Elliott, K. and Whelan, J.) Elsevier, Amsterdam.

Mazur, P., Leibo, S.P., Farrant, J., Chu, E.H.Y., Hanna, M.G. Jr, and Smith, L.H. (1970) Interactions of cooling rate, warming rate and protective additive on the survival of frozen mammalian cells, *CIBA Foundation Symposium on the Frozen Cell* (eds. Wolstenholme, G.E.W. and O'Connor, M.), J & A Churchill, London: 69–88.

Mazur, P., Rigopoulos, N. and Cole, K.W. (1982) Contribution of unfrozen fraction and of salt concentration to the survival of slowly frozen human erythrocytes: influence of cell concentration, *Cryobiology*, **19**: 679.

McCarthy, D.F. (1977) *Essentials of soil mechanics and foundations* , Reston Publishing Company, Reston, Virginia.

McDonald, J.E. (1953) Homogeneous nucleation of supercooled water drops, *J. Met.*, **10**: 416–33.

McGaw, R.W. and Tice, A.R. (1976) A simple procedure to calculate the volume of water remaining unfrozen in a freezing soil, *Proc. 2nd Conf. on Soil-Water Problems in Cold Regions*, Hydrol. Div., American Geophys Union, Washington.

McGrath, J.J. (1981) Thermodynamic modelling of membrane damage, In *Effects of low temperature on biological membranes* (eds; Morris, G.J. and Clarke, A.) Academic Press, New York: 335–77.

McRoberts, E.C. (1978) Slope stability in cold regions, In *Geotechnical engineering in cold regions* (eds. Andersland, O.B. and Anderson, D.M.) McGraw-Hill, New York: 282–404.

Menegalli, F.C. and Calvelo, A. (1979) Dendritic growth of ice crystals during the freezing of beef, *Meat Science* **3**: 179–98.

Meryman, H.T. (1966) (ed) *Cryobiology*, Academic Press, New York.

Michel. B. (1971) Winter regimes of rivers and lakes, US Army Cold Regions Research and Engineering Laboratory, Mono. III Dla.

Michel, B. (1984) Comparison of field data with theories on ice cover progression in large rivers, *Can. J. Civ. Eng.*, **11**: 798–814.

Michel, B. and Drouin, M. (1981) Backwater curves under ice cover of LaGrande River, *Can. J. Civ. Eng.*, **8** (3): 351–63.

Michel, B. and Ramseier, R.O. (1971) Classification of river and lake ice, *Can. Geotech. J.*, **8**: 36–45.

Michelmore, R.W. and Franks, F. (1982) Nucleation rates of ice in undercooled water and aqueous solutions of polyethylene glycol, *Cryobiology*, **19** (2): 163–71.

Miksch, E.S. (1969) Solidification of ice dendrites in flowing supercooled water, *Trans. Metallurgical Society of AIME*, **245**: 2069–72.

Milankovitch, M. (1941) Kanon der Erdbestrahlung und seine Andwendurig auf das Eiszeiton Problem, Royal Serb. Acad., Belgrade, Spec. Publ. **133**: 1–633.

Miller, E.E. (1980) Similitude and scaling of soil-water phenomena, In *Applications of Soil Physics* (ed. Hillel D.) Academic Press, New York.

Miller, R.D. (1966) Phase equilibria and soil freezing, In *Permafrost: Proc. Int. Conf.*, National Acad. Sci., Washington: 193–7.

Miller, R.D. (1973) Soil freezing in relation to pore water pressure and temperature, *2nd Int. Conf. on Permafrost*, Nat. Acad. Sci, Washington: 344–352.

Miller, R. D. (1978) Frost heaving in non-colloidal soils, *Proc. 3rd Int. Conf. Perm*, Nat. Res. Co. Canada, Ottawa, Publ. 16529: 707–13.

Miller, R. D. (1980) Freezing phenomena in soils, In *Applications of Soil Physics* (ed Hillel, D.) Academic Press, New York.

Milne-Thomson, L. M. (1968) *Theoretical Hydrodynamics*, 5th edn, MacMillan, New York: 172.

Minnaert, M. (1954) *The Nature of Light and Colour in the Open Air*, Dover Publications, New York.

Minsk, L. D. (1977) Ice accumulation on ocean structures, US Army Cold Regions Research and Engineering Laboratory, Report 77–17.

Mitchell J. M. Jr. (1977) The changing climate, In *Energy and Climate*, Studies in Geophysics, Nat. Acad. Sci., Washington, 51–8.

Mohr, W. P. and Stein, M. (1969) Effect of different freeze–thaw regimes on ice formation and ultrastructural changes in tomato fruit parenchyma tissue, *Cryobiology* **6**: 15–31.

Moretti, P. M. and Kays, W. M. (1965) Heat transfer to a turbulent boundary layer with varying free-stream velocity and varying surface temperature – an experimental study, *Int. J. Heat Mass Transf.*, **8**: 1187–202.

Morgenstern, N. R. and Nixon, J. F. (1971) One-dimensional consolidation of thawing soils, *Can. Geotech. J.*, **8** (4): 558–65.

Morris, G. J. (1980) Plant cells, In *Low Temperature Preservation in Medicine and Biology* (eds. Ashwood-Smith, M. J. and Farrant, J.), Pitman Medical, Tunbridge Wells: 253–84.

Morrison, D. and Cruikshank, D. P. (1982) The outer solar system. In *The New Solar System* (eds. Beatty, J. K., O'Leary, B. and Chaikin, A.,) 2nd edn, Cambridge University Press, Cambridge: 167–76.

Mossop, S. C. (1976) Production of secondary ice particles during the growth of graupel by riming, *Q. J. Roy. Met. Soc.*, **102**: 45–57.

Mossop, S. C. and Hallett, J. (1974) Ice crystal concentration in cumulus clouds: influence of the drop spectrum, *Science*, **186**: 632–3.

Muller, F. (1976) On the thermal regime of a high-arctic valley glacier, *J. Glaciol.*, **16**: 119–33.

Mulligan, J. C. and Jones, D. D. (1976) Experiments on heat transfer and pressure drop in a horizontal tube with internal solidification, *Int. J. Heat Mass Trans.*, **19**: 213–19.

Murray, W. D. and Landis, F. (1959) Numerical and machine solutions of transient heat-conduction problems involving melting or freezing, *J. Heat Trans.*, **81**: 106–12.

Nakaya, U. (1950) *Snow Crystals*, Harvard University Press, Cambridge, Mass.

Nakaya, U. (1956) Properties of single crystals of ice revealed by internal melting, SIPRE (now Cold Regions Research and Engineering Laboratory) Res. Rep. 13.

Nakaya, M., Hanajima, M. and Muguruma, J. (1958) Physical investigations on the growth of snow crystals, *J. Fac. Sci. Hokkaido Univ. Ser. II*, **5**: 87–118.

Nei, T. (1954) Freezing process of yeast cells, Part I, *J. Agri. Chem. Soc. Japan*, **28**: 91–4.

Nei, T. (1960) Effects of freezing and freeze-drying on micro-organisms, In *Recent Research in Freezing and Drying* (eds: Parkes, A. S and Smith, A. V.) Blackwell, Oxford: 78–86.

Nei, T. (1963) In *Culture Collections: Perspectives and Problems* (ed: Martin, S. M.) University of Toronto Press, Toronto: 74–6.

Norrie, D.H. and DeVries, G. (1973) *The Finite Element Method: Fundamentals and Applications*, Academic Press, New York.

Norrish, K. and Rausell-Colom, J.A. (1962) Effect of freezing on the swelling of clay minerals, *Clay Minerals Bull.*, **5**: 9–16.

Nye, J.F. (1967) Theory of regelation, *Phil. Mag.*, **16** (144): 1249–66.

Nye, J.E. and Frank, F.C. (1969) Hydrology of the inter-granular veins in a temperate glacier, Int. Assoc. Sci. Hydrology, Pub. 95, Cambridge.

O'Callaghan, M.G., Cravalho, E.G. and Huggins, C.E. (1982) An analysis of the heat and solute transport during solidification of an aqueous binary solution, *Int. J. Heat Mass. Transfer*, **25**: 553–73.

O'Neal, D.L. and Tree, D.R. (1984) Measurement of frost growth and density in a parallel plate geometry, *Trans. ASHRAE.*, **2A**: 278–90.

O'Neill, K. (1983) The physics of mathematical frost heave models: a review, *Cold Regions Sci. and Tech.*, **6**: 275–91.

Okihara, T., Kanamori, M., Kamimura, A., Hamada, N., Matsucda, S., Buturlia, J. and Miskolczy, G. (1980) Design, testing and shipboard evaluation of a heat pipe de-icing system, *Proc. 15th Thermophysics Conf.*, AIAA, Colorado.

Oleinik, O.A. (1960) A method of solving the general Stefan problem (in Russian) *Dokl. Akad. Nauk. SSSR*, **135**: 1054–7.

Olien, C.R. (1961) A method of studying stresses occurring in plant tissue during freezing, *Crop Science*, **1**: 26–8.

Olien, C.R. (1971) A comparison of desiccation and freezing as stress vectors, *Cryobiology*, **8** (3): 244–8.

Olien, C.R. (1974) Energies of freezing and frost desiccation, *Plant Physiol.*, **53**: 764–7.

Onik, G., Cobb, C., Diamond, D., Steele, G., Cady, B., Kane, R., Rubinsky, B. and Porterfield, B. (1987) Cryosurgery for unresectable hepatic tumours, *Proc. 24th Meeting*, Society for Cryobiology, Edmonton.

Ono, A. (1969) The shape of riming properties of ice crystals in natural clouds, *J. Atmos. Sci.*, **26**: 138–47.

Openchowski, S. (1883) Sur l'action localisée du froid appliqué à la surface de la région corticale du cerveau, *Compt. Rend. Soc. de Biol.*, **5**: 38–43.

Osterkamp, T.E. (1978) Frazil ice formation: a review, *J. Hyd. Div. ASCE*, **104**: 1239–55.

Outcalt, S.I. (1976) A numerical model of ice lensing in freezing soils, *Proc. 2nd Conf. on Soil-Water Problems in Cold Regions*, Hydrol. Div., American Geophysical Union, Washington: 63–74.

Outcalt, S.I. and Carlson, J.H. (1975) A coupled soil thermal regime surface energy budget simulator, *Proc. Conf. on Soil-Water Problems in Cold Regions*, Hydrol. Div., American Geophysical Union, Washington.

Palosuo, E. (1961) Crystal structure of brackish and freshwater ice, Int. Union Geod. Geophys. Gen. Assemb. (Snow & Ice) IASH Pub 54: 9–114.

Paré, A., Carlson, L.E., Bourns, M. and Karin, N. (1987) The use of an additive in sprayed sea water to accelerate ice structure construction. *Proc. Int. Symp. on Cold Regions Heat Transfer, Edmonton*, ASME, New York: 123–9.

Paterson, W.S.B. (1981) *The Physics of Glaciers*, 2nd edn, Pergamon Press, Oxford.

Pearcey, T. and Hill, G.W. (1956) The accelerated motion of droplets and bubbles, *Aust. J. Physics*, **9**: 19–30.

Pegg, D. E., Diaper, M. P., Skaer, H. leB. and Hunt, C. J. (1984) The effect of cooling rate and warming rate on the packing effect in human erythrocytes frozen and thawed in presence of 2 M Glycerol, *Cryobiology*, **21**: 491–502.

Peierls, R. (1929) Zur Kinetischen Theorie der Warmeleitung in Kristallen, *Ann. Phys.* **3**: 1055–101.

Péwé, T. L. (1983) The periglacial environment in North America during Wisconsin time, In *The Late Pleistocene* (ed. Porter, S. C.) University of Minnesota Press, Minneapolis: Ch 9.

Pfannkuch, H. O. (1962) Contribution à l'étude des deplacements des fluides miscibles dans un milieu poreux, Inst. Francais du Petrole, Paris.

Philberth, K. and Federer, B. (1971) On the temperature profile and the age profile in the central part of cold ice sheets, *J. Glaciol.*, **10** (58): 3–14.

Philip, J. R. (1978) Water on the earth. In *Water: Planets, Plants and People*, Aust. Acad. Sci, Canberra: 35–59.

Pippard, A. B. (1957) *Elements of Classical Thermodynamics*, Cambridge University Press, Cambridge.

Pitter, R. L. (1977) A re-examination of riming on thin ice plates, *J. Atmos. Sci.*, **34**: 684–5.

Pitter, R. L. and Pruppacher, H. R. (1974) A wind tunnel investigation of freezing of small water drops falling at terminal velocity in air *Q. J. Roy. Met. Soc.*, **99**: 540–50.

Planck, M. (1926) *Treatise on Thermodynamics*, 3rd english edn (Trans. A. Ogg) Dover Publications, New York.

Pohlhausen, K. (1921) Zur naherungsiveisen Integration der Differentialgleichung der laminaren Grenzschicht. *Z. angew. Math. Mech.*, **1**: 252–68.

Pollack, J. B. (1982a) Atmospheres of the terrestrial planets. In *The New Solar System* (eds. Beatty, J. K., O'Leary, B. and Chaikin, A.,) 2nd edn. Cambridge University Press, Cambridge: 57–60.

Pollack, J. B. (1982b) Titan, In *The New Solar System* (eds. Beatty, J. K., O'Leary, B. and Chaikin, A.,) 2nd edn, Cambridge University Press, Cambridge: 161–6.

Poots, G. (1962) On the application of integral methods to the solution of problems involving the solidification of liquids initially at fusion temperature, *Int. J. Heat Mass Transf.*, **5**: 525–31.

Porter, K. R. (1987) Structural organization of the cytomatrix. In *Organization of cell metabolism* (eds. Welch G. R. and Clegg J. S.) Plenum Press, New York: 9–14.

Porter, K. R., Berkele, M. and McNiven, A. (1983) The cytoplasmic matrix, *Mod. Cell. Biol.*, **2**: 259.

Portnov, J. G. (1962) Exact solution of freezing problem with arbitrary temperature variation on fixed boundary, *Sov. Phys. Doklady*, **7**: 186–9.

Pounder, E. R. (1965) *The Physics of Ice*, Pergamon Press, New York.

Powrie, W. D. (1973) Characteristics of fluid foods and their behaviour during freeze-preservation. In *Low Temperature Preservation of Foods and Living Matter*, Marcel Dekker, New York: 242–81.

Prandtl, L. (1957) *Fuhrer durch die Stromungslehre*, F Vieweg, Braunschweig.

Pratt, E. J. (1958) *The Collected Poems of E. J. Pratt*, 2nd edn (ed. Frye, N.), MacMillan Company of Canada, Toronto: 214.

Preobrazhenskii, L. Y. (1973) Estimate of the content of spray drops in the near-water layer of the atmosphere, *Fluid Mechanics – Sov Res* **2**: 95–100.

Pringle, M. J. and Chapman, D. (1981) Biomembrane structure and effects of temperature. In *Effects of low temperature on biological membranes*, (eds. Morris, J. and Clarke, A.) Academic Press, New York: 21–37.

Pruitt, W. O. (1960) Animals in the snow, *Scientific American*, **202** (1): 60–8.

Pruitt, W. O. (1978) *Boreal Ecology*, Edward Arnold, London.

Pruppacher, H. R. and Klett, J. D. (1980) *Microphysics of clouds and precipitation*, D. Reidel Pub, Dordrecht.

Quamme, H. A. and Gusta, L. V. (1987) Relationship of ice nucleation and water status to freezing patterns in dormant peach flower buds, *Hort. Sci.*, **22** (3): 465–7.

Quamme, H., Weiser, C. J and Stushnoff, C. (1973) The mechanism of freezing injury in xylem of winter apple twigs, *Plant Physiol.*, **51**: 273–7.

Ragle, R. H. (1963) Formation of lake ice in a temperature climate, US Army Cold Regions Research and Engineering Laboratory, Hanover, Res. Rep. 107.

Rapatz, G. L. and Luyet, B. J. (1960) Microscopic observations on the development of the ice phase in the freezing of blood, *Biodynamica* **8**: 195–239.

Rapatz, G. L., Menz, L. J. and Luyet, B. J. (1966) Anatomy of the freezing process in biological materials, In *Cryobiology* (ed. H. T. Meryman) Academic Press, New York: 139–62.

Rasmussen, R. M., Levizzani, V. and Pruppacher, H. R. (1984a) A wind tunnel and theoretical study of the melting behaviour of atmospheric icing particles II: a theoretical study for frozen drops of radius $< 500\,\mu m$, *J. Atmos. Sci.*, **41**: 374–80.

Rasmussen, R. M., Levizzani, V. and Pruppacher, H. R. (1984b) A wind tunnel and theoretical study of the melting behaviour of atmospheric icing particles III: experiment and theory for spherical ice particles of radius $> 500\,\mu m$, *J. Atmos. Sci.*, **41**: 381–8.

Raymond, C. F. and Harrison, W. D. (1975) Some observations on the behaviour of the liquid and gas phases in temperate glacier ice, *J. Glaciol.*, **14**: 213–33.

Richardson, M. (1975) *Translocation in plants*, 2nd edn, Edward Arnold, London.

Ring, R. A. (1980) Insects and their cells, In *Low temperature preservation in medicine and biology* (eds. Ashwood-Smith, M. J. and Farrant, J.) Pitman Medical, Tunbridge Wells: 187–218.

Ringwood, A. E. (1978) Water in the solar system, In *Water: Planets, Plants and People*, Aust. Acad. Sci., Canberra: 18–34.

Robe, R. Q. (1980) Iceberg drift and deterioration, In *Dynamics of Snow and Ice Masses* (ed. Colbeck, S. C.) Academic Press, New York, Ch. 4.

Roberts, C. G. D. (1974) *Selected Poems of Sir Charles G. D. Roberts*, (ed. Pacey, D.) McGraw-Hill Ryerson Ltd, Toronto: 38.

Robin, G. deQ. (1955) Ice movement and temperature distribution in glaciers and ice sheets, *J. Glaciol.*, **2** (18): 523–32.

Roedder, E. (1967) Metastable superheated ice in liquid-water inclusions under high negative pressures, *Science*, **155**: 1413–17.

Rose, H. E. (1945) (1) An investigation into the laws of flow of fluids through beds of granular material, (2) The isothermal flow of gases through beds of granular materials, (3) On the resistance coefficient –

Reynolds number relationship for fluid flow through a bed of granular
material, *Proc. Inst. Mech. Eng.*, **153:** 141–61.

Rosenhead, L. (ed) (1963) *Laminar Boundary Layers*, Oxford University Press,
Oxford: 292 *et seq.*

Rossini, F. D., Wagman, D. D., Evans, Wm.H., Levine, S. and Jaffe, I., (1952)
Selected values of chemical thermodynamic properties, US Nat. Bur. Stds.
Circular 500: 126–8.

Rubinsky, B. (1986) Cryosurgery imaging with ultrasound, *Mechanical
Engineering,* Jan: 48–52.

Rubinsky, B., Onik, G., Lee, C. and Bastacky, J. (1987) The mechanism of
damage during hepatic cryosurgery, *Proc. 24th Meeting*, Society for
Cryobiology, Edmonton.

Rumer, R. R. (1969) Resistance to flow through porous media, In *Flow through
Porous Media* (ed. De Wiest, D. J. M.) Academic Press, New York: 91–108.

Sakai, A. (1979) Freezing avoidance mechanism of primordial shoots of conifer
buds, *Plant and Cell Physiol.*, **20** (7): 1381–90.

Sakai, A. (1982) Freezing tolerance of shoot and flower primordia of coniferous
buds by extraorgan freezing, *Plant and Cell Physiol*, **23** (7): 1219–27.

Saline water conversion engineering data book (1965) M. W. Kellogg Company
for US Govt. Dept. Interior, Washington.

Salt, R., W. (1952) The influence of food on cold hardiness of insects, *The
Canadian Entomologist*, **84–85**: 261–9.

Salt, R. W. (1956) Cold-hardiness of insects, *Proc. 10th Int. Cong. Entom.
Montreal*, **2:** 73–7.

Salt, R. W. (1959) Role of glycerol in the cold-hardening of *bracon cephi*
(Gahan) *Can. Journ. Zool.*, **37:** 59–69.

Salt, R. W. (1961) Principles of insect cold-hardiness, *Ann. Rev. Entomology*, **6:**
55–74.

Samarskii, A. A. and Moiseynko, B. D. (1965) An economic continuous
calculation scheme for the Stefan multi-dimensional problem (in Russian) *Zh.
Vychisl. Mat. Mat. Fiz.*, **5:** 816–27.

Santeford, H. S. (1976) Effects of a snow pillow on heat and vapour transport in
the black spruce permafrost environment, *Proc. 2nd Conf. on Soil-Water
Problems in Cold Regions*, Hydrol. Div., American Geophysical Union,
Washington.

Schaefer, W. J. (1950) The formation of frazil and anchor ice in cold water,
Trans. Am. Geophys. U. **31** (6): 885–93.

Schlichting, H. (1968) *Boundary layer theory* 4th edn, McGraw-Hill, New York:
143.

Schmidt, W. F. (1965) Water droplet impingement prediction for engine inlets
by trajectory analysis in a potential flow field sample problem, Boeing Co.
D3–6961–1.

Schneider, H. W. (1978) Equation of the growth rate of frost forming on cooled
surfaces, *Int. J. Heat Mass Transf.*, **21:** 1019–24.

Schneider, S. H. and Gal-Chen, T. (1973) Numerical experiments in climatic
stability, *J. Geoph. Res.*, **78:** 6182–94.

Schnell, R. C. and Vali, G. (1972) Atmospheric ice nuclei from decomposing
vegetation, *Nature*, **236:** 163–5.

Schohl, G. A. and Ettema, R. (1986) Theory and laboratory observations of
naled ice growth, *J. Glaciol.*, **32** (11): 168–77.

Scholander, P. F., Van Dam, L., Kanwisher, J. W., Hammel, H. T. and

Gordon, M. S. (1957) Supercooling and osmoregulation in arctic fish, *J. Cell Comp. Physiol.*, **49**: 5–24.

Schumann, T. E. (1938) The theory of hailstone formation, *Q. J. Roy. Met. Soc.*, **64**: 3–21 (17).

Schwartz, G. J. and Diller, K. R. (1983) Osmotic response of individual cells during freezing. I. Experimental volume measurements, *Cryobiology*, **20**: 61–77.

Schwerdtfeger, P. (1963a) The thermal properties of sea ice, *J. Glaciol*, **4** (36): 789–807.

Schwerdtfeger, P. (1963b) Theoretical derivation of the thermal conductivity and diffusivity of snow, IASH, **61**: 75–81.

Seban, R. A. (1960) The influence of free stream turbulence on the local heat transfer from cylinders, *Int. J. Heat and Mass Trans.*, **82**: 101–7.

Shamsundar, N. and Sparrow, E. M. (1975) Analysis of multi-dimensional conduction phase change via the enthalpy model, *J. Heat Trans.*, **97**: 333–40.

Shaw, R. J. (1984) Progress toward the development of an aircraft icing analysis capability, Tech. Memo. 83562, NASA Cleveland.

Shaw, R. J. and Richter, G. P. (1985) The UH-1H helicopter icing flight test program: an overview, *Proc. 23rd Aerospace Sci. Meeting*, AIAA.

Sherwood, T. K. Pigford, R. L. and Wilke, C. R. (1975) *Mass Transfer*, McGraw-Hill Book Co, New York.

Shibani, A. A. and Ozisik, M. N. (1977) Freezing of liquids in turbulent flow inside tubes, *Can. J. Chemical Eng.*, **55**: 672–7.

Shimizu H. (1970) Air permeability of deposited snow, *Low Temp. Sci., Ser., A* **22**: 1–32.

Shumskii, P. A. (1964) *Principles of Structural Glaciology*, (trans. Kraus, D.) Dover Publications, New York.

Silvares, O. M., Cravalho, E. G., Toscano, W. M. and Huggins, C. E. (1975) The thermodynamics of water transport from biological cells during freezing, *J. Heat Trans.* **97**: 582–8.

Single, W. V. and Marcellos, H. (1974) Studies on frost injury to wheat. IV Freezing of ears after emergence from the leaf sheath, *Aust. J. Agric. Res.* **25**: 679–86.

Slaughter, C. W. (1982) Occurrence and recurrence of aufeis in an upland taiga catchment, *Proc. 4th Can. Perm. Conf.*, Nat. Res. Co. Canada, Ottawa: 182–8.

Smith, R. V., Edmonds, D. K., Brentari, E. G. F. and Richards, R. J. (1964) Analysis of frost phenomena on a cryo-surface, In *Advances in Cryogenic Engineering*, Academic Press, New York: **9**: 88–97.

Somorjai, G. A. (1968) Mechanism of sublimation, *Science*, **162**: 755–60.

Squire, V. A. (1979) Dynamics of ocean waves in a continuous sea ice cover, Ph.D. thesis, University of Cambridge.

Sroka, M. (1972) Design of an atmospheric icing tunnel, M.Sc., thesis, University of Alberta.

St. Maur, G. (1983) *Odyssey Northwest*, Boreal Institute for Northern Studies, Edmonton, OP 18.

Stallabrass, J. R. (1970) Methods for the alleviation of ship icing, NRC Canada, Dept. Mech. Eng. Report MD–51.

Stallabrass, J. R. (1980) Trawler icing, a compilation of work done at National Research Council (NRC). Nat. Res. Co. Canada, Mech. Eng. Rep. MD–56.

Stallabrass, J. R. and Lozowski, E. P. (1978) Ice shapes on cylinders and rotor blades, NATO Panel X Symposium on Helicopter Icing, London.

Stefan J. (1891) Ueber die theorie der eisbildung, inbesondere uber die eisbildung im polarmeere, *Ann. Phys. Chem. Neue Folge*, **42**: 269–86.

Steponkus, P.L. (1984) Role of the plasma membrane in freezing injury and cold acclimation, *Ann. Rev. Plant Physiol*, **35**: 543–84.

Steponkus, P.L., Stout, D.G., Wolfe, J. and Lovelace, R.V.E. (1985) Possible role of transient electric fields in freezing-induced membrane destabilization, *J. Membrane Biol.*, **85**: 191–8.

Steponkus, P.L., Wolfe, J. and Dowgert, M.F. (1981) Stresses induced by contraction and expansion during a freeze-thaw cycle: a membrane perspective. In *Effects of low temperature on biological membranes* (eds. Morris, G.J. and Clarke, A.) Academic Press, New York: 307–22.

Stewart, M. and Vigers, G. (1986) Electron microscopy of frozen-hydrated biological material, *Nature*, **319** (6055): 631–6.

Storey, K.B. (1983) Metabolism and bound water in over-wintering insects, *Cryobiology*, **20**: 365–79.

Storey, J.M. and Storey, K.B. (1985a) Triggering of cryoprotectant synthesis by the initiation of ice nucleation in the freeze tolerant frog *Rana sylvatica*, *J. Comp. Physiol.*, **156B**: 191–5.

Storey, J.M. and Storey, K.B. (1985b) Freeze tolerance in the grey tree frog *Hyla versicolor*, *Can. J. Zool.*, **63**: 49–54.

Storey, K.B. and Storey, J.M. (1986) Freeze tolerant frogs: cryoprotectants and tissue metabolism during freeze/thaw cycles, *Can. J. Zool.*, **64**: 49–56.

Storey, K.B. and Storey, J.M. (1988) Freeze tolerance in animals, *Physiol. Rev.*, **68** (1): 27–84.

Sugawara, M., Seki, N. and Kimoto, K. (1983) Freezing limit of water in a closed circular tube, *Warme-und Stoffubertragung*, **17**: 187–92.

Swanson, C.P. (1964) *The Cell*, 2nd edn, Prentice-Hall, Englewood Cliffs, New Jersey.

Szilder, K., Gates, E.M. and Lozowski, E.P. (1986) Measurement of the average convective heat transfer coefficient around rough cylinders, *Proc. Int. Symp. Cold Regions Heat Transfer* ASME, New York: 143–7.

Takeuchi, D.M., Johnson, L.J., Callander, S.M. and Humbert, M.C. (1983) Comparison of modern icing cloud instruments, Contractor rep. 168008, NASA Cleveland.

Tammann, G.T. (1925) *States of aggregation*, Van Nostrand, New York, Ch. 9.

Tarzia, D.A. (1981–2) Una revision sobre problemas de frontera movil y libre para la ecuacion del calor. El problema de Stefan, *Mathematicae Notae*, **29**: 147–241.

Taylor, G.S. and Luthin, J.N. (1976) Numeric results of coupled heat-mass flow during freezing and thawing *Proc. 2nd Conf. on Soil-Water Problems in Cold Regions*, Hydrol Div., American Geophysical Union, Washington: 155–72.

Taylor, M.J. and Pegg, D.E. (1983) The effect of ice formation on the function of smooth muscle tissue stored at −21°C or −60°C, *Cryobiology*, **20**: 36–40.

Thomas, L.J. and Westwater, J.W. (1963) Microscopic study of solid-liquid interfaces during melting and freezing, *Chem. Eng. Prog. Symp. Ser.*, **59** (41): 155–64.

Thomason, S.B., Mulligan, T.C. and Everhart, J. (1978) The effect of internal solidification on turbulent flow heat transfer and pressure drop in a horizontal tube, *J. Heat Transf.*, **100**: 387–94.

Tiller, W.A. (1958) Principles of solidification, In *Liquid Metals and Solidification*, ASM, Cleveland: 276–312.

Toscano, W.M., Cravalho, E.G., Silvares, O.M. and Huggins, C.E. (1975) The thermodynamics of intra-cellular ice nucleation in the freezing of erythrocytes, *J. Heat Transf.*, **97**: 326–32.

Tsytovich, N.A. (1958) Bases and foundations on frozen soil, Highw. Res. Board Spec. Rep. 58. NAS-NRC Washington (original published in 1958 by Academy of Sciences USSR, Moscow).

Tsytovich, N.A. (1975) *The mechanics of frozen ground* (ed. Swinzow, G.K.) Scripta Book Company/McGraw-Hill Book Company, New York.

Tyndall, J. (1858) On some physical properties of ice, *Proc. Roy. Soc.*, **9**: 76–80.

Tyutyunov, I.A. (1964) *An Introduction to the Theory of the Formation of Frozen Rocks*, (trans: Rast, N.) Pergamon Press, Oxford.

Ubbelohde, A.R. (1965) *Melting and crystal structure*, Oxford University Press, Oxford: 7 and 224.

Ueda, T. and Penner, E. (1978) Mechanical analogy of a constant heave rate, *Proc. Int. Symp., on Frost Action in Soils*, (eds. Jacobsson, A., Anderson, D. and Pusch, R.) Div. Soil Mech., University of Lulea, Lulea, Sweden.

Untersteiner, N. (1957) Glazial-meteorologische Untersuchungen im Karakorum, II, Warmehaushalt, *Arch. Meteor. Geophys. Bioclimatol. Ser. B*, **8**: 137–71.

Van Everdingen, R.O. (1982) Management of groundwater discharge for the solution of icing problems in the Yukon, *Proc. 4th Can. Perm. Conf.*, Nat. Res. Co. Canada, Ottawa: 212–26.

Van Fossen, G.J., Simoneau, W.A., Olsen, W.A. and Shaw, R.J. (1984) Heat transfer distributions around nominal ice accretion shapes formed on a cylinder in the NASA Lewis Research Tunnel, AIAA paper 84–0017.

Vanier, C.R. and Tien, C. (1968) Effect of maximum density and melting on natural convection heat transfer from a vertical plate, *Chem. Eng. Prog. Symp. Serv. 82*, **64**: 240–54.

Venkatesh, S., El Tahan, M. and Mitten, V.T. (1985) An Arctic iceberg deterioration field study and model simulations, *Ann. Glac*, **6**: 195–9.

Voller, V. and Cross, M. (1983) An explicit numerical method to track a moving phase change front, *Int. J. Heat Mass Trans.*, **26** (1): 147–50.

Von Karman, T. (1921) Uber Laminare und turbulente Reibung, *Z. angew, Math. Mech.*, **1**: 233–52.

Vowinckel, E. and Orvig, S. (1971) Synoptic heat budgets at three polar stations, *J. Appl. Meteorol.*, **10**: 387–96.

Wadhams, P. (1973) The effect of a sea ice cover on ocean surface waves, Ph.D. thesis, University of Cambridge.

Wadhams, P. (1981) The ice cover in the Greenland and Norwegian seas, *Rev. Geophys. Space Phys.*, **19** (3): 345–93.

Wadhams, P. (1988) The underside of Arctic sea ice imaged by sidescan sonar, *Nature*, **333** (6169): 161–4.

Wadhams, P., Kristensen, M. and Orheim, O. (1983) The response of Antarctic icebergs to ocean waves, *J. Geophys. Res.*, **88** (C10): 6053–65.

Wadhams, P., Lange, M.A. and Ackley, S.F. (1987) The ice thickness distribution across the Atlantic sector of the Antarctic Ocean in midwinter, *J. Geophys. Res.*, **92** (C13): 14535–52.

Wakahama, G., Kuroiwa, D. and Goto K. (1977) Snow accretion on electric wires and its prevention, *J. Glaciol*, **19**: 479–87.

Washburn, A.L. (1947) Reconnaisance geology of portions of Victoria Island and adjacent regions, *Arctic Canada, Geol. Soc. America, Mem*, **22**.

Washburn, A.L. (1973) *Periglacial processes and environments*, Edward Arnold, London.

Watt, B.K. and Merrill, A.L. (1963) Composition of foods, *Agriculture Handbook 8*, US Dept. Agric.

Weeks, W.F. and Ackley, S.F. (1982) The growth, structure and properties of sea ice, US Army Cold Regions Research and Engineering Laboratory Mono. 82–1.

Weeks, W.F. and Ackley, S.F. (1986) The growth, structure and properties of sea ice, In *The Geophysics of Sea Ice* (ed. Untersteiner, N.) Plenum Press, New York.

Weeks, W.F. and Mellor, M. (1978) Some elements of iceberg technology, In *Iceberg Utilization* (ed. Husseiny A.A.) Proc. 1st Int. Conf. on iceberg utilization for fresh water production, Pergamon Press, New York: 45–98.

Weertman, J. (1968) Comparison between measured and theoretical temperature profiles of the Camp Century, Greenland, borehole, *J. Geophys. Res.* **73** (8): 2691–700.

Weickmann, H.K., Katz, V. and Steele, R. (1970) AgI sublimation on contact nucleus? *Proc. 2nd Nat. Conf. Weather Mod.*, Amer. Met. Soc., Boston: 332–76.

Weller, G. (1967) The effect of absorbed solar radiation on the thermal diffusion on Antarctic fresh-water ice and sea ice, *J. Glaciol.*, **6**: 859–78.

Weller, G. (1968) The annual energy transfer above and inside Antarctic blue ice, *IASH*, **79**: 417–28.

Wexler, H. (1958) Geothermal heat and glacial growth, *J. Glaciol*, **3**: 420–5.

Whitaker, S. (1972) Forced convection heat transfer correlations for flow in pipes, past flat plates, single cylinders, single spheres and for flow in packed beds and tube bundles, *AIChEJ*, **18**: 361–71.

White, J.E. and Cremers, C.J. (1981) Prediction of growth parameters of frost deposits in forced convection, *J. Heat Transf.*, **103**: 3–6.

Wiegand, K.M. (1906) Some studies regarding the biology of buds and twigs in winter, *Bot. Gaz.*, **41**: 373–424.

Williams, P.J. (1964a) Experimental determination of the apparent specific heats of frozen soils, *Geotechnique*, **14** (2): 133–42.

Williams, P.J. (1964b) Unfrozen water content of frozen soils and soil moisture suction, *Geotechnique*, **14** (3): 231–46.

Williams, P.J. (1966) Downslope soil movement at a sub-arctic location with regard to variations with depth, *Can. Geotech. J.*, **3**: 191–203.

Williams, J. (1975) The influence of snowcover on the atmospheric circulation and its role in climatic change, *J. Appl. Meteor.*, **14**: 137–52.

Wolosewick, J.J. and Porter, K.R. (1979) Microtrabecular lattice of the cytoplasmic ground substance: artifact or realty? *J. Cell Biol.*, **82**: 114–39.

Wood, G.R. and Walton, A.G. (1970) Homogeneous nucleation kinetics of ice from water, *J. Appl. Phys.*, **41**: 3027–36.

Woodruff, D.P. (1968a) Morphology of the solid/liquid interface during melting of a binary alloy, *Phil. Mag.*, **17** (146): 283–94.

Woodruff, D.P. (1968b) Morphology of the solid/liquid interface during melting, *Phil. Mag.*, **18** (151): 123–7.

Woodruff, D.P. (1973) *The Solid-Liquid Interface*, Cambridge University Press, Cambridge.

Yen, Y.C. and Galea, F. (1969) Onset of convection in a water layer formed continuously by melting ice, *Phys. Fluids* **12** (3): 509–16.

Zachariassen, K. E. (1985) Cold tolerance in insects, *Physiological Rev*, **65** (4): 800–32.

Zakrzewski, W. P. (1986) Icing of ships, Part I: Splashing a ship with spray, NOAA Tech. Mem. ERL PMEL – 66.

Zarling, J. P. (1987) Approximate solutions to the Neumann problem, *Proc. Int. Symp. on Cold Regions Heat Transf.*, ASME, New York: 47–54.

Zerkle, R. D. and Sunderland, J. E. (1968) The effect of liquid solidification in a tube upon laminar flow heat transfer and pressure drop, *J. Heat Transf.*, **90**: 183–90.

Zienkiewicz, O. C. (1977) *The Finite Element Method*, McGraw-Hill, New York.

Zubov, N. N. (1979) *Arctic Ice*, Engl. transl. Airforce Geophysics Laboratory, Hanson, Mass, Rep. AFGL–TR–79–0034.

Index

a-axis growth, 62, 106, 113–15, 127, 130, 142, 201, 252, 306, 350, 360–1
ablation (see also melting) 250, 365, 370, 372
  zone, 249
acceleration parameter, 158
acclimated stem, 334–5
acclimatization (see hardening)
accretion,
  on aircraft, 178, 192, 205–6, 211, 213
  on carburettors, 17, 225–6
  decay, 373
  dry, 185, 188, 200–2, 204, 208–9, 213, 371
  on power lines, 17, 221, 224–5, 372
  with splashing, shedding, 185–6, 208
  wet, 185, 187–8, 200–2, 208, 210, 213, 223
  zone, 187–92, 202, 217
activation energy,
  barrier, 59
  for melting, 360
  for self diffusion, 358
active (layer) zone, 81, 272, 274, 374
additive growth, 122, 124, 134, 249, 347
adhesion efficiency, 188, 195–6
adsorption coefficient, 170
adsorption inhibition, 340
advective flux, 190–1
aeroglaciology, 167, 370
aerothermodynamic coupling, 199, 225
aggregation, 255
air, bubbles and grain, 128
air conditioning, 175, 221, 226
albedo, 127, 230
  of clouds, 235
  of desert, 235
  of ice, 133, 235
  of snow, 132–3, 229–30, 235, 245
  of soil, 235
  of vegetation, 300
  of water, 235
algae, 112, 141, 301, 316
amoeba, 301, 316–17
amorphous ice (or vitreous ice), 53, 168, 307
  density of, 53
anchor ice, 177, 166
  movement of, 117
  thickness, 117
anions, 303
annular growth, 154, 160–1
Antarctic, 123, 230, 242, 244, 247, 369
Antarctic ice, 118
Antarctic iceberg, 369
anti-freeze, 373
  peptide, 340
anti-icing (see structural icing control)
aquatic animals, 340
aquatic plants, 166, 329
aqueous solution, 38–40
  diffusion in, 104
  growth of ice in, 103, 107, 302, 304
Aristotle, 20, 167
Arrhenius equation, 312, 326
Arrhenius plots, 332
astronomical time scale, 1, 229
atmosphere, 298
atmospheric evolution, 4
atmospheric ice, 13–14, 371
atmospheric plants, 330
Aufeis (see naled)
avalanches, 374, 378

bacteria, 270, 313, 316–17, 345
  nucleation-active, 334
basal plane (face), 168, 170, 177, 182, 340, 316
  definition of, 50–1
  linear growth rates of, 171, 194
  melting of, 360
Beer's law, 127

Belgica antarctica, 338
Benard cells, 139
binary solution (see also aqueous solution) 104, 307, 326
biochemical reactions, 310–12, 382
bioglaciology, 298, 310, 342–3, 380
bisophere, 256, 300, 380
biota, 301
biot number, 70, 143, 150, 216, 228, 319, 372
biotic limits, 329
Bjerrum defects, 52
    and dislocations, 52
    formation of, 51–2
    mobility of, 52
Blasius problem, 144
blood, 302
    plasma, 305
blue-grass algae, 313
Boltzmann constant, 59, 170
Boltzmann distribution, 59
bottom ice, 123, 141
bottom melting, 365–6
bottom sediment, 129
boundary layer,
    flow, 137–9, 142, 144, 147, 158, 207, 228
    theory, 83
Bracon cephi, 337
break up, 13, 366, 367
    of lakes, 367
    mechanical, 366, 368–9
    of rivers, 367
    of sea ice, 367–8
brine,
    channel, 301
    pocket, 128, 267, 372

c-axis (optic axis),
    definition of, 50–1
    growth parallel to, 62, 113–14, 174, 201, 340, 360, 363, 370
Callisto, ice on, 5
capillary-induced pressure, 36–7, 246–7
capillary regime, 247
cambium 333
carburettor icing, 225–6
Carnot engine, 293
cations, 303
cell,
    dehydration, 335, 382
    dimensions, 320
    geometry, 324
    human, 312
    membrane, 302, 310, 313, 316, 318, 324, 326, 331–2, 382
    nucleolus, 315

nucleus, 303, 315
    survival, 323–4, 331
    suspension, 326, 342
    volume, 320, 327–9
    wall, 312–13
chemical potential, 34–7, 39, 44, 57–8, 109–11 268, 305, 319–20
chill injury, 332
Chlorella emersonii, 331
chloroplast, 314–15
Clausius–Clapeyron equation, 31, 154, 179, 290, 320, 353, 377
climatic effects, 126, 230, 238, 247, 250, 274, 281
closed cavity ice, 276–7
cloud cover, effect of, 240
cloud,
    cumulonimbus, 182
    droplet distribution, 180–2
    optical phenomena in, 14
cluster (see Embryo)
coherence length, 303, 310
collection efficiency, 210, 213, 223
collision efficiency, 184, 188, 201, 216
    of an aerofoil, 206, 208
    of snow crystals colliding with drops, 196
    of snow crystals colliding with snow crystals, 195
comets, ice in, 6–7
compressibility coefficient, 261, 285
condensation (see heat transfer)
conduction, thermal (see thermal and heat transfer)
conductive flux, 190
conservation of mass (see also continuity equation), 26
consolidation, 375–6
consolidation coefficient, 261
constitutional supercooling, 105–6, 112, 115, 141, 301, 306, 346, 350, 363–4, 366
constitutional superheating, 364
continuity equation, 27, 249, 261, 264, 285
convection, 66, 125, 129, 134, 237, 364
cortex, 332–5
cosmic dust, 2
critical (cluster) embryo, 58–9, 61, 104, 174, 326, 356, 358
crushed ice, 346
cryobiology, 298
cryodemulsification, 346
cryosphere, 243
cryosurgery, 343–4
    for brain tumor, 344
    for cancer tissue, 344
    hepatic, 344
    opthalmic, 344

cryosurgical probe, 344
cryopreservation of, 341
  corneas, 343
  heart valves, 343
  human skin, 343
  kidney, 343
  pancreas, 343
  parathyroid, 343
  teeth, 343
  veins, 343
cryoprotectant, 319, 326–7, 341–3, 345,
    347, 383
cubic ice (see also ice Ic), 53, 168
cuticle, 337
cytomatrix, 310, 318
cytoplasm, 310, 113–16, 318–19, 326, 342

D-defect, 52
Darcy's law, 260, 263
defect, 52
  and dislocations, 359–60
dehydration, 318, 331, 333, 335, 337, 342,
    344, 351, 382
de-icing (see structural icing control)
degree of,
  compaction, 258
  saturation, 247
  supersaturation, 172
  wetness, 246
  wetting, 59, 326
degree-day index, 81, 232, 234, 274
dendrite,
  branching, 117, 306
  breakage, 161
  growth, 106–8, 112, 114, 134–5, 141,
    152–4, 159, 173, 174–5, 182, 289, 309,
    349–50, 371
  matrix, 142
  melting, 360, 370
  tip velocity, 136
density of,
  firn, 242
  ice, 53, 242
  soil, 282
  snow, 242
deposition, 167–9, 172, 178, 238, 361
depth hoar, 243, 248, 276
desiccation, 300
de-vitrification, 343, 384
differential scanning calorimetry, 318
differential thermal analysis, 318, 334, 339
diffusion, 40–4, 261, 319
  binary, 43
  coefficient, 41, 261, 358
  coupled, 41–3, 105, 302, 326, 357, 382
  of electricity, 41

of heat, 41, 174
and irreversibility, 40
of mass, 41, 43
and mobility, 309
and phenomenological coefficients, 41–3
self, 358
surface, 174, 244
thermal, 135
vacancy, 61
volume, 244
of water vapour, 174
of water in water, 41
diffusional growth, of snow crystals, 194
diffusivity, 43, 55, 77, 262
dimethyl sulphoxide (DMSO), 331
dirac delta function, 92
discretization, 87
dislocation, 361
  and defects, 359–60
  density of, 362
  and kinks, 360
  screw, 62, 170, 340
  step, 170, 360
dispersion, 261–3
dispersion coefficient, 262
dispersivity, 262
double layer, 268–9, 271
drag,
  coefficient of, 182, 198
  on a hailstone, 201
droplet,
  collisions, 181
  freezing, 104, 308
  inertia, 183
  size distribution, 180–2, 205, 215, 218, 221
  trajectory, 182, 184, 207–8, 213, 215,
    216, 223
dry accretion, 185, 188, 200–2, 204, 208–9,
    213, 371
dry snow zone, 249
DSC method, 318
DTA method, 318, 334, 339
Dufour effect, 43, 270

earthworm, 301
Eckert number, 190
effective specific heat (see specific heat,
    apparent)
electric charge, 34, 42
electrical conductivity, 42
electrochemical potential, 42
electromagnetic heating, 385
electro-osmosis, 270
electrostatic potential, 34, 42
embryo, 56–8, 61, 318
  critical, 58–9, 61, 104, 174, 326, 356, 358

emissivity (see emittance)
emittance, 66, 236
endoplasmic reticulum, 315, 317
energy,
 of dissociation 51–2
 flux, 25, 77
 internal, 27, 76
 principle, 25, 76
energy balance, 188, 190–1, 202, 205, 217,
  227, 263, 352
 global, 230
engulfment, 108–11, 268, 317, 332
enthalpy,
 and heat capacity, 78
 and latent heat, 78
 of soil, 263, 287
entropy,
 flux, 26
 generation, 26, 28, 41, 44, 356–7
 principle, 26, 41, 357
equilibration, 323, 328–9, 342, 382
equilibrium,
 chemical, 218, 319, 343
 general criterion for, 28
 mechanical, 267
 metastable, 44, 47, 56–8, 154, 167, 175,
  178, 215, 266–7, 318–9, 323, 356, 370,
  382
 of phases with curved interfaces, 37
 stable, 28–30, 36–7, 44, 47, 57, 106, 108,
  167, 175, 177–8, 204, 266–7, 272, 290,
  310–11, 319, 321–4, 328, 340–1, 348,
  356
 thermal, 319–20
 unstable, 28, 44–5, 56, 58
erythrocyte, 303, 310, 325–6
Eurosta solidaginis, 337
eutectic point, 32–3, 311, 323–4, 382
eutectic temperature, 33, 107, 307, 325
evaporative flux, 190
evaporation (see heat transfer)
evaporative icing, 226
evapotranspiration, 231, 240, 300
extracellular fluid, 302, 310, 319–20, 328,
  338, 350
extracellular freezing, 350–1
extracellular ice, 325–6, 332, 335, 343, 345,
  349, 351, 353, 381–4
extracellular water, 334, 384
extraorgan freezing, 335–6

facetting, 360–2
fast ice, 141
finite difference approximations, 88
finite element geometry, 94
finite element matrix, 99–100

fire fighting and ice, 226
firn, 242–8, 369
firnification, 248
first order phase change, 31
flake ice, 120
'flakice', 140
flocs, 117
floes, 117–18, 122–3
flow separation, 148, 158, 163, 372
fogs, 218, 221
 droplet size distribution of, 182
Fourier's law, 41, 64, 238
Fourier number, 60, 351
frazil ice, 13, 112, 121, 166, 226, 340
 collision breeding of, 117
 discoid, 116
 formation of, 116
 fracture of, 116
 hazard, 13
 matrix, 118
 multiplication of, 116
 needle, 116
 production of, 116, 122
 and sea ice, 130, 141, 301
 seeding mechanisms of, 116, 122
 shapes of, 116
 sintering of, 117
 size of, 116
 splinter of, 117
free energy (see Gibbs free energy)
freeze,
 avoidance, 336
 concentration, 104, 114, 267, 307,
  310–12, 319–20, 322, 335, 337, 342,
  344–7, 350, 382
 damage, 349, 351
 dehydration, 331
 desalination, 104
 drying, 345, 350–1
 inhibition, 340
 injury, 332
 resistance, 330
freezer burns, 351
freeze-induced electric potential, 333
freeze-resistant (see ice resistant)
freeze-thaw cycle, 8, 18, 20, 131, 230, 346,
  364, 373–5, 377–8, 380
freeze-tolerant (see ice-tolerant)
freeze-up, 13
freezing,
 around pipe, 70
 of drops, 70, 239, 308
 of egg yolk, 347
 of embryos, 342
 fraction, 188–9, 202
 fringe, 281, 285, 288–93, 377

freezing, (*contd*)
  front, 288–9
  of fruit, 345, 349
  of meat, 348
  of milk, 346, 348
  of pools and ponds, 112–13
  protocol, 325
  of sea water, 121, 139
  of solutions, 103, 108, 346
  of sphere, 69
  of suspensions, 103, 108, 346
  of vegetables, 345, 349
  zone, 107–8
freezing front, 288, 297, 329
freezing index (see degree-day index)
freezing point, 31, 47
  depressant, 220
  depression, 40, 48, 110–11, 307, 310,
    317–18, 327, 336–7, 340, 360, 373
freezing rain, 215, 221, 223, 225
freezing temperature range, 48
freezing time of sphere, 70, 308–9
frogs, 340–1
frost, 13, 177, 226–8, 334
  layer, 178, 370
  matrix, 176–7, 188, 371
  penetration, 232, 281, 284, 288–9
frost creep, 378
frost polygons, 380
frost-resistant (see ice-resistant)
frost-tolerant (see ice-tolerant)
frostbite, 341
Froude number, 120
frozen soil, 13
frozen spray, 13–14
frozen zone, 281, 285–6, 290
funicular regime, 247

galactic formations, 1
gamma function, 205
Ganymede, ice on, 5
gelation, 124, 134, 347
geliflocculation, 116–17, 119, 121, 160
gelifluction, 378–79
geoglaciology, 229, 310, 374, 378
geological time scale, 380
geothermal gradient, 129, 240, 374
geothermal heat flux, 250–1
Gibbs equation, 34
Gibbs–Duhem equation, 35, 247
Gibbs free energy (see Gibbs function)
Gibbs function, 28–9, 34–6, 58
Gilpin bands, 158–9
glaciation in the Universe, 2
  in comets, 6, 7
  on cosmic dust, 2

  as ice II, V and VI, 6
  on moons, 5
  on planets, 3, 4–6
  in 'snowballs', 6
glacier, 17
  ablation zone of, 249
  base, 250
  densification of, 248
  dry snow zone of, 249
  in northern hemisphere, 11
  percolation zone of, 249
  snow cover on, 249
  superimposed ice on, 249
  temperature, 245
  wet snow zone of, 248
glass transition (see vitrification)
glaze ice, 185, 207–8, 210–12, 233, 289
Golgi body, 315
Graetz problem, 155, 164
Grashof number, 165
graupel, 198, 200, 204
grease ice, 121
Greenland, 242, 244, 247–8
ground ice classification, 275
groundwater and ice, 19, 122, 272, 276,
    280, 377–8

H_2O, on planets, 3–4, 6
habitat, 141
habits of crystals, 169, 172, 174
  and growth rates, 171, 194
  and corners, 173
  and edges, 173
  effect of,
    electric fields on, 174
    foreign molecules on, 174
    temperature on, 196
  hexagonal, 20
  primary features of, 172–4, 176
  prismatic, 173
  secondary features of, 172–4
  and sectored plates, 173
  and sheaths, 173
haemolymph, 388, 340
hail, 13, 17, 192, 198, 204
hailstone, 198, 202
  ablation of, 371
  and capillary action, 200
  dry growth of, 201–2
  energy balance of, 202–3
  fall mode of, 199
  with lobes, 201
  origin of, 200
  stability of, 200
  terminal velocity of, 202
  wet growth of, 201–2

hailstone, (*contd*)
  wind tunnel studies of, 201
Haines jump, 267
hair frost (see needle crystal)
haloes, 14
hanging dam, 124, 130, 140
hardening,
  cold, 336, 339
  frost, 336
heat balance, 124, 127, 216, 231, 234, 239, 245, 272, 274
heat conduction equation, 72, 88, 128, 356
heat flux, 26, 41, 190
heat generation, 77
heat of transport, 42
heat of wetting, 259, 283
heat transfer,
  coefficient of, 66–7, 126, 134, 137–40, 142–3, 146, 162–3, 210, 228, 372
  by condensation, 172, 178, 228, 300
  by conduction, 41, 113, 174, 261
  by evaporation, 113, 126, 203, 252, 261, 300
  by radiation, 113, 126–7, 231, 252, 261, 357
  by sublimation, 230, 244, 261, 265, 353, 361, 370
heat valve effect, 241
heaving (see also ice segregation), 18, 255, 280, 281, 289
heterogeneous nucleation, 59–60, 113, 134, 192, 252, 266, 270, 307, 318, 323, 326, 334–6, 343, 346, 382
hexagonal bi-pyramids, 168
hexagonal plates, 171
hexagonal prisms, 168, 171, 176
hexagonal symmetry, 20, 168
hibernation, 336–7
history, and ice, 20
hoarfrost, 17, 19, 160, 175–6, 186, 215
homogeneous (or spontaneous) nucleation, 57, 59, 163, 178, 182, 185, 318, 326, 344, 336, 343, 358, 382
homoiotherm, 336, 341
Hopper structure, 173
humus, 270–1
hydrates, 43
hydraulic conductivity, 257–8, 260, 285, 296–7, 342, 375
hydration, 310
hydrogen bonding, 50, 174, 303, 327
hydroglaciology, 102, 364
hydrophilic behaviour, 109, 271
hydrophobic behaviour, 271, 303
hydrosphere, 256, 298
hysteresis effects, 357, 377

IAH classification, 198
ice,
  albedo of, 133, 235
  and Inuit, 23
  in history, 20
  in poetry, 20–3
  perceptions of, 21–2
ice Ih (hexagonal ice), 20, 357
  axes of, 50–1
  bond angle in, 51
  bonding energy, 51–2
  crystal structure of, 20, 51
  defects in, 51
  density of, 53
  growth of, 194
  ploycrystalline, 113, 144, 154–5, 158, 160–1, 175, 245, 361–4
  proton positions in, 50–2
  specific heat of, 54
  thermoelectric properties of, 52
ice Ic (cubic ice), 53, 168
  density of, 53
ice polymorphs, 6, 52–3
ice ages, 7–12, 380
  and Antarctica, 11
  astronomical theory of, 8–11
  and continental drift, 10
  and Gondwana, 10–11
  and historical record, 12
  early, 10
  effect of: axis tilt on, 8; climate on, 9; eccentricity on, 8, 229; precession on, 9; spin on, 8
  geological evidence for, 10
  interglacial periods of, 9, 11, 230, 242
  and Little Ice Age, 12
  and South Pole, 10–11
  present, 10–11
ice cap, 13, 242
ice content, 232
ice covers (see also sea ice), 118, 230
ice cream, 347
ice crystal (see also habits of crystals and snow crystal), 14
  growth of, 194
  size destribution of, 307
  stellar, 360
ice dam (veil), 163
ice fibres (see needle crystal)
ice forming nuclei (see nucleant)
ice fraction, 29, 204, 254, 265, 285
ice-free zone, 155
ice jam, 13, 120, 123, 367
ice lens, 292, 297, 376
ice pellet, 17, 198, 200
ice point temperature, 31, 40, 47

ice pressure, 292, 294
ice-resistant insects, 338–9
ice-resistant plants, 334–5
ice segregation (see also segregated ice), 267, 277, 280–1, 289–97, 331, 335, 342, 376
ice sheet, 13, 141, 229–30, 242, 248
    and Ice Ages, 10–11
    temperature distribution in, 13
ice shelf, 242, 368
ice stream, 17, 368
ice-tolerant insects, 338–9
ice-tolerant plants, 335
ice wedge, 277, 376, 379–80
iceberg, 13
    calving, 18, 368–9
    deterioration of, 369; convection effects on, 369; radiation effects on, 369; thermal shock effects on, 369; wave effects on, 369
    meltwater, 369
    metamorphosis, 369
    tabular, 12, 369
    and the *Titanic*, 22
    weathered, 368
icicle, 14, 220
icing (see naled)
icy mantles, 2, 5–6
impact icing, 226
impurities,
    dissolved, 103
    particulate, 103
interstellar medium, 2
insect,
    adult, 337
    egg, 337
    larva, 337
    pupa, 337
insolation (see also solar radiation), 234
integral technique, 83, 85
interface,
    curvature of, 110–11
    equation, 60, 64, 66, 71–7, 74, 77, 89, 93, 125, 128, 137, 144, 227, 377
    faceted, 361–3, 370
    non-faceted, 360, 363
    rough, 61, 360, 363, 366
    shape of, 154, 360–1, 363, 366
    smooth, 61
    temperature of, 84
    velocity of, 60, 76
interfacial energy balance (see also interface equation), 60, 158, 365
interfacial instability, 75, 106, 117, 156–8, 161, 163–4, 177
interfacial tracking, 89

interglacial, 9, 11, 230, 242
internal melt figures, 360
internal sublimation, 360
interpolation function, 94–5
interstitial defect, 52
intracellular fluid, 302, 310, 385
intracellular freezing, 318, 334
intracellular ice, 318, 322, 324, 329, 332, 338, 340, 342, 344–5, 349–51, 381
intracellular nucleation, 319, 325–6
intracellular sap, 315
intracellular supercooling, 381
intracellular water, 318, 320, 322–3, 335, 349, 383
intrusive ice, 277–8, 280–1
inversion temperature, 53, 129, 136–9, 153, 164–5
ionization defects,
    and dislocations, 52
    formation and migration of, 52
    mobilities of, 52
irradiation, 234–5

Kepler, 20
killing point, 334
kinetic flux, 190
Kirchoff's law, 236

L-defect, 52
lake ice, 113, 129
    decay, 364
lake water, 103
lamellar crystal, 268
lamellar space, 268
Landau transformation, 81, 89–90
language, literature and ice, 20–3
latent heat,
    of evaporation, 302
    distributed, 92–3, 128, 264, 285
    flux, 190
    of fusion, 31, 53–4, 64, 232, 266, 283
    of sublimation, 52, 54–5, 351
latitude, effect of, 230
lattice energy, 54
leading edge, 145
leaves, 270
ledge energy, 170
lenticel, 334
leucocyte, 303, 312
Lewis number, 105–6, 115, 363–4
linear growth rate, 171
liquidus curve, 32, 106, 307, 310–11, 383
lithosphere, 256
Lobelia teleki, 230
lone-pair orbitals, 49, 51
long-wave radiation, 127

lymph, 302
lysosome, 315

Mach number, 208
macromolecules, 307, 310
magpie, 301
marine icing (see also accrection), 213,
    215, 217
Mars, ice on, 4–5
mass fraction, 128
mass penetration, 105
mass transfer coefficient, 190
mass wasting, 378
mean volume radius, 186
median volume diameter, 216
melt (or Tyndall) figures (see thaw
    figures)
melting,
    effect of solar radiation on, 362
    intergranular, 362, 366, 373
    pressure, 291–2
    point, 318, 361
melt pond, 135, 301, 365
meltwater, 245–8, 277, 331, 366
membrane,
    cell, 302, 318
    dehydration, 318, 333
    destabilization, 332–3
    disruption, 333, 345
    folds, 313, 315
    permeability, 320, 323, 325–6, 383
    pore, 310, 318
    semi-permeable, 38, 43, 314, 319
    shrinkage, 318
    thermotropic transition in, 332
    thickness, 313, 315
mesophyll, 332, 335
metabolism, 334
metamorphosis, 122, 127, 231, 242–3, 301,
    369
metastability, 42–4
micelle, 303, 307, 346
micrometeorological effects, 126
milk, 346, 348
microtrabecular lattice, 310, 318
microvilli, 316
mitochondria, 303, 314, 317
molality, 39–40
molar volume, 321
mole fraction, 39
molecular diffusion coefficient, 261
molecular diffusivity, 262
muscle (see also myosystems), 301, 311,
    345
mush frost (see needle crystal)
mushy ice, 112

mycoplasma, 313
myosystems, 345, 347, 349

nacelle icing, 213
naled, 229, 242, 252–3, 378
    and climate, 18–19
    and culverts, 252
    dimensions, 252
    and groundwater, 122, 252
    origin of, 252
    and river ice, 252
    toe, 254
natural convection, 129, 134, 364
needle (ice) crystal, 242, 254–5, 267, 280
Newton's law of cooling, 324
nilas, 121
nivation, 378
normalization, 60, 78–9, 104, 227, 328,
    376
northern ecology and ice, 19
nucleation,
    agent, 160
    catalyst, 337–9
    effect of solutes on, 104
    efficiency of, 192
    by frost, 334
    heterogeneous, 59–60, 113, 134, 192,
        252, 266, 270, 307, 318, 323, 326,
        334–6, 343, 346, 382
    homogeneous, 57, 59, 163, 178, 182, 185,
        318, 326, 334, 336, 343, 358, 382
    inhibition, 340
    and innoculation, 326
    in a pore, 264–7
    by snow, 116
nucleants, 178, 192–3, 222, 326, 387
nucleation rate of ice crystals in water, 59,
    116
numerical bondary conditins, 89, 100
numerical convergence, 86
numerical formulation,
    explicit, 88, 91, 98
    implicit, 88, 92, 99
numerical stability, 86, 88
Nusselt number, 138, 145, 165, 180

oceanic heat flux, 123, 133
offshore structure icing 213, 218–21
Onsager's reciprocal relation, 41
open cavity ice, 275
optic axis (see *c*-axis)
optical phenomena in clouds, 14
order of magnitude, 78, 251
organ permeability, 342
organ preservation, 342
organelle, 316, 332, 349

organism response,
  effect of hoar on, 20
  effect of freeze-thaw on, 20
osmotic,
  dehydration, 341
  pressure, 39, 44, 317, 320, 328, 333, 366, 383
  stress, 338, 343
overburden pressure, 292, 296
overwintering, 337

pancake ice, 117–18, 121–2
parhelion, 14
patterned ground, 18, 229, 380
Peclet number, 263
pellets (or sleet), 17
pendular regime 247
peptide antifreeze, 340
percolation zone, 249
permafrost, 229, 380
  aggradation of, 241
  alpine, 274
  continuous, 273
  discontinuous, 273
  distribution of, 273–4, 300
  on Mars, 4
  table, 252, 278, 374, 380
perturbation solution, 81
phagocytosis, 316
phase diagram,
  of binary solution, 104
  of H₂O, 30
  modified, 306–7
  of sea water, 33–4
  of sodium chloride solution, 32
phloem, 332–3
phonons,
  diffusion of, 55
  and thermal conductivity, 55
photosynthesis, 314
phytosystems, 345, 348–9
piezometric head, 260–1, 263, 287
pingo, 18, 229, 279–80
pinocytosis, 316
pipe ice, 151–61, 355, 372
pipkrake (see needle crystal)
planets, ice on, 3–6, 9
planetesimals, 2
plant,
  cell, 314–15
  leaf, 333
  root, 330–3
  stem, 333
  survival, 20
plasmalemma, 335, 349
plasmodesmata, 314

plate ice, 119–20, 122
Pluto and ice, 6
pneumatic de-icing devices (see structural
  icing control)
poetry and ice, 20–3
poikilotherm, 336, 339–40
polar ice, 4
polycrystalline ice, 144, 154–5, 158, 160–1,
  175, 245, 361–4
polynyas, 141
pond water, 103
pore,
  diameter, 260
  geometry, 260, 284
  nucleation of, 289
  pressure, 260, 277–81, 300, 375–7
  volume, 260
potential flow, 207–8
Poynting effect, 38
Prandtl number, 143, 163
pressure,
  difference at curved surface, 38
  ice, 292, 294
  osmotic, 39, 44, 317, 320, 328, 333, 366, 383
  overburden, 292, 296
  pore, 260, 277–81, 300
primary crystallization, 306
primary heave, 255, 280
primary ice, 119–121
primordial medium, 1
prism face, 168, 177, 361
  linear growth rate of, 171, 194
propeller icing, 211
proton, mobility, 269
protoplasm, 333
protoplasmic streaming, 333
Pterostichus brevicornis, 337
pyramid function, 94–101

radiation (see also heat transfer),
  albedo, 127
  balance, 127, 236–7, 360
  from ice, 66
  long-wave, 127
  short-wave, 127
  solar, 127, 229, 235, 365–6
radiative flux, 190
radiosity, 234, 236
Raoult's law 39, 320
Rayleigh instability, 135, 138–9, 364
Rayleigh number, 136, 139, 180, 228
recovery factor, 190, 209
recrystallization,
  in organs, 381, 384–5
  in snow, 244

reflectance or reflectivity (see albedo)
refrigerator, 176
regolith, 4, 272
reindeer, 301
re-laminarization, 148, 158
residual error, 95–101
reticulated ice pattern, 281
reversibility, 356, 361
Reynolds number, 141–2, 148–9, 156, 158, 162–4, 180, 183, 199–200, 207, 228, 284
Reynolds transport theorem, 27, 84
rime (see also accretion and hoarfrost), 207–8, 210, 212, 223
    feathers, 186
    needles, 194
    simulated, 185
    of snowcrystals, 200
rind, 121
river ice, 113
    decay, 364
river water, 103
rock, frozen, 13, 256
roots, 270
rotor icing, 211–12
roughness, 209, 238, 300
ruled surface, 31, 48, 57, 167
runback (water), 192, 217

Saccharomyces cerevisiae, 322–3, 345
saline ice (see also sea ice), 128
Salmonella, 345
saltation of frazil, 123
salt plumes, 139
saturation ratio, 179
saturation vapour pressure, 172
Saturn's rings, ice in, 6
scales, 78–9, 250–1, 376
Schmidt number, 180
Schumann–Ludlam limit, 187
Scolytus multistriatus, 337
screw (or spiral) dislocations, 62, 170, 340
sea biota, 112
sea ice,
    Arctic, 131–3
    crystallography of, 142
    decay of, 364
    environment, 301
    first year, 366
    hazard, 13
    melting of, 365
    model of, 131
    multi-year, 365
    properties of, 128
    and radiation, 128
    salinity of, 128

second year, 366
    smooth, 147
    snow cover over, 131–3
    variations in, 130–3
    young, 366
sea temperature, 216
sea water, 31, 103
secondary crystallization, 306
secondary growth, 172–4
secondary heave, 281–9
secondary ice, 121
sedimentation, 123
Seebeck coefficient, 42
Seebeck effect, 42
segregated ice, 277, 280–1
segregation criterion, 294
segregation potential, 295–7
segregation temperature, 296
Seiraphera diniana, 337
separation (see flow separation)
sensible heat flux, 190
shed water, 202–3, 205, 217, 317
Sherwood number, 180
shipboard ice (see also marine icing), 16–17, 192, 215, 218
shorefast ice, 119–22
short-wave radiation (see also solar radiation), 127
sill ice, 277–8
similarity solution, 81–3, 108
sintering, 196, 245
skim ice, 113–14, 119–22
sleet (ice pellets), 17, 223
slumping, 18, 374, 378
slush, 118, 122, 124, 140, 154, 160, 246, 252, 254, 365
slush balls, 117
slushflow, 378
small hail particle (see ice pellet)
snow (see also snowcover and snowfall),
    blizzard, 17
    blowing, 242–3
    decay of, 365
    ice, 122–3
    melting of, 365
    metamorphosis of, 244–6, 249, 301
    permeability of, 243
    porosity of, 243–4
    wet, 223
'snowballs', 6
snow blizzard, 17
snowcover,
    albedo of, 132, 230
    and capillary action, 122, 247
    decay of, 365
    energy exchange in, 127

snowcover, (*contd*)
  funicular regime in, 247
  global climate related to, 13
  insulating effect of, 19, 300, 330, 375
  metamorphosis of, 122, 127, 231, 242–3,
    301
  moisture in, 238
  pendular regime in, 247
  on sea ice, 131–3
  and superimposed ice, 122
  and taiga, 19
  thermal resistance of, 125
  and tundra, 19
  tunnelling in, 301
snow crystal 14–15
  equilibrium shape of, 244
  melting of, 365, 370
  models of 197
  modes of fall of, 198–9
  observations of, 18
  prismatic refraction of, 18
  riming of, 194
  shape of, 15, 20
  symmetry of, 15
  terminal velocities of, 197
snow drift, 18, 378
snowfall, 122, 230
snowflake (see also snow crystals),
  aggregation of, 194
  rimed, 194, 371
snow ice, 365
snow pellet (see graupel)
snowpack (see snowcover)
soft hail particle (see graupel)
soil,
  aggregation, 257
  capillarity of, 300
  clayey, 258–9, 270–1
  density of, 282
  fragments, 257
  freezing fringe in, 281, 285, 288–93
  freezing front in, 288–9
  frost susceptible, 331
  frozen, 13
  frozen zone in, 281, 285–6, 290
  gravelly, 277
  heat capacity of, 282–3
  hydraulic conductivity of, 257–8, 260,
    285
  interconnections in, 257, 259–60
  mineral, 258–60, 271
  morphology of, 258–9
  organic, 257, 259, 270–1
  particle sizes of, 256, 258
  permeability of, 258–9, 264, 271, 288
  porosity of, 257, 259–61, 282, 284

residual, 257
sandy, 277
shrinkage of, 268, 270–1
silty, 270
specific heat of, 282
stress in, 375, 377
thawing of, 375–80
thermal conductivity of, 283
transported, 257
unfrozen zone in, 281, 285–6, 293
void ratio of, 282
solar radiation, 229
  absorption of, 366, 370
  and snowmelt, 365
  spectrum of, 235
solar system, ice in 2–7
Soret effect, 43, 270
species principle (see conservation of
  mass), 26
specific heat, 53
  apparent, 78, 92–3, 128, 283
  of crystalline ice, 54
  isobaric, 53
  isometric, 53
  of soil, 282–3
  of supercooled water, 54
  variable, 78, 92
specific surface area, 270
spinodal, 58
spiral dislocation, 62, 170, 340
  growth of, 170
  step, 170
splintering, 194
spongy ice, 201, 204–5, 218
spontaneous (homogeneous) nucleation, 46
spray, 214, 216–18, 220
  icing, 226
spring flood, 252
squirrel, 301
stagnation flow, 144, 146, 209
stalk ice (see needle crystal)
Staphylococcus, 345
state of the sea, 215
Stefan, 20, 63
Stefan equation (solution), 65, 67, 72, 74,
  227, 233
Stefan number, 73–4, 79–81, 84–5, 91, 125,
  135, 142–3, 147, 154, 185–6, 192–3,
  215, 246, 276, 329, 344, 353, 372, 374
Stefan problem (simplified), 64–71, 124
Stefan–Boltzmann constant, 236
stellar crystal, 306
stem, 270
step (see also dislocation),
  collection distances of, 170
  effects on linear growth rates, 173

step, (*contd*)
  heights of, 170
  migration distance on, 170
  propagation of, 170
  velocity of, 168–70
stick-slip behaviour, 160
Stokes' law, 183
stomata, 334
storage temperature, 383
storage time, 383
Streptococcus, 345
structural icing, (see also accretion) 17, 371, 373
structural icing control,
  anti-icing, 373–4
    by chemical methods, 220, 373
    by thermal methods, 373
  de-icing, 373–4
    by mechanical methods, 373
    by thermal methods, 220, 373
subcooling, 48
sublimation (see heat transfer)
supercooling, 48
superheat ratio, 79, 143, 165
superheated ice, 358
superheated water, 000
superimposed ice, 122, 249
  zone, 249
supersaturation, 174–5
surface energy, (see surface tension)
surface ripples, 150
surface transition, 148
surface tension, 34, 58, 61, 110, 179, 254, 260, 281, 289, 377
suspensions, growth of ice in, 103, 302, 311
symmetry, 356, 358, 377

tap water, 103
Taylor–Goertler waves, 147
temperate glacier, 245, 248
temperature,
  of comets, 7
  critical, 267
  datum freezing, 46–7
  in deep space, 1
  of earth, 7
  and entropy, 26
  global, 7, 230, 274
  in the ground, 240
  homologous, 53
  of maximum density, 53
  protocol, 342, 381
  spontaneous freezing, 46, 48, 59, 113, 307, 340
stable equilibrium deposition, 177
stable equilibrium freezing, 46–7, 59, 64,

113, 121, 125, 129, 134, 140, 147, 153–5, 159, 161, 178, 185, 191, 218, 247, 266–7, 281, 305, 307, 310, 318, 329, 340, 346–9, 362, 370, 372, 381, 384
triple point, 47, 167, 243

tension crack ice, 277, 279
terminal velocities, 198, 223
  of snow flakes, 195
thaw,
  figures, 359–60, 362
  holes, 365
thaw–consolidation ratio, 376–7
thawing of (see also melting),
  organs, 384
  soil, 374
  tissue, 384
thawing index (see degree-day index)
thermal conductivity, 41–2, 53, 55, 68, 110, 127–8, 228, 232, 283–4
thermal contraction, 368
thermal diffusion, 135
thermal diffusivity, 55, 77
thermal erosion, 368, 371
thermal fluctuations, 56
thermal growth, 124, 126, 131
thermal history, 355
thermal penetration, 105–6
thermal properties, 53–5, 281–6, 374
thermal regime, 19, 249–50
thermal shock, 193
thermodynamics,
  classical, 24
  First Law of, 25
  Second Law of, 26
thermodynamic variables, 34
thermoelectric effect,
  theory of, 42
thermokarst, 380
thermophysical properties, 53
throttle icing, 226
Titan, ice on, 6
Tollmien–Schlichting waves, 147
topography, effect of, 230, 234
tonicity, 311
tortuosity, 262, 284, 351
translocation, 233
triple point of water, 31, 47, 57, 167, 176–8, 192–3, 228, 243, 265, 358, 370
triple T curves, 308–9, 318
tropospheric cooling, 230
trumpet curve, 240–1
tubules, 315
tundra, 231
turbine icing, 225–6

turbulence effects, 116, 196, 238, 366–7
turgor, 315
turnover, 115, 121, 367
two-factor hypothesis, 342
Tyndall (or melt) figures (see thaw)

undercooling, (see supercooling)
underturning, 120
unfrozen water, 188, 202, 204, 220, 231,
    235, 268–70, 276, 279, 281, 285–9,
    310–11, 330–1, 348–9, 372, 380
unfrozen zone, 281, 285–6, 293
unicellular bacterium, 312
unicellular organism, 316–17, 329–30
universal gas constant, 40
Uranus' moons and ice, 6

vacancy,
    defect, 52
    diffusion, 61
vacuole, 314–15, 349
valence orbitals, 49, 51
van't Hoff factor, 179
vascular system, 303, 334, 342–4, 382
vegetation, effect of, 230–1, 234, 238, 240
vein ice, 277
vessel-generated spray, 213–16, 218
virus, 313
viscous dissipation, 190
vitreous humour, 305
vitreous ice (see amorphous ice)
vitrification, 49, 307–8, 311, 318, 327, 343,
    382–4
void ratio, 282
volume diffusion, 244
volume fraction, 260, 283
volumetric ice content, 264

wake capture, 196
water,
    atmospheric, 299
    biological, 299–300, 380
    chillers, 163
    content, 205–6, 311
    distribution on Earth, 299, 380
    in ice caps, 299–300
    in humans, 301–3
    in lakes and rivers, 299

flickering cluster model of, 269, 302, 358
hard sphere model of, 56
mobility of, 260, 272, 302
in the oceans, 299–300
origin on Earth, 298–9
quasi-crystalline model of, 56, 302
turnover of, 300–1
transport of molecules in, 60
in the Universe, 298
water content, 214, 221, 285, 288, 301
water molecule, 49–51
    bond angle in, 49, 51
    bond length of, 50
    bonding orbitals of, 49
    modes of vibration of, 51
    hydrogen atoms in, 49–51
    hydrogen bonding in, 50
    lone-pair orbitals in, 49, 51
    polar nature of, 268, 303
    valence bonding in, 49
    valence orbitals in, 49, 51
water spray, 213
water turnover, 300–1
wave equation, 356
waves, 115
weathering, 256–7, 374
wedge flow, 144
weighting function, 95
weighting residual, 95
wet snow, 223
    zone, 249
white ice (see snow ice)
wind, effect of, 231, 238
wind-blown spray, 182, 205
woody plants, 335
work,
    chemical, 34, 77, 286
    displacement, 34, 75–6, 286
    dissipation, 77, 249, 251
    electrical, 34, 77
    surface tension, 34, 286
work flux, 26
wound, 334
Wulff's theorem, 000

xylem, 332–3, 335

yeast, 311